Integrated Wireless
Propagation Models

Integrated Wireless Propagation Models

William C. Y. Lee, Ph.D.

David J. Y. Lee, MBA, Ph.D.

New York Chicago San Francisco
Athens London Madrid
Mexico City Milan New Delhi
Singapore Sydney Toronto

Integrated Wireless Propagation Models

1 2 3 4 5 6 7 8 9 0 DOH 1 2 0 9 8 7 6 5 4

ISBN 978-0-07-183751-4
MHID 0-07-183751-5

This book is printed on acid-free paper.

Sponsoring Editor
Michael McCabe

Editing Supervisor
Stephen M. Smith

Production Supervisor
Pamela A. Pelton

Acquisitions Coordinator
Amy Stonebraker

Project Manager
Raghavi Khullar,
Cenveo® Publisher Services

Copy Editor
Bruce Owens

Proofreader
Nikhil Roshan,
Cenveo Publisher Services

Indexer
Robert Swanson

Art Director, Cover
Jeff Weeks

Composition
Cenveo Publisher Services

About the Authors

William C. Y. Lee, Ph.D., honorary dean of the School of Advanced Communications, Peking University, and formerly chief scientist of Vodafone Plc, is one of the original pioneers who developed wireless technology, and the Lee propagation point-to-point model in 1978, at Bell Labs. World-renowned for his development of commercial CDMA technology, Dr. Lee is a technologist, innovator, teacher, and writer. He is an honorable professor at Beijing University of Aeronautics and Astronautics, Southwest Jiao Tong University (Chengdu), and Taiwan National Chiao Tung University. Dr. Lee has written four prominent books and been granted more than 40 patents in the area of mobile communications. In 2014, he received the IEEE Vehicular Technology Society's Hall of Fame Award, the society's most prestigious award.

David J. Y. Lee, MBA, Ph.D., is a senior software development manager at Cisco Systems. Before that, he was a mobile solutions manager at Cisco. Dr. Lee was previously the vice president of product marketing and operations at Exio Communications and the director of strategic technology at Vodafone. He was also the RF director for Vodafone Romania, responsible for the nationwide planning, deployment, and optimization of the GSM network. Dr. Lee participated in many network optimization and merger and acquisition activities while with Vodafone in the United States, Europe, and Asia. He has had 10+ patents granted and 50+ papers published. Dr. Lee also worked at Motorola and Bell Labs.

Contents

(A listing of abbreviations and acronyms is available at www.mhprofessional
.com/iwpm.)

Preface

Ever since I created the macrocell prediction model in 1977 at Bell Labs, many people wanted to know the details of the model. AT&T had held it as proprietary and did not disclose it. Afterward, I created the microcell model in 1988. Both models are briefly described in my previous books, but the whole models were not fully disclosed at that time because of Pactel's proprietary interest. In 1991, David Lee worked for me at Pactel (the company was renamed AirTouch in 1994), making many field measurements in the different countries of Pactel's markets for deploying the desired cellular systems. In the meantime, the merit of these two models as a tool in deploying the systems in those countries has been shown. Starting in 1995, the two of us had worked on the in-building (picocell) model, which used mostly empirical data. In 2008, David asked me if I wanted to write a book on the Lee model. He would do all the preparation work for the book. From his hard work collecting all the necessary material, this book resulted.

Chapter 1 introduces all the terms and describes the natural phenomena in the mobile communications environment. Chapter 2 introduces the macrocell models that have been created by others. They are the most popular area-to-area models used in the industry. Chapter 3 introduces the point-to-point prediction models in the macrocell models. There are two models. The Lee model is for short distances (<10 miles), and the Longly–Rice model is for long distances (>10 miles). Chapter 4 introduces the microcell models. The Lee microcell model is introduced in the first part of the chapter. The near-in distance used in microcell prediction is clearly defined based on the equations derived in my earlier published books. A lot of empirical data collected from different areas, in both domestic and foreign markets, were used to verify the Lee model.

Chapter 5 introduces the Lee in-building model for both picocell and femtocell. The newly defined close-in distance for the indoor environment is derived. Many earlier papers regarding the Lee in-building model published by David and me have been modified in this book, such as in the case of "same floor" and especially in the case of "interfloor," shown in Sec. 5.5.3. In Chap. 6, the integration of the three Lee models is described. David has made a software tool to plot a prediction signal strength chart covering the three different Lee models.

In this book, many other prediction models are included with their merits for readers to understand besides the Lee models. Also, the integration of the three Lee models—macrocell, microcell, and in-building (picocell and femtocell)—is displayed and used for future planning of an overall cellular system in a specific area. Users should understand that a good prediction model can be accepted if its standard deviation of prediction error is within ±8 dB. In small cells, the prediction error should be ±5 dB.

Measuring signal strengths in a large area is costly. Therefore, we may depend on many prediction charts on a specified coverage generated roughly by area-to-area models. Each chart represents a specified environment. Then the Lee model derives a point-to-point model using a terrain map and the antenna effective height to fine-tune the signal strength. The point-to-point prediction model is useful when cell size is large. When cells become small, measuring signal strength in a small area is considerably easier and less expensive; we can collect more empirical data in a particular confined area. Thus, the prediction model becomes more accurate based on the empirical data.

How was the Lee model created? In 1976, engineers at Bell Labs were discussing the deployment of a cellular system in an area without doing a massive measurement in that area. Then I wrote an internal memorandum about my point-to-point prediction based on Okumura et al.'s legendary paper in 1968, which was the first paper to provide the ground work for an area-to-area model. At that time, I had to prove to Bell Labs engineers that the tool was working. Initially, in 1976, I had requested engineers from the Tri-State Telephone Team (which consisted of New York Bell, New Jersey Bell, and Connecticut Bell) to assist me because they could provide me with the measured data that I wanted in their territory. Also, they had the manpower to generate the coarse terrain charts I wanted. First, I had used the commercial 50,000:1 scale topographical maps printed by the U.S. Defense Mapping Agency. Each map was roughly 5 by 8 miles (8 by 13 km). The contour lines were in 20-ft increments. To make plotting a terrain contour along a mobile path on a map easier, we generated grid maps. Each grid was about one-half of a square kilometer. Each map had 18 by 24 grids. The Tri-State engineers used an eyeball average of the terrain altitudes in each grid. Each grid had one value. Each map had 432 (18 by 24) values, and one in each grid. It was a tedious job. After the grid maps were generated, we started to pick the routes in the Tri-State territory for getting the measured data. Every planned route on a particular map was drawn, crossing the grids where the route laid to make the signal strength measurements on that route. The altitudes along the route were plotted according to the eyeball average altitudes. Once the transmitter antenna was set up with its known location, height, power, and gain, as well as the height and gain of the mobile unit, the Tri-State engineers passed information on to me. I was the one to use hand calculations according to the eyeball values in the corresponding grids on the terrain contour map to plot the predicted signal strength curve along the mobile path and gave that to the Tri-State engineers before the measurement. The next day, the Tri-State engineers took the measured data on the particular planned route and compared these with my predicted values. I never went out and participated in taking measurements. Therefore, it was a totally independent process between the measurement and prediction.

The process went on for more than a year, and the predicted values were very good compared with the measured data in general. Not only was my prediction tool good, but the eyeball average of altitudes also turned out to be a good method. After more than a year of doing hand calculations to predict signal strengths in many different Tri-State areas, I was like a computer. At that time, I was young and had a lot of energy, and I was not afraid of my colleagues teasing me.

Finally, Bell Labs decided to use my tool in 1978. The eyeball average proved that the coarse grid contour maps were adequate. Then the 1° by 1° tape scaling 250,000:1 was purchased from the Defense Mapping Agency Topographic Center (DMATC).

A software tool was created using the 1° by 1° tape for inputting the terrain data and my prediction tool for system planning. It was called ACE, later renamed ADMS. The planning engineers at AT&T and later the Baby Bells were trained to use the tool in their cellular markets starting in 1983. On October 30, 1979, Mr. C. S. Phelan, patent attorney of Bell Labs, wrote me a letter (printed on page 20 of *Lee's Essentials of Wireless Communications* [McGraw-Hill, 2001]) and acknowledged my contribution from a Bell Labs internal memorandum, "A New Mobile Radio Propagation Model," Case 39445-7, March 30, 1979, written by me. The use of the terrain contour data maps with the effective antenna height gain to predict the signal strength at any known location was the invention of my point-to-point prediction model. Bell Labs did not want to patent this model but rather wanted to keep it for internal use.

In 1980, Hata wrote a paper on his area-to-area model and I was one of the reviewers. Also, I was the associate editor of *IEEE Transactions on Vehicular Technology* to accept his paper. Later, the Hata model was adopted by CCIR. I am very glad that it became an internationally well-known model. Readers may not know that I have been working in this field for 50 years. If you are interested in the past, I have many stories to tell and most of them are good ones.

This is my fourth book that has been published by McGraw-Hill. I hope this book will provide readers with clear guidelines to further implement the propagation models in future 4G or 5G systems.

William C. Y. Lee, Ph.D.

Acknowledgments

In preparing this book, we have drawn on many models that are related to the Lee model. We should apologize for not including all the models in our book because of page limitations. We are thankful for the kind advice of many scholars—Prof. Bingli Jiao of Beijing University, China; Prof. Lajos Hanzo of the University of Southampton, UK; Prof. Ping-Zhi Fan of Southwest Jiaotong University, China; Prof. Yi-Bing Lin of National Chiao Tung University, ROC; Dr. Joseph Shapira of Netvision Inc., Israel; and Prof. J. R. Cruz of the University of Oklahoma—during the preparation of this book.

We also should thank Mr. Sherman Yang for his assistance in proofreading the manuscript. Mr. Steve Chapman, Senior Publisher, and Mr. Michael McCabe, Senior Editor, at McGraw-Hill, and Ms. Raghavi Khullar and Ms. Kritika Kaushik, Project Managers at Cenveo Publisher Services, have helped us publish this book. Without their encouragement and patience, this book would not have been published.

Finally, we would like to thank our wives for giving us the time to write and finish this book. Also, this book may inspire our children and grandchildren

     Bill's two grandchildren, Alex and Sophia
     David's son, Richard

to follow our steps in their future careers.

William C. Y. Lee, Ph.D.
David J. Y. Lee, MBA, Ph.D.

Integrated Wireless Propagation Models

CHAPTER 1

Introduction to Modeling Mobile Signals in Wireless Communications

1.1 Why Write This Book?

Almost 50 years ago, mobile communications research was born at Bell Laboratories;[1] the major task at that time was to focus on understanding the natural phenomenon of a mobile signal's propagation along a transmission path. This is similar to diagnosing a disease in medicine. Then the treatment process follows. If the diagnosis is accurate, then the treatment for the disease will usually be more effective. Therefore, in this book, we try to put together diagnostic methods and model the natural phenomenon to give the reader an overall view of all the tools that have been used to diagnose signal propagation. Afterward, signal strength can be predicted from many different models, which are based on the natural phenomenon.

1.2 Differences Between Free Space Communications and Mobile Communications in Propagation

There are three major phenomena created by mobile signals:[2]

1. Because the mobile unit is always surrounded by human-made structures and buildings or is inside the buildings, many multipath waves may be created by the reflections of the transmitting signal, causing the resultant strength of the signal to be different at different locations. Also, the reflected signals do not arrive at the mobile unit simultaneously. Thus, the time delay between the first wave and the last wave causes the distortion among the received digital symbols. This is called intersymbol interference (ISI) in the data transmission.

2. Because the mobile unit moves from one location to another location along the travel path, the signal strength received at the mobile unit from a radio path at each location is different. These variations in signal strength vary along the radio paths and are called short-term signal fading. The fading rate can be fast or slow, depending on the mobile speed. Also due to the mobile speed, the frequency of the signal carrier will be shifted. It can be higher or lower, depending

1

on the direction of the mobile unit traveling toward or away from the transmitting antenna of the base station. This is called the Doppler shift.

3. Due to the mobile unit always moving on the ground, the ground contour affects the received signal at the mobile unit and makes the average signal strength vary up or down. It is found that the average signal follows the lognormal distribution and is called long-term fading. The signal along the radio path will be visualized up and down at its mean value of the signal. Besides, the signal propagation experiences a huge path loss due to the signal reception in the mobile environment. It is a 40-dB-per-decade loss observed from this mobile environment instead of a 20-dB-per-decade loss observed from free space propagation.

1.3   Treatment of Mobile Signals

1. To deal with the multipath (short-term) fading signals, we use diversity techniques to reduce the fading: spacing diversity, polarization diversity, angle diversity, field component diversity, and so on.[3] Sometimes, the spread spectrum, such as frequency hopping or CDMA, also reduces the fading.[4]

2. For dealing with the lognormal (long-term) fading, we first have to model the signal receptions based on different natural environments.[4,5] Based on the size of the cell, there are three popular models: macrocell, microcell, and in-building (picocell). In each of the model environments, we may encounter different man-made structures, such as metropolitan, suburban, rural area, and open land. There are different landscapes, such as water surface, foliage, hills, and so on. Once we can model those particular environments, signal strength can be easily predicted. Second, by using the output from a predicting model working in an area of implementing a mobile system, we can engineer our system equipment with a desired signal coverage in that area.

3. Once we understand the signal coverage in an area, we can use many techniques to improve the reception of the signal. One very promising technique is MIMO.[6] The techniques of improving signal reception or increasing the spectrum efficiency are not covered in this book. However, in Sec. 2.18 the on-body model is applied in the MIMO system.

1.4   History of Developing the Lee Model[7]

In 1974, the first designing stage of the analog system AMTS (Advanced Mobile Telephone Service) had already been completed.[8] Bell Labs was starting to search for ways to implement the AMTS system in the field. At that time, the cost of each cellsite[1] was over U.S. $1 million. How many cellsites are needed for an adequate signal-strength coverage in a given deployed area? To answer this question, the Lee model was developed. At that time, all the propagation models were area-to-area models. Developing a point-to-point model for predicting the signal strength along any mobile traveling route in an area was a challenge. The Lee model was created for the point-to-point prediction, and its accuracy needed to be verified before determining whether it would be accepted. Bell Labs first suggested using printed DMA (Defense Map Agent)

---

[1]The term *cellsite* was coined by F. H. Blecher at Bell Labs. The cellsite is a base station with many system control features.

50,000:1 scale topographic maps. Each map size was roughly 5×8 square miles. The contour lines were in 20-ft increments. The maps were purchased from local map stores in Morristown, New Jersey. The verification project was carried out in a joint effort with Bell Labs and the TriState Team (New York Bell, New Jersey Bell, and New England Bell). First, to make an elevation-grid map from a 5×8-sequence-mile topographic map, the map was divided into 18×24 grids. Each grid is about one-half square kilometer. Because several terrain elevation contour lines run through each grid, an eyeball average of the terrain elevation lines in each grid representing the elevation of that grid was proposed. Then the elevation-grid maps were generated. The project team members were responsible for providing the elevation-grid maps in their selective geographical area with the measurement data available in such corresponding areas. Bell Labs took the elevation-grid maps received from the team members and applied them to the Lee model, then the predicted signal strengths on any particular streets on the maps were obtained. Use of the predicted outcome to compare with the measured data showed a good match between the measured and predicted data. The eyeball average values turned out to be good enough to be used for the Lee model prediction. Nevertheless, the eyeball average process for getting the grid maps and the hand calculation of the Lee model were labor intensive. The alternative was to purchase the $1° \times 1°$ map tape of the DMATC (Defense Map Agency Topographic Center). It had 1200×1200 data grids. The size of each grid was 3 arc-second $\times$ 3 arc-second. The altitude value in each grid was stored in the tape from a 250,000:1 digitized map. Bell Labs took an average of 30 altitudes from the surrounding 30 grids (5×6 grids) to get an average altitude for each grid. Using this average altitude at each grid to compare with the value obtained from the eyeball average at the same grid, the result was that the differences were very small. Therefore, the average altitudes from DMA $1° \times 1°$ tape were accepted to make the process less labor intensive. Bell Labs programmed the Lee model and stored the DMA tapes into the computer. The implementation of the Lee model using computer computation was called ACE (Advanced Coverage Estimate) in 1978. The name of ACE has changed to ADMS (Area Deployment of Mobile System). All the Baby Bells deploying their cellular systems in 1983 were using the ADMS tool. The Lee model[9] went through many phases of revision at PacTel and AirTouch. The Prediction Tool called Phoenix was based on the Lee model and was used internally in all the Pactel and, later, AirTouch markets, both domestic and international, such as in Germany, the United Kingdom, Italy, Portugal, Romania, Spain, Korea, and Japan. Lee's books[2,4,5] described the model very briefly. In 1985, some portion of Ericsson's early internal prediction model used the Lee model.

There are many commercial prediction tools that also used the Lee model.[9-13] In 2008, the Federal Communications Commission (FCC) and the International Telecommunication Union (ITU) paid a lot of attention to the Lee model.[14,15] This model has not been patented. Therefore, it is worthwhile to be documented in this book.

1.5   Basic System Operations[16]

In every wireless system, an antenna is used to radiate and receive electromagnetic energy. The antenna acts as a transducer between the system and free space is referred to as the air interface. While a comprehensive treatment of the subject of antennas is beyond the scope of this book, it is helpful to understand how they operate and how they can affect mobile signal strength. For many types of antennas, we can estimate the gain from the physical dimensions (size and shape) or from knowledge of the antenna beam widths.

When a mobile unit travels along a street, the mobile signal received can be predicted. This book discusses the tools for predicting the mean signal level within different areas, such as in rural, suburban, and urban environments, and with many different characters, including the details of the terrain, building clutter and the extent of foliage, and over the water along the radio path. Some factors are more important than others. The prediction is only a statistical estimate, as the measured data can vary about the predicted mean as the mobile moves around within a small area.

The mobile signal is a fading signal. We can separate the fading signal into two parts. The first part is that the mean signal level is varied following lognormal statistics with a standard deviation, depending on the nature of the local environment. The variation of the local mean signal level is called long-term (lognormal) fading. The second part is that, on top of the slow variation of the local mean signal, there is a rapid and deep variation known as a short-term (or Rayleigh) fading, caused by multipath propagation in the immediate vicinity of the mobile. This follows Rayleigh statistics over fairly short distances. There are other important characteristics of the signal channel, such as noise and interference. When a wideband channel is used, issues of frequency-selective fading and ISI arise.

When a mobile communication system was first deployed in the 1980s, a powerful transmitter was used on a high site to cover a large cell with few signal channels. Since the natural radio spectrum is limited, we need to balance the requirements of area coverage and system capacity during system planning. The AMPS system lowered the transmitter power and applied the frequency reuse scheme to effectively increase both the area coverage and the system capacity at the same time. The prediction tools can help engineers optimize system performance.

The provision of a wide area coverage will always involve the development of an infrastructure of radios and/or line links to connect a number of base stations via one or more control points so that the nearest base station to any mobile can be used to send messages to and receive from that mobile. Structuring a national network entirely using radio links is clearly complicated and costly. An alternative is to use the public telephone network (PTN), which is already available. If the PTN is used as the backbone infrastructure to connect the base stations together, then there are many connection points between the base stations and the fixed network. Each cellsite (base station) has to cover only a small area. This in itself is a major step toward achieving much greater frequency reuse in a cellular system with many cellsites. Moreover, in principle, a mobile within the coverage area of any cellsite needs to connect itself to the facilities of the national and international telephone networks.

The potential of this deployment strategy was realized from the start, but before any systems could be implemented some major issues in the field had to be solved from the signal strength prediction tools, and the tools have to be updated continuously as new issues arise.

1.6   Mobile Radio Signal: Fading Signal

1.6.1   Conditions of Mobile Signal Reception

1.6.1.1   In a Standstill Condition

The received mobile signal strength at one location can be strong, but just moving a wavelength away from its location, the signal can be very weak. This is because the height of the mobile antenna is close to the ground level, and usually the mobile signal cannot be a line-of-sight signal received from the base station. Two or more wave paths of a signal would be created by reflection, refraction, or diffraction from the surroundings and arrive at the mobile receiver antenna at different times.

The combination of these wave paths generates a resultant signal that can be very different from the original transmitted signal in both amplitude and phase while arriving at the mobile. Its received signal is also different from each location every multiple of a half wavelength apart on a road.

1.6.1.2   In a Moving Condition

When a mobile unit is traveling along a street, climbing a hill, or moving inside a building, the mobile antenna receives a signal, strong and weak, along the path. This is called signal fading. Mobile signal fading is hard to handle in mobile communication reception. If we can predict the received mobile statistically (i.e., long-term fading) before the system is deployed, then we can plan a highly efficient system in the field. The time consumed, the labor, and the cost for deploying a system will be reduced. Otherwise, the cut-and-try method in the field has to be used. In addition, the prediction tool also helps engineers find reasons that signal strength is unexpectedly high or low.

1.6.1.3   Obstruction Condition

There are two obstruction conditions that the received mobile signal may experience:

1. Obstructed or nonobstructed by human-made structures or a forest:
   - Nonobstructed by a human-made structure or forest—it is in a LOS condition.
   - Obstructed by a human-made structure or forest—it is in a non-line-of-sight (NLOS) condition.
2. Obstructed or nonobstructed by the terrain:
   - Nonobstructed by terrain—it is in a nonshadow region.
   - Obstructed by terrain—it is in a shadow region.

1.6.1.4   Direct Path and Diffraction Path

1. Direct path—A radio path is not obstructed by the terrain. There are LOS path and NLOS paths.
2. Diffraction path—A radio path is obstructed by the terrain in a shadow region.

1.6.2   Types of Signal Fading

Different types of fading[17] depend on the natural properties of the original transmitted signal and related parameters, such as path loss, bandwidth, symbol period, and frequency.

Various types of scattering and multipath phenomena, which can cause severe signal fading, are attributable to the mobile-radio communications medium and also affect mobile-radio fading signals. Two types of fading, long term and short term, compound the mobile-radio signal fading as mentioned earlier.

In short-term fading, the types of fading can be further classified as follows:

1. Based on the time-delay spread affected in the received signal:
   Flat fading and frequency-selective fading.

2. Based on the bandwidth of the signal:
   The fading of the received mobile signal can be reduced dependent on the basis of the bandwidth of the signal. The CDMA signal has less fading than the narrow-band signal.

3. Based on the mobile speed and the impulse response rate:
   Fast fading and slow fading.

4. Based on the mobile speed only:
   Flat fading and time-selective fading caused by Doppler spread.

Before we describe the different types of fading, the attributes of the fading signal should be discussed first.

1.6.3   Attributes of Signal Fading

1.6.3.1   Local Mean

The signal strength $r(t)$ or $r(x)$, shown in Fig. 1.6.3.1, is the actual received signal level in dB. Based on what we know about the cause of signal fading,[2,17] the received $r(t)$ can be artificially separated into two parts by its cause, long-term fading $m(t)$ and short-term fading $r_0(t)$ when measured in time as

$$r(t) = m(t) \cdot r_0(t) \tag{1.6.3.1}$$

or when measured in distance as

$$r(x) = m(x) \cdot r_0(x) \tag{1.6.3.2}$$

where $m(t)$ or $m(x)$ is the long-term fading.

We know the long-term fading is the average of the fading signal as the dotted curve shown in Fig. 1.6.3.1. It is also called a local mean since, along the long-term fading, each value corresponds to the mean average of the field strength at each local point. The estimated local mean $\hat{m}(x_1)$ at point x_1 along x-axis can be expressed mathematically as

$$\hat{m}(x_1) = \frac{1}{2L} \int_{x_1-L}^{x_1+L} r(x)\, dx = \frac{1}{2L} \int_{x_1-L}^{x_1+L} m(x) r_0(x)\, dx \tag{1.6.3.3}$$

Assume that $m(x_1)$ is the true local mean; then, if we choose the length properly at point x_1 in Fig. 1.6.3.1, the estimated local and the true local mean should be equal:

$$m(x = x_1) = \hat{m}(x = x_1) \qquad x_1 - L < x_1 < x_1 + L \tag{1.6.3.4}$$

Therefore, let $\hat{m}(x_1)$ approach $m(x_1)$ in Eq. (1.6.3.3); then, the following relation must hold:

$$\frac{1}{2L} \int_{x_1-L}^{x_1+L} r_0(x)\, dx \rightarrow 1 \tag{1.6.3.5}$$

The length L has been determined after fully understanding the statistical characteristics of short-term fading $r_0(x)$. Lee was the first one to study the length of L[18] in 1974 based on the signal received at 800 MHz. From the comparison of the standard deviation σ of sample mean $\hat{m}(x)$ with different length L, the length $2L = 40\,\lambda$ is desirable because the standard deviation σ reduced to 1 dB at $L = 40\,\lambda$. If the length $2L$ is shorter than $40\,\lambda$, the average output would retain some portion of Rayleigh fading. The length

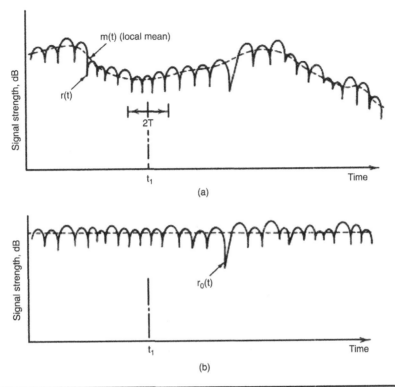

FIGURE 1.6.3.1   A mobile radio signal fading representation. (*a*) A mobile signal fading. (*b*) A short-term signal fading.

cannot be greater than 40 λ; otherwise, the excessive length of averaging would smooth out the local mean information, which it is not supposed to do. In practice, because at a low-frequency transmission the one wavelength is very long in distance, L in the 20 λ to 40 λ ranges is acceptable. Also, to calculate the local mean from a tape of measured data, we have to first digitize the measured data and then be sure that every 50 samples cover a length of 40 λ to get an average value of local mean.[19]

1.6.3.2 Time-Delay Spread

As we know, the direct path is always the shortest path between transmitter and receiver if such a path exists. Therefore, any other path is longer than the direct path. Thus, any other signal that traveled other than the direct path must use more time to reach the receiver. That is how delay happens. The multipath signal(s) will not align with the direct signal, and the cumulative signal will be smeared in time. This is called *delay spread* and is illustrated in Fig. 1.6.3.2.1.

Some metrics are used to represent the multipath delay spread effects of a mobile radio signal:

Mean excess (average) delay: a weighted average (first moment) of the power delay profile (magnitude squared of the channel impulse response).

RMS (root mean square) delay spread: the RMS value (second moment) of the power delay profile, denoted by s_t.

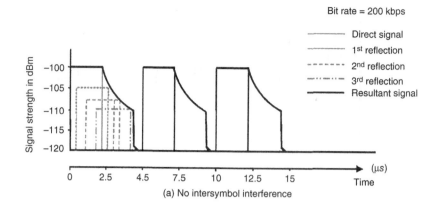

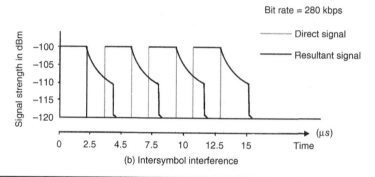

Figure 1.6.3.2.1 Illustration of the effect of multipath delay spread on received symbols. (a) Resultant signal with three reflections. Delay spread = 4 µs, T_b = 5 µs, R_b = 200 kbps. No ISI. (b) Resultant signal with three reflections. Delay spread = 4 µs, T_b = 3.73 µs, R_b = 280 kbps. ISI occurs.

Delay window: the width of middle portion of the power delay profile that contains a certain percentage of energy in the profile.

Delay interval: the length of time between the two farthest separated points of where the impulse response drops to a given power level.

The typical ranges for these parameters are summarized as follows:

Parameter	Urban	Suburban
Mean delay time d	1.5–2.5 µs	0.1–2.0 µs
Corresponding path length	450–750 m	30–600 m
Maximum delay time (–30 dB)	5.0–12.0 µs	0.3–7.0 µs
Corresponding path length	1.5–3.6 km	0.9–2.1 km
Range of delay spread Δ_l	1.0–3.0 µs	0.2–2.0 µs
Mean delay spread		

It is desirable to have the maximum delay spread be small relative to the symbol interval $T = 5$ μs (also, the data rate is $1/T = 200$ kbps) of a digital communication signal so that the received signal cannot be affected by the delay spread, as shown in Fig. 1.6.3.2.1(*a*). When the maximum delay spread is greater than the symbol interval, ISI occurs, as shown in Fig. 1.6.3.2.1(*b*). A correlated requirement is that the coherence bandwidth (which is the function of inversed time-delay spread) be greater than the signal bandwidth. Coherence bandwidth is defined as the bandwidth over which the channel can be considered flat with linear phase. A flat frequency response with linear phase implies no signal distortion.

1.6.3.3 Coherence Bandwidth

Coherence bandwidth is a statistical measurement of the range of frequencies over which the channel can be considered invariant, or "flat." If the signal bandwidth is less than the coherence bandwidth, $B < B_c$, then the medium channel is considered narrowband or flat (flat fading). Otherwise, it is called a wideband channel (frequency-selective fading). Flat fading causes the amplitude of the received signal to vary, but the spectrum of the signal remains intact. The use of flat fading is to determine the allowable outage time under the fading and also to use the Rayleigh probability density function (PDF) to determine the required fade margin to meet the requirement. For selective fading, some modulations are relatively tolerant of frequency dropouts, whereas in other cases an equalizer may be used.

One definition of coherence bandwidth is that if the amplitude correlation is at 0.5, then the coherence bandwidth is approximately[2]

$$B_C = \frac{1}{2\pi\Delta_\tau} \qquad (1.6.3.3.1)$$

Since an exact relationship between coherence bandwidth and delay spread does not exist, Eq. (1.6.3.3.1) is a reasonable estimate. In general, spectral analysis techniques are required to realize the exact impact of a time-varying multipath on a particular transmitted signal.[20,21]

1.6.3.4 Doppler Spread

The time-dispersive nature of a radio channel can be described by Delay spread and coherence bandwidth. Consider a mobile moving at a constant velocity v along a path length d between points a and b while it receives signals from a remote source P, as illustrated in Fig. 1.6.3.4.1. The difference in a path length traveled by a wave from source P

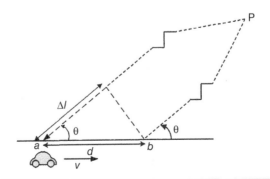

FIGURE 1.6.3.4.1   Illustration of Doppler spread effect.

to the mobile at points a and b is $\Delta l = d \cos\theta = v\Delta t \cos\theta$, where Δt is the time required for the mobile to travel from a to b and the incident angle θ of the incoming wave is assumed to be the same at points a and b since the source is assumed to be far away. The phase change in the received signal due to the difference in path lengths is therefore

$$\Delta\phi = \frac{2\pi\Delta l}{\lambda} = \frac{2\pi v\Delta t}{\lambda}\cos\theta \qquad (1.6.3.4.1)$$

where λ is the wavelength. The apparent change in frequency due to the motion of the mobile, or Doppler shift, is given by f_d, where

$$f_d = \frac{1}{2\pi}\frac{\Delta\phi}{\Delta t} = \frac{v}{\lambda}\cdot\cos\theta \qquad (1.6.3.4.2)$$

The Doppler shift shown in Eq. (1.6.3.4.2) is a function of the mobile velocity and the incident angle between the direction of motion of the mobile and the direction of arrival of the wave. It can be seen from Eq. (1.6.3.4.2) that if the mobile is moving toward the direction of arrival of the wave, the Doppler shift is positive (i.e., the apparent received frequency is increased), and if the mobile is moving away from the direction of arrival of the wave, the Doppler shift is negative (i.e., the apparent received frequency is decreased). Multipath components from a continuous-wave signal that arrive from different directions contribute to Doppler spreading of the received signal, thus increasing or decreasing the signal bandwidth.

If the baseband signal bandwidth is much greater than f_d, the effects of Doppler spread are negligible at the receiver. This is a slow-fading channel.

1.6.3.5   Coherence Time

Coherence time is the time duration over which two received signals have a strong amplitude correlation in a Rayleigh fading environment. The coherence time and maximum Doppler spread are inversely proportional to each other. That is,

$$T_C \approx \frac{1}{f_m} \qquad (1.6.3.5.1)$$

Coherence time T_C is used to characterize the time-varying nature of the frequency dispersiveness of the channel in the time domain. When the reciprocal bandwidth of the baseband signal (e.g., symbol interval) is greater than the coherence time of the channel, the channel will vary in time during the transmission of the signal, and distortion at the receiver occurs. If the coherence time is defined as the time over which the amplitude correlation coefficient is equal or greater than 0.5, then the coherence time is found as[22]

$$T_C = \frac{9}{16\pi f_m} = 0.179/f_m \qquad (1.6.3.5.2)$$

where $f_m = v/\lambda$, which is the maximum Doppler shift. Equation (1.6.3.5.2) is used in analog communication. In digital communications, the coherent time is taking a square root of multiplication of Eqs. (1.6.3.5.1) and (1.6.3.5.2). That is,

$$T_C = \sqrt{\frac{9}{16\pi f_m^2}} = \frac{0.423}{f_m} \qquad (1.6.3.5.3)$$

Comparing Eq. (1.6.3.5.3) with Eq. (1.6.3.5.2), the coherence time for digital communications is longer than that for analog communication because of the nature of code sequences.

We may calculate their coherence time T_C when two signals are arriving at different times. For example, for a vehicle traveling 60 mph at an 800-MHz carrier, the value of $T_C = 2.5$ ms for the analog communication is found from Eq. (1.6.3.5.2). This means that when the symbol rate is greater than $\frac{1}{T_C} = 400$ bps, the channel will not cause distortion due to the motion. In a digital transmission system, the coherence time $T_C = 7.6$ ms is found from Eq. (1.6.3.5.3). The symbol rate $\frac{1}{T_C}$ must exceed 132 bps in order to avoid distortion due to frequency dispersion.

The two signals are arriving within the coherence time, but distortion could still occurs due to multipath delay spread if it exceeds symbol duration T_s.

1.6.4 Flat Fading

When the mobile radio transmission medium, sometimes called radio channel, has a constant gain and linear phase response over a bandwidth B_H, which is greater than the bandwidth of the transmitted signal B_s, $B_H > B_s$, shown in Fig. 1.6.4.1, then the spectral characteristics of the transmitted signal are conserved at the received end. It is called flat fading in a multipath radio channel. However, the strength of the received signal varies with time due to the fluctuation from the natural causes. The characteristics of a flat fading channel are illustrated in Fig. 1.6.4.1, where T_0 is the arrival time.

Figure 1.6.4.1 shows that the received signal $r(t)$ varies in gain, but the spectrum of the transmission is conserved. In a flat fading channel, the time-delay spread is less than the symbol interval, as shown in Fig. 1.6.3.2.1(a). The channel impulse response $h(t)$, which is from a single delta function $\delta(T)$ sending at the transmitter, has a bandwidth B_H greater than the bandwidth of the signal B_s. Flat fading channels are also known as narrowband channels since the bandwidth of the signal is narrow compared to the channel impulse response bandwidth of flat fading. The distribution of the instantaneous signal envelope variation of flat fading channels is the Rayleigh distribution. The Rayleigh distribution will be introduced later.

When a transmitted signal suffers a flat fading, its spectral characteristics are conserved if both

$$B_S \ll B_H \qquad (1.6.4.1)$$

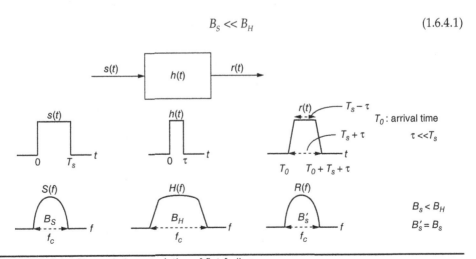

Figure 1.6.4.1   Channel characteristics of flat fading.

and

$$T_s \gg \Delta_\tau \qquad\qquad (1.6.4.2)$$

where B_s is the bandwidth of the transmitted signal, T_s is the symbol interval (e.g., reciprocal bandwidth), and Δ_τ and B_H are the delay spread and the bandwidth of the fading channel, respectively.

1.6.5 Signal Fading Caused by Time-Delay Spread: Frequency-Selective Fading

Basically, based on multipath time-delay spread, the fading could be divided into two groups: flat fading and frequency-selective fading. Frequency-selective fading is when the coherence bandwidth of the channel is larger than the bandwidth of the signal. Coherence bandwidth is defined as within such a bandwidth if the signal is frequency insensitive with a linear phase, as mentioned previously. If the coherence bandwidth of the channel is smaller than the bandwidth of the signal, it is called flat fading. In this case, the same magnitude fading would happen to all signals with different frequency components.

Frequency-selective fading is called such when two signals with different frequencies received, after having gone through a radio channel, have different fading characteristics. The condition of creating frequency-selective fading is different from that of flat fading. When the bandwidth of a received signal is greater than the bandwidth of a radio channel in which a constant gain and linear phase response are preserved, then the received signal is spread in time and is distorted; thus, its fading characteristics change and differ from flat fading. The distorted signal causes the time dispersion of its transmitted symbols and introduces the ISI. To avoid ISI, the symbol rate has to be calculated so that the symbol interval is longer than the delay spread, as shown in Fig. 1.6.3.2.1(*a*). The characteristics of a frequency-selective fading channel is depicted in Fig. 1.6.5.1.

For frequency-selective fading, when the symbol interval of the transmitted signal is greater than the time-delay spread, the spectrum $S(f)$ of the signal has a bandwidth B_s, which is greater than the coherence bandwidth B_C of the channel, $B_s > B_C$.

Frequency-selective fading channels are wideband channels since $B_s > B_C$. In summary, a transmitted signal is suffering in a frequency selective fading if

$$B_S > B_C \qquad\qquad (1.6.5.1)$$

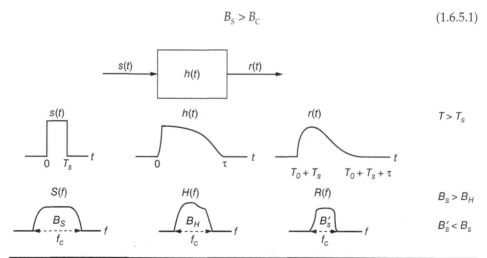

Figure 1.6.5.1 Characteristics of frequency-selective fading channel.

and

$$T_S < \Delta_\tau \qquad (1.6.5.2)$$

For a conservative guideline, we may set $T_S < 0.1\,\Delta_\tau$ to be the condition of having a frequency-selective fading. In some types of modulation, T_S of the signal can be larger than $0.1\,\Delta_\tau$ before causing the frequency-selective fading.

1.6.6   Fading Signal Caused by Doppler Spread

Based on Doppler spread, fading can be divided into two categories: fast fading and slow fading. Fast fading and slow fading are classified based on how rapidly the transmitted baseband signal changes compared to the rate of the channel impulse response changes. If the former is at a rate much slower than the latter, the channel may be assumed to be a fast-fading channel. Otherwise, it is assumed to be a slow-fading channel. Therefore, the velocity of the mobile unit causes a baseband signal through the channel at the terminal faded. The velocity determines whether a signal undergoes fast fading or slow fading.

1.6.6.1   Fast Fading

When the coherence time (T_C) of the channel is smaller than the symbol interval (T_S) of the transmitted signal, a fast fading channel is formed. In a fast-fading channel, the frequency is dispersed due to Doppler spreading. Therefore, the fast-fading channel is also called the time-selective fading channel. A transmitted signal suffers fast fading if

$$T_S > T_C \qquad (1.6.6.1.1)$$

and

$$R_S < f_d = 2V/\lambda \qquad (1.6.6.1.2)$$

where R_S is the data rate of the signal. When Eq. (1.6.6.1.2) is met, the random FM noise occurs, and the signal is distorted.[23] As we can see from Eq. (1.6.6.1.2), it is rare to use a very low data rate that is less than the Doppler frequency. Therefore, the fast-fading case is not of interest here.

1.6.6.2   Slow Fading

In a slow-fading channel, the coherence time is much longer than the symbol interval of the signal $s(t)$, and the data rate of the signal is much faster than the Doppler spread frequency, as shown below. Therefore, the condition of slow fading is

$$T_S \ll T_C \qquad (1.6.6.2.1)$$

and

$$R_S \gg f_d = 2V/\lambda \qquad (1.6.6.2.2)$$

In this case, no random FM noise can distort the signal,[21] and the channel is considered static over the time of transmission. The velocity of the mobile unit, the carrier frequency, and the signal symbol rates are the three factors that determine whether a signal suffers fast fading. In general, we are always dealing with the slow-fading case.

Therefore, in general, flat fading and frequency-selective fading are the slow fading also in the category of the short-term fading signal. The short-term fading signal is not used to predict the propagation path loss models, as will be mentioned in the next section.

1.6.7   Short-Term and Long-Term Fading Signal

Short-term fading, or simply fading, is used to describe the rapid fluctuation of the amplitude of a radio signal over a short period of time or travel distance. Long-term fading is obtained by averaging a long piece of a short-term fading signal. Short-term fading is usually observed from the measurement data caused by multipath waves in the field, while long-term fading smooths out the real-time signal variation in the short-term fading and preserves the slow variation of the fading caused by the contours of terrain, hills, forests, and buildings. Therefore, the long-term fading data are created from the short-term fading data by averaging them to achieve a local mean.[18] The statistics of long-term fading provides a way to compute an estimate of the path loss as a function of distance and other factors.

1.7   Co-Channel Interference Created from the Frequency Reuse Scheme

The frequency reuse concept was first proposed by Doug Ring at Bell Labs in 1957.[8] Lee has shown the implementation of frequency reuse,[24] which serves as a guideline for designing cellular systems. Lee has also concluded in general that the frequency reuse factor K has to be 7 for a requirement of $C/I = 18$ dB when the path loss exponent $\gamma = 4$. In this section, a simulator is introduced for further studying the relationship between frequency reuse factor K and the path loss exponent γ. Theoretically, $\gamma = 40$ dB per decade. However, this path loss exponent value γ varies due to different human-made environments. The frequency reuse can generate many tiers of co-channel cells. For $K = 3$, the first tier has six co-channel interfering cells, and the second tier has 12 co-channel interfering cells. For $K = 7$, the first tier has six co-channel interfering cells, and the second tier also has 12 co-channel interfering cells. Sectorization is way to reduce co-channel interference. The system may need to first calculate the tolerable co-channel interfence that the system can allow then decide what frequency reuse factor number should be used. From what we can obtain from the prediction tool, the optimized frequency reuse factor varies with path loss exponent as well as with sectorization. The tool also can calculate the relative distribution of the co-channel interference from the first and higher tiers. The effect of power control on C/I can also be obtained from the prediction tool.

1.7.1   Basic Concepts[24]

Frequency reuse is the core concept of the cellular mobile radio system. In a frequency reuse system, users in different cells may simultaneously use the same frequency channel. Thus, the system increases capacity and improves spectrum efficiency.

Frequency reuse factor K defines the frequency reuse pattern (may not be unique). It is the number of cells in a frequency reuse cluster. The frequency reuse patterns for $K = 4, 7$ are shown in Fig. 1.7.1.1.

The frequency reuse distance D is the minimum distance between two cells that use the same frequency, that is, the co-channel cells. It can be determined from

$$D = \sqrt{3K} \cdot R \tag{1.7.1.1}$$

where R is the radius of the cell. Reuse of an identical frequency channel in different cells will create co-channel interference that has to be eliminated with various techniques. Otherwise, it can become a major problem. The co-channel interference can be

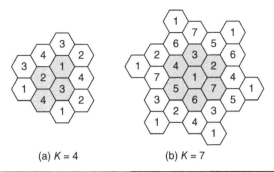

(a) $K = 4$ (b) $K = 7$

Figure 1.7.1.1   Frequency reuse patterns.

reduced by increasing the distance D, which is increasing the value K. The co-channel interference depends on many other factors, such as the number of co-channel cells in the vicinity of the center cell, the type of geographic terrain contour, and antenna height. The prediction tool should take care of this issue. In this section, we focus mainly on the effect of sectorization and the path loss exponent, which characterizes the geographical terrain.

According to the general propagation path loss formula, the received power at distance d from the transmitting antenna can be written as

$$P(d) = P_0 \cdot d^{-\gamma} \tag{1.7.1.2}$$

where γ is the path loss exponent. The received power can be represented as a signal carrier or an interference. Thus, we can write the carrier to interference ratio C/I as follows:

$$\frac{C}{I} = \frac{P_0 \cdot R^{-\gamma}}{\sum_{k=1}^{Z} I_k} = \frac{R^{-\gamma}}{\sum_{k=1}^{Z} D_k^{-\gamma}} \tag{1.7.1.3}$$

where D_k is the distance from the kth co-channel interferer to the receiver in the center cell and Z is the total number of interference co-channel cells. We can see from Eq. (1.7.1.3) that the C/I ratio depends on the number of interfering cells Z and the path loss exponent γ.

The interference from the first tier is the most dominant, and we will verify this later. If we consider only the contribution from the first tier, then $Z_1 = 6$. Since D_k is not much different from frequency reuse distance D in the first tier, we let $D_k = D$. Equation (1.7.1.3) then becomes

$$\frac{C}{I} = \frac{R^{-\gamma}}{6D^{-\gamma}} = \frac{1}{6} \cdot \left(\frac{D}{R}\right)^{\gamma} = \frac{1}{6}(3K)^{\frac{\gamma}{2}} \tag{1.7.1.4}$$

In an urban area, $\gamma = 4$, and the C/I requirement for analog communication is 18 db, which converts to a linear value of 63. Then we have our reuse factor:[2]

$$K = \frac{1}{3} \cdot \sqrt{6\left(\frac{C}{I}\right)} = \frac{1}{3} \cdot \sqrt{6 \cdot 63} = 6.5 \approx 7 \tag{1.7.1.5}$$

This gives a seven-cell reuse pattern for an urban area.

We also studied the case with more than one interfering tier and created an interference ratio of I_k of kth tier to I_1 of the first interfering tier, $\frac{I_k}{I_1}$ for $k = 2, 3, 4$. The center cell of the first-tier co-channel cells is the same center cell of the second- and third-tier co-channel cells. Thus, starting from Eq. (1.7.1.3), the C/I of the center cell from only the first-tier co-channel interfering cells for the frequency reuse factor $K = 7$ (i.e., a total bandwidth will be divided evenly in a cluster of seven cells) is

$$C/I_1 = Z_1 \cdot (D_1/R)^4 = 6\,(D_1/R)^4 \qquad\qquad (1.7.1.6)$$

where Z_1 is the number of co-channel interfering cells in the first tier. From Eq. (1.7.1.1),

$$D_1/R = \sqrt{3K} = 4.7 \qquad \text{for } K = 7 \qquad\qquad (1.7.1.7)$$

The distance D_k from kth-tier co-channel interfering cell to the center cell is

$$D_k = k \cdot D_1 \qquad\qquad (1.7.1.8)$$

and Z_k is the number of total co-channel interfering cells at the kth tier:

$$Z_k = k \cdot Z_1 \qquad\qquad (1.7.1.9)$$

Then the co-channel interference for the kth-tier co-channel cells is

$$C/I_k = Z_k\,(D_k/R)^4 = k\,Z_1\,(k\,D_1/R)^4 \qquad\qquad (1.7.1.10)$$

Thus, I_k/I_1 can be found from Eqs. (1.7.1.6) and (1.7.1.10) as

$$I_k/I_1 = k^{-5} \qquad\qquad (1.7.1.11)$$

$$\frac{I_2}{I_1} = 2^{-5} = 0.0312 = -15 \text{ dB} \qquad\qquad \text{(interference by the second tier)}$$

$$\frac{I_3}{I_1} = 3^{-5} = 0.004115 = -24 \text{ dB} \qquad\qquad \text{(interference by the third tier)}$$

The total interference (first tier and second tier) versus the first-tier interference is

$$(I_1 + I_2)/I_1 = 1 + 0.0312 = 1.0312 = 0.133 \text{ dB} \qquad\qquad (1.7.1.12)$$

The total interference (first and second and third tier) versus the first-tier interference is

$$(I_1 + I_2 + I_3)/I_1 = 1 + 0.0312 + 0.004115 \approx 1.0353 = 0.15 \text{ dB} \qquad (1.7.1.13)$$

From the results from Eqs. (1.7.1.12) and (1.7.1.13), we can conclude that the co-channel interference caused by the second and third tiers is very small and can be neglected.

As frequency reuse factor K increases, the frequency reuse distance D increases, and the interference is reduced. The cell size becomes insignificant larger compared with the reuse distance D for large K. Then we can approximate each cell by its center. Usually, the ratio of interference caused by the second tier and the first tier is the largest.

One way to reduce co-channel interference is to divide the cell into sectors and a use directional antenna in each sector, as shown in Fig. 1.7.1.2. A directional antenna provides a very low gain in the back lobe and will interfere with the cells only in the front (shown in Fig. 1.7.1.2). Thus, the interference caused by the co-channel cells to the center cell will be reduced by more than one-half.

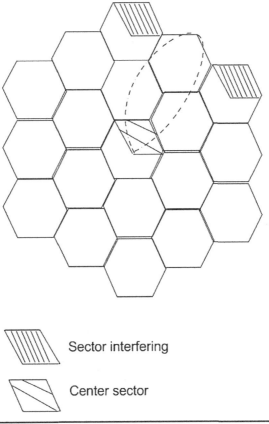

Sector interfering

Center sector

FIGURE 1.7.1.2   Cell sectorization.

In this section, we will study the effect of sectorization and the path loss exponent on the C/I ratio and the frequency reuse factor K. The C/I received from both the forward and the reverse link are calculated. The impact of power control on C/I is also studied.

1.7.2   Simulation Model

The simulator supports a general propagation model with different path loss exponents. Ideal hexagonal cells are used. Multiple sectors, such as three sectors, Lee's microcell,[25-27] and six sectors, with the antenna used in each sector, can be specified by the user. A real antenna pattern is used in the simulation (Fig. 1. 7.2.1). Every cell or sector has the same parameters (ERP, antenna pattern, and so on) except for the antenna direction. Each cell or sector has one mobile, and that mobile is distributed randomly in the cell or sector with a probability density $1/A$, where A is the area of cell or sector. All mobiles have the same ERP when transmitting with omnidirectional antennas if power control is absent. Each cell or sector uses the same kind of antenna for transmission and reception. Only the first interfering tier is used to calculate the C/I ratio, although our simulator is capable of doing multiple interfering tiers. The contribution of higher tiers is compared with the first tier to show that they are negligible.

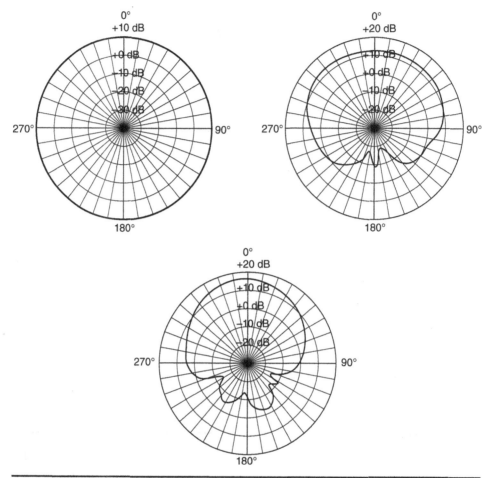

FIGURE 1.7.2.1   Antenna patterns used for simulation—omni and three and six sectors.

Our simulator is run in both forward and reverse link for different path loss exponents (γ = 2, 2.5, 3, 3.5, 4), different sectorization schemes (three sectors, Lee's microcell,[25] and six sectors), and different frequency reuse factors (K = 3, 4, 7, 12). Different real antennas are used in different sectorization schemes (shown in Fig. 1.7.2.1). The impact of power control in the reverse direction is also studied.

1.7.3   Simulation Result

The rule of thumb is that C/I at the center cell improves as the path loss exponent or number of sectors increases. Also, the reverse link C/I ratio increases with the power control added. This property has been known for a long time, but our simulation results provide the exact data on the improvement of the system performance for the first time.

For γ = 4, we can see from Fig. 1.7.3.1 that we need K = 7 to provide a coverage of 99 percent or more area in order to satisfy the criterion C/I > 18 db. This verifies Lee's theory.[25] Also, we find that for a three-sector cell, Lee's microcell, and a six-sector cell,

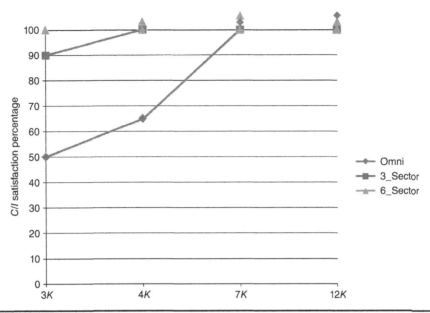

FIGURE 1.7.3.1   The satisfaction percentage of coverage (for $C/I = 18$ dB) versus K, forward link without power control.

we need only $K = 4$ for providing a coverage of 100 percent satisfaction. If we lower the C/I requirement, say, a coverage of 94 percent satisfaction, then we can even have $K = 3$. This tells us that for a reasonable satisfaction of coverage, we can have $K = 3$ (or even lower) for using a sectorized cell.

For considering a reverse link, we studied the cases with and without power control. The use of power control essentially improves the C/I ratio at the cell boundary of the center cell, where the C/I ratio is the worst. So we can see a significant improvement of the C/I in the cell. Also, because of the power control, the C/I ratio is the same at the base regardless of where the mobile is in the cell. This is the reason that we indicate only 0 percent or 100 percent coverage, as shown in Fig. 1.7.3.2. In order to have 100 percent coverage, the omni-cell system needs $K = 4$ and others need only $K = 3$, as depicted in Fig. 1.7.3.2.

Figure 1.7.3.3 is the satisfaction percentage of coverage versus K due to the reverse link transmission without power control. Comparing Fig. 1.7.3.2 with Fig. 1.7.3.3, one can find that for an omni-cell system, the coverage satisfaction percentage of 68 percent in Fig. 1.2.3.3 becomes of 100 percent in a three-sector cell system in Fig. 1.7.3.2 for $K = 4$ and $C/I = 18$ dB.

Another interesting thing we found is that the coverage satisfaction percentage is basically the same for both the forward and the reverse link without power control (comparing Figs. 1.7.3.1 and 1.7.3.3). Thus, the analysis used for the forward link should be applied for the reverse link.

The co-channel interference is unchanged at a mobile in a cell on a reverse link. However, for calculating the forward link, the co-channel interference varies, depending on the location of the mobile in the cell. We take the location with the worst interference in calculating $\frac{I_2}{I_1}$ and shown in Fig. 1.7.3.4.

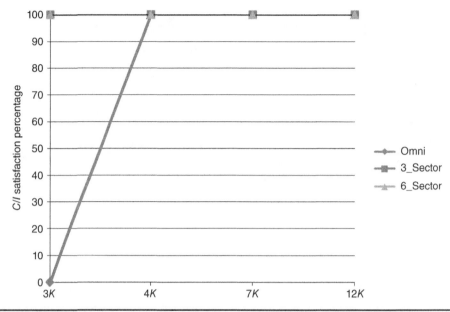

FIGURE 1.7.3.2 The satisfaction percentage of coverage (for C/I = 18 dB) versus K, reverse link with power control.

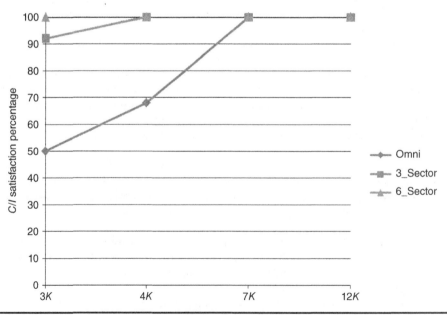

FIGURE 1.7.3.3 Interference percentage versus K, reverse link without power control.

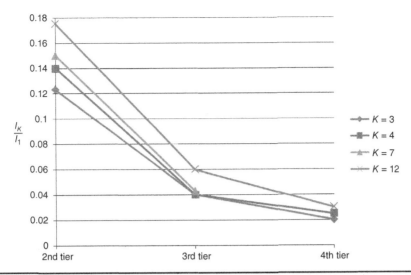

Figure 1.7.3.4   Forward link path loss exponent $\gamma = 4$, omni antenna (worst case).

On a reverse link without power control, the worst interference positions for Lee's microcell system with omni-antennas are at the vertex of the hexagon. For three-sector cells and six-sector cells, they are at the center of the cell. The ratio of $\frac{I_2}{I_1}$ for the omni-cells with path loss $\gamma = 4$ is shown in Fig. 1.7.3.5. The $\frac{I_2}{I_1}$ ratio for the three-sector cells with $\gamma = 4$ is shown in Fig. 1.7.3.6. The $\frac{I_2}{I_1}$ ratio remains almost the same regardless of the difference of frequency reuse factors and sectorized or unsectorized cells, as shown in these two figures.

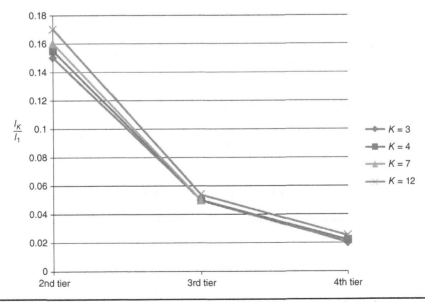

Figure 1.7.3.5   Reverse link without power control path loss exponent $\gamma = 4$, omni-antenna.

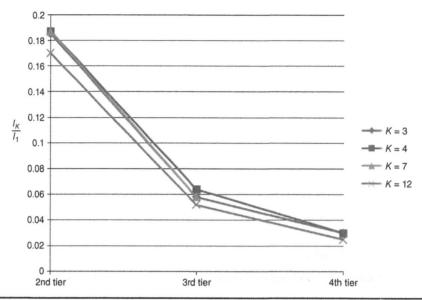

Figure 1.7.3.6   Reverse link without power control path loss exponent = 4, three sectors.

A simulator was built to give an exact C/I analysis. From this analysis, we can see that the coverage satisfaction percentage for a given C/I varies with frequency reuse factor K and path loss exponent γ. This is an important characteristic in designing a cellular system. Specifically, for a coverage satisfaction percentage of 94 percent, we need $K = 3$ only if a sectorized-cell system is used. Also, we can see that the first interfering tier contributes the most to interference. We also give a conservative approximation of the *interference ratio* $\frac{I_k}{I_1}$ for $k = 2$ and 3 (second and third tiers), one from a mathematical calculation and another from the simulation with uniformly distributed mobiles in the reverse link. We find that power control in the reverse link can improve C/I ratio satisfaction up to 45 percent.

1.8   Propagation Fading Models

1.8.1   Rayleigh Fading Model—Short-Term Fading Model

All statistical characteristics that are not functions of time are called "first-order statistics." For example, the average power, mean value, standard deviation, pdf, and cumulative probability distribution are all first-order statistics.

1.8.1.1   The CDF of $r_e(t)$ from an *E* Field Signal

Assume that the *E* field of the signal $s(t)$ has two components, real and imaginary $E = X_1 + jY_1$, and X_1 and Y_1 are Gaussian with zero mean and variance one. Then the envelope of the signal is

$$r = (X_1^2 + Y_1^2)^{1/2} \tag{1.8.1.1}$$

The pdf of $r(t)$ can be expressed as[4]

$$p(r) = \frac{r}{\sigma^2} \exp\left(-\frac{r^2}{2\sigma^2}\right) \tag{1.8.1.2}$$

and the cdf of $r(t)$, the probability that r is less than level A, is

$$p(r \le A) = \int_0^A \frac{r}{\sigma^2} \exp\left(-\frac{r^2}{2\sigma^2}\right) dr = 1 - \exp\left(-\frac{A^2}{2\sigma^2}\right) \tag{1.8.1.3}$$

where the mean square of r is

$$E[r^2] = E[X_1^2] + E\{Y_1^2\} = 2\sigma^2 \tag{1.8.1.4}$$

It is also called average power, and $\sqrt{2}\,\sigma$ is called RMS. Letting $A = \sqrt{2}\,\sigma$ in Eq. (1.8.1.3), we find that

$$p(r \le \sqrt{2}\,\sigma) = 63\% \tag{1.8.1.5}$$

This means that 63 percent of the signal is equal or below average power (also equal or below its RMS). The function of Eq. (1.8.1.3) is plotted on Rayleigh paper, as shown in Fig. 1.8.1.1.

The mean value of r is

$$E[r] = \int_0^\infty r p(r) dr = \int_0^\infty \frac{r^2}{\sigma^2} \exp\left(-\frac{r^2}{2\sigma^2}\right) dr = \sqrt{\frac{\pi}{2}}\sigma \tag{1.8.1.6}$$

1.8.2   Log-Normal Fading Model—Long-Term Fading Model

A lognormal distribution is used for studying a long-term fading. It is a continuous probability distribution of a random variable x whose logarithm is normally distributed in probability theory. This model is used to quantify the distribution of multiple reflections and diffractions rays between a transmitter and a receiver. The lognormal PDF can be given as

$$p(x) = \frac{1}{x\sqrt{2\pi\sigma^2}} \exp\left\{-\frac{[\ln(x) - m]^2}{2\sigma^2}\right\} \tag{1.8.2.1}$$

where m is the median value and σ is the standard deviation of the corresponding normal distribution. From the measurement data, we take the local mean average (see Sec. 1.6.3.1) to average out the short-term fading; the remaining data are called long-term fading, which follows the lognormal fading model shown in Eq. (1.8.2.1). All the prediction models are predicting the long-term fading signal. The long-term fading signal is obtained from averaging the sampled data.[19] Therefore, the true long-term fading signal has to be treated very carefully, as described in the next section.

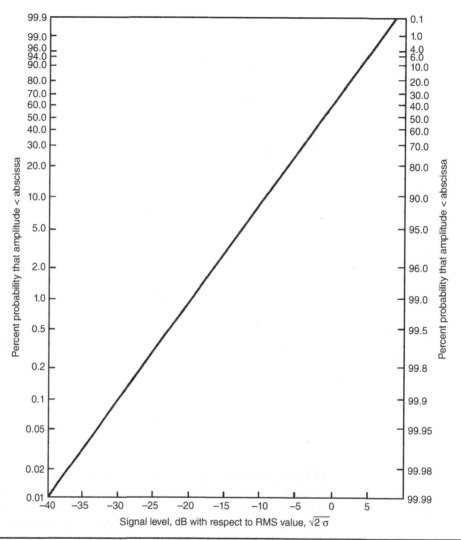

FIGURE 1.8.1.1   Statistical properties of a Rayleigh-fading signal.

1.8.3   Estimating Unbiased Average Noise Level

Usually, the sampled noise in a mobile environment contains high-level impulses that are generated by the ignition noise of the gasoline engine.[28,29] Although the level of these impulses is high, the pulse width of each impulse generally is very narrow (see Fig. 1.8.3.1). As a result, the energy contained in each impulse is very small and should not have any noticeable effect on changing the average power in a 0.5-s interval.

However, in a normal situation, averaging a sampled noise is done by adding up the power values of all samples, including the impulse samples, and dividing the sum by the number of samples. This is called the *conventionally averaged noise power* and is denoted n_c.

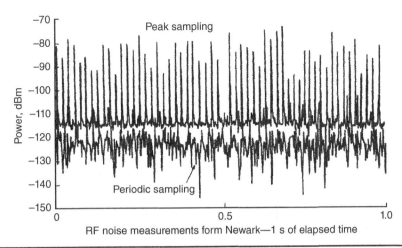

FIGURE 1.8.3.1   Environmental noise trace.

In this case, these impulse samples dominate the average noise calculation and result in a mean value that is not representative of the actual noise power. Use of this value affects the design requirements of the system (signaling and voice). Hence, before the following new technique was introduced, it was not known why there was no correlation between BER and the signal-to-noise ratio measured in certain geographic areas. In a new statistical method, the average noise is estimated by excluding the noise impulses while retaining other forms of interference. This technique is compatible with real-time processing constraints.

1.8.3.1   Description of the Method[30,31]

A counter in the mobile unit counts the instantaneous noise measurements that fall below a preset threshold level X_t and sends a message containing the number of counts n to the database for recording. From the database data, we can calculate the percentage of noise samples x_i below the present level X_t,

$$P(x_i \leq X_t) = \frac{n}{N} \qquad (1.8.3.1)$$

where in our case N is the total number of samples. Once we know the percentage of noise samples below level X_t, we can obtain the average "noise" X_0 exclusive of the noise spikes from the Rayleigh model. Furthermore, the level X_t can be appropriately selected for both noise and signal measurements because both band-limited noise and mobile radio fading follow the same Rayleigh statistics.

1.8.3.2   Estimating the Average Noise X_0

For a Rayleigh distribution (band-limited noise), the average noise power should be exclusive of the noise spikes; X_0 can be obtained from

$$X_0 = 10 \log \left\{ -\frac{1}{\ln \left[1 - P(x_i \leq X_t)\right]} \right\} + X_t \quad \text{dBm} \qquad (1.8.3.2)$$

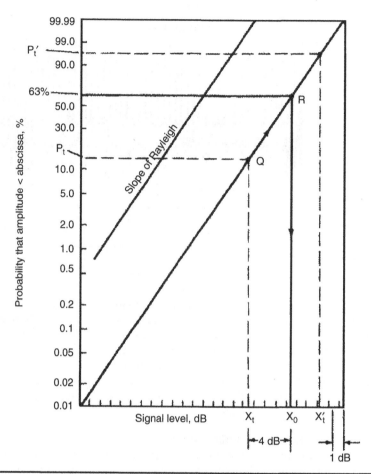

FIGURE **1.8.3.2**   Technique of estimating average noise.

This technique can be illustrated graphically using Rayleigh paper. Since $P(x_i \leq X_t)$ is known for a given X_t, we can find a point P_t on the paper, as illustrated in Fig. 1.8.3.2. Through that point, we draw a line parallel to the slope of the Rayleigh curve and meet the line of $P = 63\%$, which is the average power lever (see Eq. (1.8.1.5)). This crossing point corresponds to the X_0 level (unbiased average power in decibels over 1 mW, dBm).

Example 1.8.3.1   If a total number of samples is 256 and 38 samples are below a level -119 dBm, then the percentage is

$$\frac{38}{256} = 15\%$$

Draw a line at 15 percent and meet at Q on the Rayleigh curve. Assume that the X_t is -119 dBm. The X_0 is the average power because 63 percent of the sample is below that level. Then

$$X_0 = X_t + 4 \text{ dB} = -119 + 4 = -115 \text{ dBm}$$

Example 1.8.3.2   The total number of samples is 256. Three noise spikes are 20 dB above the normal average. Find the errors, using the following two methods. Compare the results.

Use geometric average method. Let the power value of each sample (of 253 samples) after normalization be 1; that is, the average is 1. Then the measured average of 256 samples, including three spikes, is

$$N_0 = \text{Measured average} = \frac{\sum_1^{253} x_i + 100 \sum_1^3 x_i}{256} = 2.16 \qquad (\text{assume } x_i = 1)$$

$$= 3.3 \text{ dB} \quad \text{above the true average}$$

Statistical average method

$$63\% \text{ of samples} = 256 \times 0.63 = 161 \text{ samples}$$

This means that 161 samples should be under the average power level. Now three noise spikes added to the 161 samples increases the number of samples to 164:

$$\frac{164}{256} = 64\%$$

The power levels at 63 and 64 percent show almost no change. Typical data averaging using the geometric and statistical average methods is illustrated in Fig. 1.8.3.3. The corrected value is approximately −118 to −119 dBm based on the statistical average. The geometric average method biases the average value and causes an unacceptable error, as shown in the figure.

1.8.4   Rician Distribution

The Rician fading model is a stochastic model made by S. O. Rice for studying radio propagation when the signal arrives at the receiver by several different paths and

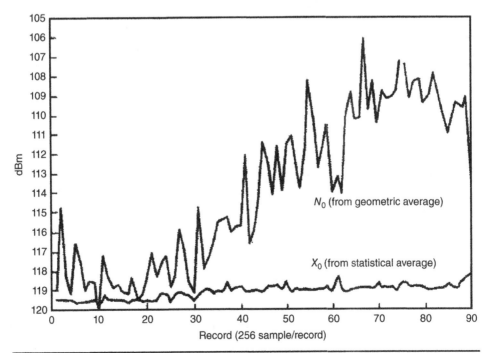

FIGURE 1.8.3.3   An illustration of comparison of N_0 with X_0.

at least one of the paths is dominant and can be treated as a nonfading signal component under an LOS propagation condition. At the reception, the envelope of the signal is affected by adding a dc component to the random multipath.

The Rician distribution's pdf can be given by

$$p(r) = \begin{cases} \dfrac{r}{\sigma^2} e^{-\frac{(r^2+A^2)}{2\sigma^2}} I_0\left(\dfrac{Ar}{\sigma^2}\right) & \text{for } (A \geq 0, r \geq 0) \\ 0 & \text{for } (r < 0) \end{cases} \qquad (1.8.4.1)$$

The parameter A denotes the dc amplitude of the dominant signal, and I_0 (*) is the modified Bessel function of the first kind and zero order. The Rician distribution is often described in terms of a parameter α:

$$\alpha(\text{dB}) = 10 \log \frac{A^2}{2\sigma^2} \quad \text{dB} \qquad (1.8.4.2)$$

α is the ratio of the deterministic signal power and the average power of the multipath signal.

When the dominant signal A becomes weak, as seen from Eq. (1.8.4.1), the composite signal starts to resemble a noise signal. When the dominant signal fades away, the Rician distribution degenerates to a Rayleigh distribution, as shown in Eq. (1.8.1.2).

1.9   Three Basic Propagation Mechanisms[32]

Three basic mechanisms affect radio signal propagation. They are reflection, diffraction, and scattering. When a signal encounters an obstacle, which mechanism occur, depends on the size of obstacle. If the obstacle is very large compared to the wavelength, reflection will occur, such as a radio wave impinging on something, such as the earth's surface or walls. At the place where the obstruction has occurred, the radio wave could bend or propagate around the obstacle; this is diffraction. But if the obstacle is small compared to the wavelength, scattering will occur.

1.9.1   Reflection

1.9.1.1   Introduction and Principles

The specular reflections from a smooth surface conform to Snell's law, which states that the product of the refraction index N_1 and the cosine of the grazing angle ($\cos \theta_1$) is constant along the path of a given ray of energy. This relationship is illustrated in Fig. 1.9.1.1.1.[32] Since N_1 is always the same for both incident and reflected waves, it follows that the incident and reflected wave angles are also the same.

1.9.1.1.1   Reflection Coefficient   The ratio of the incident wave to its associated reflected wave is called the *reflection coefficient*. The generalized Fresnel formulas for horizontally and vertically polarized radio waves can be used to determine the power and boundary relationships of transmitted, incident, and reflected radio waves conforming to Snell's law. These relationships are shown in Fig. 1.9.1.1.2.[29] By disregarding that part of the transmitted wave that is transmitted into the ground, the formulas can

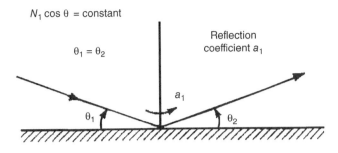

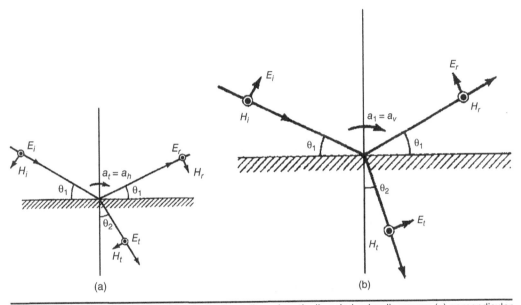

FIGURE **1.9.1.1.1**   Smooth-surface reflections.

FIGURE **1.9.1.1.2**   Wave relationships for horizontally and vertically polarized radio waves: (a) perpendicular (horizontal) polarization (may result from building reflection) and (b) parallel (vertical) polarization (may result from ground reflection).

be applied, in terms of the incident and reflected waves, to calculate the reflection coefficient:

$$a_h = \frac{\sin\theta_1 - (\varepsilon_c - \cos^2\theta_1)^{1/2}}{\sin\theta_1 + (\varepsilon_c - \cos^2\theta_1)^{1/2}} \qquad \text{(horizontal)} \qquad (1.9.1.1.1)$$

$$a_v = \frac{\varepsilon_c \sin\theta_1 - (\varepsilon_c - \cos^2\theta_1)^{1/2}}{\varepsilon_c \sin\theta_1 + (\varepsilon_c - \cos^2\theta_1)^{1/2}} \qquad \text{(vertical)} \qquad (1.9.1.1.2)$$

where

$$\varepsilon_c = \varepsilon_r - j60\sigma\lambda \qquad (1.9.1.1.3)$$

where the permittivity ε_r is the principal component of the dielectric constant ε_c, σ is the conductivity of the dielectric medium in Siemens per meter, and λ is the wavelength.

Some typical values of permittivity and conductivity for various common types of media are shown in the following table:

Medium	Permittivity ε_r	Conductivity σ, s/m
Copper	1	5.8×10^7
Seawater	80	4
Rural ground (Ohio)	14	10^{-2}
Urban ground	3	10^{-4}
Fresh water	80	10^{-3}
Turf with short, dry grass	3	5×10^{-2}
Turf with short, wet grass	6	1×10^{-1}
Bare, dry, sandy loam	2	3×10^{-2}
Bare, sandy loam saturated with water	24	6×10^{-1}

For ground reflections in a mobile radio environment, the permittivity ε_r of the dielectric constant ε_c is always large, and the angle θ_1 is always much less than 1 rad for the incident and reflected waves. When the incident wave is vertically polarized, then Eq. (1.9.1.1.2) can be applied, as illustrated in Fig. 1.9.1.1.2(b), with the result that

$$a_v \approx -1 \tag{1.9.1.1.4}$$

When the incident wave E_i is horizontally polarized, then Eq. (1.9.1.1.1) can be applied, as illustrated in Fig. 1.9.1.1.2(a), with the result that

$$a_h \approx -1 \tag{1.9.1.1.5}$$

In both of these cases, the actual value of the dielectric constant ε_c does not have a significant effect on the resultant values of a_v and a_h. It should be noted that Eqs. (1.9.1.1.4) and (1.9.1.1.5) are valid only for smooth terrain conditions, where $\theta_1 \ll 1$.

In the analysis of reflections from highly conductive surfaces of buildings and structures, the imaginary part of the dielectric constant ε_c is very large, and the following are true:

$$a_v \approx -1 \qquad \theta \text{ very small}$$
$$a_v \approx 1 \qquad \theta \text{ close to } (2n + 1)\, \pi/2 \tag{1.9.1.1.6}$$
$$\text{and} \qquad a_h \approx -1 \qquad \text{either } \theta \text{ very small or } \varepsilon_c \text{ very large} \tag{1.9.1.1.7}$$

Again, as in Eqs. (1.9.1.1.4) and (1.9.1.1.5), the actual value of the dielectric constant ε_c does not affect the resultant values of a_v and a_h in Eqs. (1.9.1.1.6) and (1.9.1.1.7).

When the incident wave E_i is vertically polarized (but horizontally with respect to the building walls) and the reflected wave is from a highly conductive surface, as in Fig. 1.9.1.1.2(a), then the reflection coefficient a_h is always approximately equal to -1, as shown in Eq. (1.9.1.1.7).

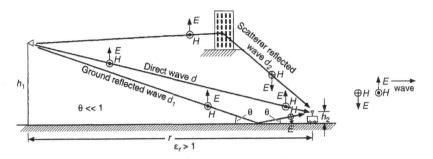

FIGURE 1.9.1.1.3   Reflections typically found in an ideal mobile radio environment.

Figure 1.9.1.1.3[29] illustrates a hypothetical situation that is typical of an ideal mobile radio environment. The surface of the terrain is smooth, and only one highly conductive building is reflecting the transmitted wave.

The received signal at the mobile unit can be calculated from the stationary properties of the energy rays by using Fermat's principle and by applying Snell's law:

$$E_r = E_i(1 - e^{-j\beta\Delta d_1} + a_h e^{-j\beta\Delta d_2})   \tag{1.9.1.1.8}$$

where Δd_1 and Δd_2 are the propagation-path differences, defined as

$$\Delta d_1 = d_1 - d$$

$$\Delta d_2 \approx d_2 - d$$

In reality, the incident angle θ is very small for the ground-reflected wave.

It is also proper to rewrite Eq. (1.9.1.1.8) in terms of time delay:

$$E_r = E_i(1 - e^{-j\omega\Delta\tau_1} + a_h e^{-j\omega\Delta\tau_2})   \tag{1.9.1.1.9}$$

where

$$\Delta\tau_i = \frac{\Delta d_i}{v_c}$$

and where v_c is a constant equal to the speed of light.

1.9.1.2   Brewster Angle

Basically, the Brewster angle is an angle at which no reflection occurs from the surface of the medium. In other words, it is the angle where the reflection coefficient $R = 0$, and the mathematical expression is given as

$$\sin\beta = \left(\frac{\varepsilon_1}{\varepsilon_1 + \varepsilon_2}\right)^{\frac{1}{2}}   \tag{1.9.1.2.1}$$

where β is the Brewster angle and ε_1 and ε_2 are the permittivity of the first and second medium, respectively.

For the case when the first medium is free space and the second medium has a relative permittivity **r**, Eq. (1.9.1.2.1) can be expressed as

$$\sin(\beta) = \frac{\sqrt{\varepsilon_r - 1}}{\sqrt{\varepsilon_r^2 - 1}} = \frac{1}{\sqrt{\varepsilon_r + 1}} \tag{1.9.1.2.2}$$

Note that the Brewster angle occurs only for vertical polarization, v i.e., E-field is in the plane of incidence, as shown in Fig. 1.9.1.1.2(a).

1.9.1.3   Ground Reflection (Two-Ray) Model

A signal transmitting from the base station to the mobile unit because the mobile is very close to the ground produces a direct wave and a reflected wave. Each wave travels in its own path, a direct path and a ground-reflected path, as shown in Fig. 1.9.1.1.3. The received power P_r at the mobile unit can be expressed as[4]

$$P_r = P_0 \left(\frac{1}{4\pi d / \lambda} \right)^2 \left| 1 + a_v e^{-j\Delta\phi} \right|^2 \tag{1.9.1.3.1}$$

where a_v = the reflection coefficient (see Eq. (1.9.1.1.4))
  $\Delta\phi$ = the phase difference between a direct path and a reflected path
  P_0 = the transmitted power P_t + antenna gain G_t at the base + antenna gain G_m at the mobile
  d = the distance
  λ = the wavelength

Equation (1.9.1.3.1) indicates a two-wave model, which is used to understand the path-loss phenomenon in a mobile radio environment. It is not the model for analyzing the multipath fading phenomenon. In a mobile environment, $a_v = -1$ because of the small incident angle of the ground wave caused by a relatively low cell-site antenna height (see Eq. (1.9.1.1.4)).
    Thus,

$$P_r = P_0 \left(\frac{1}{4\pi d / \lambda} \right)^2 \left| 1 - \cos\Delta\phi - j\sin\Delta\phi \right|^2$$

$$= P_0 \frac{2}{(4\pi d / \lambda)^2} (1 - \cos\Delta\phi) = P_0 \frac{4}{(4\pi d / \lambda)^2} \sin^2 \frac{\Delta\phi}{2} \tag{1.9.1.3.2}$$

where $$\Delta\phi = \beta \Delta d \tag{1.9.1.3.3}$$

and Δd is the difference, $\Delta d = d_1 - d_0$, from Fig. 1.9.1.3.1:

$$d_1 = \sqrt{(h_1 + h_2)^2 + d^2} \tag{1.9.1.3.4}$$

and $$d_0 = \sqrt{(h_1 - h_2)^2 + d^2} \tag{1.9.1.3.5}$$

Since Δd is much smaller than either d_1 or d_0,

$$\Delta\phi = \beta \Delta d \approx \frac{2\pi}{\lambda} \frac{2 h_1 h_2}{d} \tag{1.9.1.3.6}$$

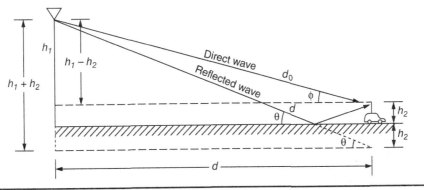

FIGURE 1.9.1.3.1   A simple model—two-ray model.

Then the received power of Eq. (1.9.1.3.2) becomes

$$P_r = P_0 \frac{\lambda^2}{(4\pi)^2 d^2} \sin^2 \frac{4\pi h_1 h_2}{\lambda d} \tag{1.9.1.3.7}$$

If $\Delta\phi$ is less than 0.6 rad, then $\sin(\Delta\phi/2) \approx \Delta\phi/2$, $\cos(\Delta\phi/2) \approx 1$, and Eq. (1.9.1.3.7) simplifies to

$$P_r = P_0 \frac{4}{16\pi^2(d/\lambda)^2} \left(\frac{2\pi h_1 h_2}{\lambda d} \right)^2 = P_0 \left(\frac{h_1 h_2}{d^2} \right)^2 \tag{1.9.1.3.8}$$

From Eq. (1.9.1.3.8), we can deduce two relationships as follows:

$$\Delta P = 40 \log \frac{d_1'}{d_1} \qquad \text{(a 40-dB-per-decade path loss)} \tag{1.9.1.3.9}$$

$$\Delta G = 20 \log \frac{h_1'}{h_1} \qquad \text{(an antenna height gain of 6 dB per octal} \tag{1.9.1.3.10}$$

where ΔP is the power difference in decibels between two different path lengths and ΔG is the gain (or loss) in decibels obtained from two different antenna heights at the cell site. From these measurements, the gain from a mobile antenna height is only 3 dB per octal, which is different from the 6 dB per octal, for h_1' shown in Eq. (1.9.1.3.10). Then

$$\Delta G' = 10 \log (h_2'/h_2) \qquad \text{(an antenna height gain only 3 dB per octal)} \tag{1.9.1.3.11}$$

Finally, the received power at a distance d from the transmitter can be expressed as

$$P_r = P_0 \frac{h_1^2 h_2^2}{d^4} = P_t G_t G_m \frac{h_1^2 h_2^2}{d^4} \tag{1.9.1.3.12}$$

The path loss for the two-ray model (with antenna gains) can be expressed in dB as

$$PL \, (\text{dB}) = 40 \log d - (10 \log G_t + 10 \log G_m + 20 \log h_1 + 10 \log h_2) \tag{1.9.1.3.13}$$

1.9.2   Diffraction

1.9.2.1   Fresnel Zones and Huygens' Principle

From the propagation of electromagnetic wave, if the wave impacts on an obstacle that is "large" compared to the wavelength in dimension, reflection usually occurs, while if it impacts on an obstacle that is more or less the same wavelength in dimension, diffraction usually occurs. Diffraction is a physical phenomenon that an electromagnetic wave can pass around an obstacle. Huygens' principle provides some insight into diffraction.[33] The Principle states that every point of the wave front can act as a source to generate a secondary wavelet when the wave front encounters a obstacle.

Let's assume that the wave front is infinite. Then if an obstacle blocks some part of the wave front, the wave front will produce some wavelets, and these wavelets that occurs around the obstacle are disturbed. The directions of these wavelet propagations may be different from the original wave front. This phenomenon is called *diffraction*.

Let's consider a scenario shown in Fig. 1.9.2.1.1 that a plane wave front moves toward the plane obstacle AA' with several holes, and these holes are relatively not large to the wavelength so that diffraction can occur. From the observation, we notice that the diffracted wavelet beyond AA' points to different directions, and in each direction the amplitude of the wavelet is different, proportional to $(1+\cos\alpha)$, and will be shown in Eq. (1.9.2.1.3) later. When the wavelet has the same direction as the original wave front ($\alpha = 0$), the amplitude reaches the maximum value 2, while when the wavelet has the opposite direction as the original wave front ($\alpha = \pi$), the amplitude reaches the minimum value 0. Therefore, the amplitude of the wavelet varies from 0 to 2, depending on the angle α from 0° to 180°, as shown in the equation $0 \le 1+\cos\alpha \le 2$.

Several facts are observed:

Diffraction still happens regardless of whether the obstacle is conductive or nonconductive.

The fields of the shadow area are not strictly zeroes because the directions of the wavelets could be different from the propagation of the original wave front, and the propagation energy could reach the shadow area via diffraction.

We can calculate the fields of the shadow area based on Huygens' principle. The amplitude of the wavelet is proportional to the angle between the directions of the original wave front and wavelets.

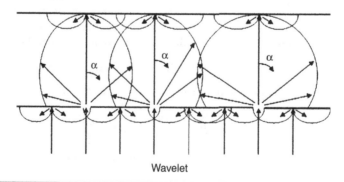

Wavelet

Figure 1.9.2.1.1   Diffraction illustrated.

1.9.2.1.1 Mathematical Presentation Consider the case of a point source located at a point P_0, oscillating at a frequency f. The disturbance may be described by a complex variable U_0 known as the complex amplitude. It produces a spherical wave with wavelength λ and wave number $\beta = 2\pi/\lambda$. The complex amplitude of the primary wave located at a distance r from P_0 is given by

$$U(r) = \frac{U_0 e^{i\beta r}}{r} \qquad (1.9.2.1.1)$$

The magnitude U decreases inversely proportional to the distance r where the wavelet traveled, and the phase changes as β times the distance r.

Using Huygens' theory and the principle of superposition of waves, the complex amplitude at a further point P is found by summing the contributions from each point on the sphere of radius r. The complex amplitude of the wavelet at P is then given by

$$U(P) = \frac{iU(r)}{\lambda} \int_S \frac{e^{i\beta r}}{r} K(\alpha)\, dS \qquad (1.9.2.1.2)$$

where S describes the surface of the sphere and r is the distance between P_0 and P. $K(\alpha)$ is the inclination factor shown in Eq. (1.9.2.1.3), and α is the angle of incidence.

The various assumptions made by Fresnel emerge automatically in Kirchhoff's diffraction formula,[34] to which the Huygens–Fresnel principle can be considered an approximation. Kirchoff gives the following expression for $K(\alpha)$:

$$K(\alpha) = -\frac{i}{2\lambda}(1 + \cos\alpha) \qquad (1.9.2.1.3)$$

K has a maximum value at $\alpha = 0$ as in the Huygens–Fresnel principle; K is equal to 0 at $\alpha = \pi$.

1.9.2.2 Knife Diffraction

1.9.2.2.1 Single-Knife Diffraction From Huygens' Principle, the fields in the shadow area are not absolutely 0. There is still some energy to reach the shadow area via diffraction. Also, Huygens' Principle helps to analyze the diffraction caused by a knife edge in a mathematical way.[33,35-37]

In electromagnetic theory, the field strength of a diffracted radio wave due to a knife edge can be expressed as

$$\frac{E_d}{E_0} = F e^{j\Delta\phi} \qquad (1.9.2.2.1.1)$$

where E_0 is the free-space electromagnetic field with no knife edge present, E_d is the diffracted wave, F is the diffraction coefficient, and $\Delta\phi$ is the phase difference with respect to the path of the direct wave.

We consider that the diffraction loss is a propagation loss, and thus it could be expressed by the original power divided by the diffracted power in decibels, which is given by[30,32-34]

$$L(v) = 20 \log\left(\left|\frac{E_d}{E_0}\right|\right) = 20 \log|F| \qquad (1.9.2.2.1.2)$$

where E_d is the diffracted filed and E_i is the incident field, while F is a function of the phase difference $\Delta\phi$, which is given by

$$F = \frac{S+0.5}{\sqrt{2}\,\sin(\Delta\phi + \pi/4)}$$ (1.9.2.2.1.3)

and the phase difference $\Delta\phi$ is

$$\Delta\phi = \tan^{-1}\left(\frac{S+0.5}{C+0.5}\right) - \frac{\pi}{4}$$ (1.9.2.2.1.4)

where C and S are the Fresnel integrals, expressed as

$$C = \int_0^v \cos\left(\frac{\pi}{2}x^2\right)dx$$

$$S = \int_0^v \sin\left(\frac{\pi}{2}x^2\right)dx$$

As we know, the diffraction loss is due to the knife edge, so the diffraction loss is relative to the diffraction parameter v, and the expression of v with terms of geometrical parameters is given by

$$v = -h'\sqrt{\frac{2(d_1' + d_2')}{\lambda d_1' d_2'}}$$ (1.9.2.2.1.5)

where λ is the wavelength and the other parameters could be found in Fig. 1.9.2.2.1.1.

In the real world, usually $d_1, d_2 \gg h$, the expression of diffraction parameter v can be simplified by

$$v \approx -h\sqrt{\frac{2(d_1 + d_2)}{\lambda d_1 d_2}}$$ (1.9.2.2.1.6)

where λ is the wavelength and the other parameters can be found in Fig. 1.9.2.2.1.1.

Notice how the integration limits in Eq. (1.9.2.2.1.4) indicate the nature of the summation of secondary sources from the top of the knife edge, with parameter v, up to infinity. The eventual result $L(v)$ is illustrated in Fig. 1.9.2.2.1.2. It can be numerically evaluated using standard routines for calculating Fresnel integrals or approximated for $v > 1$ with accuracy better than 1 dB.

$$L(v) \approx 20\log\frac{1}{\pi v\sqrt{2}} = 20\log\frac{0.225}{v} \quad v < -2.4$$ (1.9.2.2.1.7)

The approximate solutions for the different values of v can be found in Table 3.1.2.3.1 in Chap. 3.

Figure 1.9.2.2.1.3 shows an example of the Fresnel zones. By the definition of an ellipsoid, the radius of the nth zone r_k must match the following condition:

$$a+b = d_1 + d_2 + \frac{k\lambda}{2}$$ (1.9.2.2.1.8)

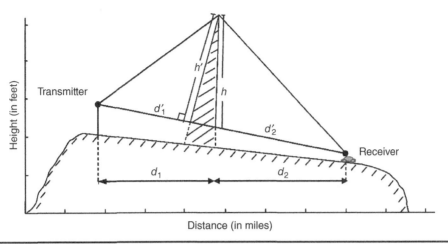

FIGURE 1.9.2.2.1.1 Geometrical distance of single knife.

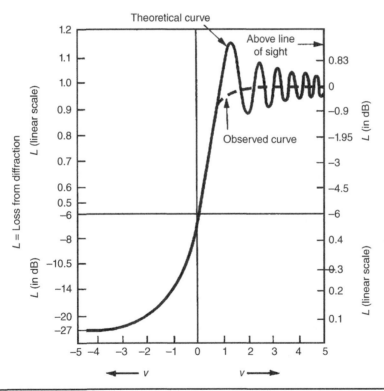

FIGURE 1.9.2.2.1.2 Diffraction loss.

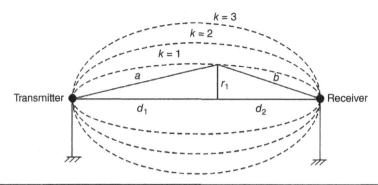

FIGURE **1.9.2.2.1.3**   Fresnel zone illustrated.

In the real world, $r_k \ll d_1, d_2$, we can thus make the expression simple by a good approximation, which is given by

$$r_k \approx h \sqrt{\frac{k\lambda d_1 d_2}{d_1 + d_2}} \qquad (1.9.2.2.1.9)$$

Substituting Eq. (1.9.2.2.1.9) into Eq. (1.9.2.2.1.5) yields

$$v \approx \frac{-h^2}{r_k} \sqrt{2k} \qquad (1.9.2.2.1.10)$$

This is a relationship between the diffraction parameter v and the number of Fresnel zones.

The main propagation energy is diffracted in the first Fresnel zone, and any obstacle outside the first Fresnel zone has little effect on the propagation. That is why the diffraction parameter could be expressed in terms of the first Fresnel zone.

1.9.2.2.2   Multiple-Knife Diffraction   We talked about the single-knife-edge diffraction above. But in the real world, it is more likely that the propagation will encounter several obstacles, especially in a hill terrain scenario. This is called *multiple-knife diffraction*. Several models aim to handle this issue.[38–40] Bullington[41] suggested that the series of obstacles be replaced by a single equivalent obstacle so that the path loss can be obtained using single-knife-edge diffraction models. This method, illustrated in Fig. 1.9.2.2.1, oversimplifies the calculations and often provides optimistic estimates of the received signal strength. In a more rigorous treatment, a wave-theory solution for the field behind two knife edges in series was derived.[38] This solution is useful and can be applied easily to predicting diffraction losses due to *two* knife edges. However, extending this to more than two knife edges becomes a mathematical problem to be solved. Many mathematically less complicated models have been developed to estimate the diffraction losses due to multiple obstructions.[39, 40]

Bullington's Equivalent Knife Edge
This algorithm replaces all the obstacles of terrain with a single knife edge. As shown in Fig. 1.9.2.2.1, the real terrain is replaced by a single "virtual" knife edge at the point of

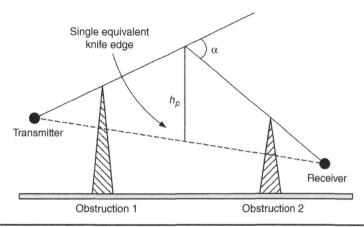

FIGURE 1.9.2.2.2.1   Single-knife-edge model.

intersection of the horizon ray from each of the terminals. Then we can use the same approach to obtain the path loss as the single knife diffraction.

This solution is easy to apply but oversimplifies the calculations and often provides optimistic estimates of the received signal strength. Therefore, it provides results that may underestimate the path loss due to the fact that an important obstacle could be ignored.

Epstein–Peterson Method

The primary limitation of using the Bullington method is that the prominent obstacles can be wiped out. To ease this limitation, the Epstein–Peterson method[42] can be used. This method computes loss caused by each obstacle by using the neighboring peaks as two ends to form a basic single knife, shown in Fig. 1.9.2.2.1.1, in calculating diffraction loss. Then, by summing up all losses from the individual obstacles (peaks), the overall path loss is obtained. As shown in Fig. 1.9.2.2.2(a), in the propagation from the transmitter to the receiver, there are two knife edges. From the figure, we see two values of diffraction parameters from two knife-edge heights, h_1 and h_2. Then the individual diffraction loss is obtained. In Fig. 1.9.2.2.2(b), there are three obstacles, 01, 02, and 03, between T (transmitter) and R (receiver).

We calculate three losses caused by the three obstacles, respectively:

$$L_{01} = f(d_1, d_2, h_1)$$

$$L_{02} = f(d_2, d_3, h_2)$$

$$L_{03} = f(d_3, d_4, h_3)$$

where L_{01}, L_{02}, and L_{03} are the individual losses caused by three obstacles, d_i is the distance, and h_i is the effective height.

Potential errors and correction method when two obstacles are located closed to each other have been discussed.[43,44] So when diffraction parameters v of both edges 01 and 02 are smaller than unity, the correction is given by

$$L = 20 \log_{10}(\mathrm{cosec}\ \alpha) \tag{1.9.2.2.2.1}$$

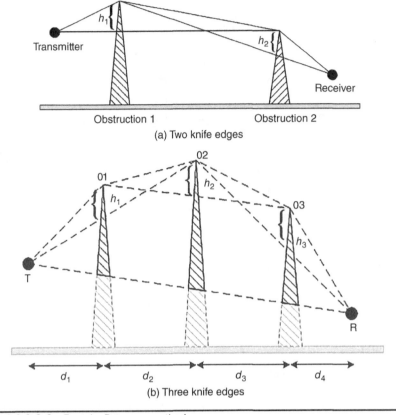

FIGURE **1.9.2.2.2.2**   Epstein–Peterson method.

where α is a spacing parameter expressed as

$$\text{cosec } \alpha = \left(\frac{(d_1 + d_2)(d_2 + d_3)}{d_2(d_1 + d_2 + d_3)} \right)^{\frac{1}{2}}$$ (1.9.2.2.2.2)

Japanese Method
The Japanese method[45] follows the same procedure as the Epstein–Peterson method, calculating losses caused by each knife edge first and then summing all to obtain the overall path loss. But it uses different parameters of distance and effective height. Figure 1.9.2.2.2.3 shows the effective transmitter heights T, T', and T'', each of which is found by projecting a ray from the top of the knife edge, passing on the top of the left neighboring knife edge, and ending at the different height at the location of the transmitter. The path loss for each knife edge is given by

$$L_{01} = f(d_1, d_2, h_1)$$

$$L_{02} = f(d_1 + d_2, d_3, h_2)$$

$$L_{03} = f(d_1 + d_2 + d_3, d_4, h_3)$$

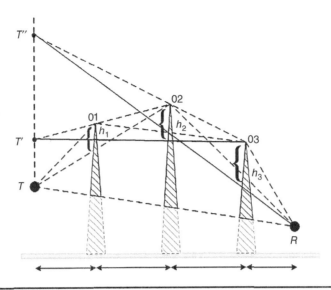

FIGURE 1.9.2.2.2.3 Japanese method.

Deygout Method

The Deygout method[39] is depicted in Fig. 1.9.2.2.2.4 for a three-knife-edge path. It calculates the v parameter for each edge alone, as if all other edges were absent. The edge having the largest value of v is named the main edge, and its loss is obtained from the standard calculation. From Fig. 1.9.2.2.2.4, the edge h_2 is the main edge. Then the additional diffraction losses for edges h_1 and h_3 are found with respect to a line joining the main edge to the terminals T and R, respectively. The total diffraction loss is the main edge loss plus two additional losses, one from edge h_1 and the other from edge h_2. The same method can be applied to the case of more than three knife-edge paths.

Estimates of path loss using this method[39] generally show fairly good agreement. However, when there are multiple edges and they are close together,[46] the results become pessimistic, and the path losses are overestimated.

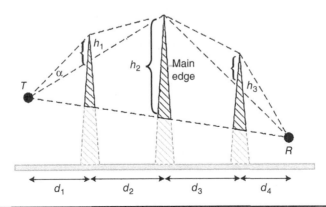

FIGURE 1.9.2.2.2.4 The Deygout diffraction construction.

1.9.3   Scattering

Scattering occurs when photons or electromagnetic waves pass through a medium, such as water or glass, and the direction of propagation changes. Scattering can be of two kinds: elastic scattering and inelastic scattering. If the scattered photon or electromagnetic wave has the same frequency as the incident wave, it is elastic scattering since the photon energy of the scattered photons is not changed. If the photon energy of the scattered photons changes because its frequency changed, it is inelastic scattering.

1.9.3.1   Elastic Scattering

There are two typical kinds of elastic scattering: Rayleigh scattering and Mie scattering. Because Mie scattering deals with spherical particles and less with wavelength and because Rayleigh scattering deals mainly with wavelength. We talk only about Rayleigh scattering here.

When the wavelength of an incident wave is equal to or larger than 10 times the radius of scattered photons, the shape of the scattered photon does not matter anymore and can be assumed as a spheroid. The principles of Rayleigh scattering state that if the incident and scattered waves have the same frequency, then the intensity of the scattered wave is related to the angle of the incident wave and inversely proportional the to biquadrate of the wavelength of the incident wave. The expression of the equation is given by

$$I = I_0 \frac{8\pi^4 N \alpha^2}{\lambda^4 R^2}(1 + \cos^2\theta) \tag{1.9.3.1}$$

where α is molecular polarizability, I is the intensity of the scattered wave, while I_0 is the intensity of incident wave, λ is the wavelength, N is the number of scatters, θ is the incident angle, and R is the distance from scatters.

From Eq. (1.9.3.1), we see that the intensity of the scattered wave is proportional to the incident angle θ and that the number of scatters is inversely proportional to the wavelength λ squared and the distance R.

1.10   Applications of the Prediction Models

1.10.1   Classification of Prediction Models

There are three different kinds of prediction models: the area-to-area model, the point-to-area model, and the point-to-point model. The descriptions are shown below.

1.10.1.1   Area-to-Area Model

This kind of prediction model can predict the received signal strength in only a general area, and the locations of both the base station and the mobile unit are not specified.

1.10.1.2   Point-to-Area Model

This kind of prediction model can predict the received signal strength in a general area with the location of the base station specified but not the location of the mobile unit.

1.10.1.3   Point-to-Point Model

This kind of prediction model can predict the received signal strength transmitted from a given base station and received by a mobile at a given point. This model is considered to have more practical use. All point-to-point models can be used as area-to-area models but not vice versa.

1.10.2   Prediction Models for Propagating in Areas of Different Sizes

1.10.2.1   Macrocell Model

The macrocell prediction model is used mainly to predict the coverage of a large cell with maximum power from a base station under government regulation. The interference from the co-channel reused cells can be calculated. The predicted carrier-to-interference ratio C/I is used to deploy a system. Macrocell prediction models are the prediction models that were deployed in early cellular systems. When planning a cellular system in a large area, it is very costly to select proper locations for base stations based on the measured data in that area. If a wrong location is selected for a base station, not only does the system result in a poor coverage but strong interference among base stations prevails as well. Therefore, this macrocell prediction model is the most valuable tool among all the other small area prediction models.

1.10.2.2   Microcell Model

The microcell prediction model is used for managing system capacity. The size of the cells is smaller, and the transmitted power is lower in a confined area. More interference problems occur, and more handoffs need to be considered. The prediction tool can be used to optimize the overlaid area between any two adjacent cells for maximizing system capacity. It is a useful tool for designing a microcell system economically on the basis of capacity.

1.10.2.3   In-Building Model (For Picocell and Femtocell)

Because more people use mobile units (cell phones) while walking inside buildings, a different prediction model has to be built to deal with this different environment. The in-building model is especially important with wireless data usages growing exponentially enabled by WLAN/WiFi. To optimize the system with best coverage and minimize interference in very packed rooms/buildings such as picocell becomes more and more critical. Femtocell is a low power cellular base station, which is targeted for in home use and small business. The in-building propagation model can be applied to both picocell and femtocell deployment.

1.10.3   Aspects for Predicting the Signal Strengths in a General Environment

1.10.3.1   Mobile Path, Radio Path, and Radial Path

Local means are calculated from the measured data while the mobile units are traveling along a road or along a certain path. This path is called the *mobile path*, as shown in Fig. 1.10.3.1.1 on the x-axis.

However, each local mean is based on a radio path x' between the base station and the mobile unit at a corresponding spot on the mobile path. Identifying radio paths is not important if we are modeling an area-to-area model. However, it is very important for modeling a point-to-point model because from the radio path, the predicted signal strength at the corresponding location of the mobile unit can be calculated.

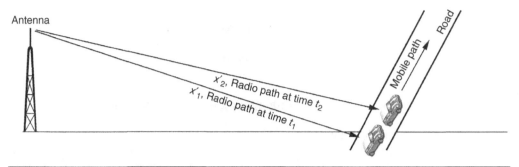

FIGURE 1.10.3.1.1 Mobile radio path.

For systematically calculating the predicted propagation loss from a prediction model on each radio path in a different angle of arrival, a group of radial paths can be artificially drawn on a contour map, the first, being a group of radial lines formed with its original at a base station. The radial lines from the base station are incremented in very small angles, say 0.25° or 0.5° on the map. Also, along each radial line on the map, there are sequential spots separated by a given incremental distance, say, 300 ft or 100 m. The predicted signal strength at each spot is obtained from the prediction model. Then, at each radial path, there are the predicted values of signal strength attached to the corresponding spots along the radial line. One does not need to know the mobile paths or the radio paths when calculating predicted signal strengths at sequential spots along the radial paths from a terrain contour map. On the radial path map, the signal strengths of all the locations of a mobile unit can be found. Therefore, the predicted signal strengths from a radial-path map are more useful in designing a cellular system, but the calculation time is lengthened.

1.10.3.2 Antenna Height Gain at Base Station and at Mobile Unit
From the two-wave propagation model in Sec. 1.9.1.3, the antenna height gains of both base station and mobile unit are 6 dB per octal. Nevertheless, from the actual measurement, the antenna height gain is 6 dB per octal at the base station but only 3 dB per octal at the mobile unit.

1.10.3.3 Reflection Coefficient of the Ground Wave
Concerning ground reflections in a mobile radio environment, the incident angle (or reflected angle) is very small because the base station antenna heights and the mobile unit antenna heights (in feet or meters) are relatively short compared with the distance between the two antennas (in miles or kilometers). The energy of the incident wave will completely reflect back from the ground regardless of the values of the permittivity of dielectric constants of the ground. However, the phase of the signal will be shifted 180° after reflecting from the ground. The principle of reflection is shown in Sec. 1.9.1.1.

1.10.3.4 Radius of Effective Local Scatterers[47,48]
The local scatterers are defined such that the sizes of the scatterers are greater than the wavelength of the operational frequency, and the heights of the scatterers are higher than the mobile antenna height. Naturally, the houses and building surrounding the mobile unit meet this definition and are local scatterers. The local scatterers surrounding

the mobile unit cause the short-term fading while the mobile is moving. How large is the area in which the effective scatterers are surrounding the mobile unit? The radius r of a group of active local scatterers cannot be measured. However, we can obtain it indirectly by comparing the measured data to a theoretical model described by Lee:[47]

$$r = (R \times BW)/2 \tag{1.10.3.4.1}$$

where R is the propagation distance from the transmitter to the receiver and BW is the beamwidth of the incoming signal.

The experimental data from 850 MHz is trying to match the theoretical curve of cross correlation for two base station antennas spacing between 0 and 50 λ and $R = 3$ mi with various BW from 0.35° to 1.2°. The close match is at an angular sector of 0.5°. The radius of effective local scatterers can be roughly estimated from Eq. (1.10.3.4.1) and turns out to be around 70 wavelengths at 850 MHz. Then we may conclude that the radius r of effective scatterers is around 50 to 100 λ variously from different operational frequencies.

1.10.3.5   Near-In Propagation Distance[49]

When the base station antenna is mounted close to the ground in an unobstructed condition, the signal propagates in free space and can maintain its exponent loss of two to only a certain distance before experiencing a higher exponent loss due to the effect of the ground. This range is called *near-in propagation distance*, d_{sf}. The derivation of d_{sf} is from Eq. (1.9.1.3.6). Let the location of the phase difference $\Delta\phi$ between the direct wave and ground-reflected wave be π:

$$\Delta\phi \approx \frac{2\pi}{\lambda}\frac{2h_1 h_2}{d} = \pi \tag{1.10.3.5.1}$$

At this location, the wave strength can still remain strong, as shown in Eq. (1.9.1.3.2). From Eq. (1.10.3.5.1), we come up the near-in propagation distance d_{sf}:

$$d_{sf} = \frac{4h_1 h_2}{\lambda} \tag{1.10.3.5.2}$$

Equation (1.10.3.5.2) will be a criterion used in the prediction model of the microcell (Chap. 4). Another defined distance called close-in propagation distance will be a criterion used in the in-building cell (Chap. 5).

1.10.4   Predicting the Interference Signals

The prediction tool has to predict not only the local means of the mobile signal but also the interference signals that would affect the mobile signal. Once the mobile signal and the interference signals are predicted, the carrier-to-interference ratio, C/I, can be calculated.

1.10.4.1   Co-Channel Interference

Based on the frequency reuse scheme deployed in the cellular system, the frequency reuse factor K is usually 3, 4, 7, or 12. For an example, $K = 7$ means a cluster of seven cells that will share the total allocated bandwidth of the cellular signals. The distance of two co-channel cells can be found from Eq. (1.7.1.1). In any value of K, there are different tiers of interfering cells. At the first tier, the number of interfering cells is $Z_1 = 6$; at the second tier, $Z_2 = 12$; and at the third tier, $Z_3 = 18$. The total number of interfering cells, Z, is

$$Z = \Sigma Z_n \quad \text{for } n = 1 \text{ to } N \tag{1.10.4.1.1}$$

Usually, we consider only the interfering cells at the first tier.

1.10.4.2 Adjacent-Channel Interference

The adjacent channels within the cluster of K cells also can create interference due to channel filter characteristics. We can figure out the adjacent signals that can leak into the mobile signal based on the slope of the channel filter. The leaked signals are the additional interference to the mobile signal. Therefore, the sectorized cells can help reduce this kind of interference.

1.10.4.3 Simulcast Interference

The data transmitted over two or more transmitters operating on the same RF carrier frequency is called *simulcast*. The advantage of using simulcast in an area-wide coverage is to simplify dispatching or to simplify area-wide, mobile-to-mobile communications. In a mobile radio telephone system, normally no simulcast interference exists. Recently, simulcast has been used in some systems to improve coverage in mobile systems.

1.11 Summary

The basic theory of mobile communications is discussed in this chapter. In the following chapters, we will first discuss the popular propagation models and then move to the Lee macrocell model, followed by the Lee microcell model and then the Lee in-building model. Chapter 6 ties all the Lee models together to present an integrated Lee model that can be applied to all mobile environments.

References

1. Bell Laboratories. "High Capacity Mobile Telephone System Technical Report," Submitted to the Federal Communications Commission, December 1971.
2. Lee, W. C. Y. *Mobile Communications Engineering: Theory and Applications*. 2nd ed. New York: McGraw-Hill, 1998.
3. Ibid, Chap. 9.
4. Lee, W. C. Y. *Wireless and Cellular Telecommunications*. 3rd ed. New York: McGraw-Hill, 2006.
5. Lee, W. C. Y. *Mobile Communications Design Fundamentals*. 2nd ed. New York: John Willey & Sons, 1993.
6. Foschini, G. J. and Gans, M. J. "On Limits of Wireless Communications in a Fading Environment When Using Multiple Antennas." *Wireless Personal Communications* 6 (1998): 311–35.
7. Lee, W. C. Y. *Lee's Essentials of Wireless Communication*. New York: McGraw-Hill, 2001.
8. Young, W. R. "Introduction, Background and Objectives—AMPS." *Bell System Technical Journal* 58 (January 1979): 1–14.
9. IEEE VTS Committee on Radio Propagation. "Lee's Mobile." *IEEE Transactions on Vehicular Technology,* (February 1988): 68–70.
10. Evans, Greg, Joslin, Bob, Vinson, Lin, and Foose, Bill. "The Optimzation and Application of the W. C. Y. Lee Propagation Model in the 1900 MHz Frequency Band," *IEEE Transactions on Vehicular Technology* (March 1997): 87–91.
11. Alotarbi, Faihan D., and Ali, Adel A. "Tuning of Lee Path Loss Model Based on Recent RF Measurement in 400 MHz Conducted in Riyadh City, Saudi Arabia." *Arabian Journal for Science and Engineering* 33-1B (April 2008): 145–52.

12. Kostanic, Ivica, Rudic, Nikola, and Austin, Mark. "Measurement Sampling Criteria for Optimization of the Lee's Macroscopic Propagation Model." 48th IEEE International Conference on Vehicular Technology, VTC 98. Vol. 18–21, May 1998, 620–24.

13. Kostanic, I., Guerra, I., Faour, N., Zec, J., and Susani, M. "Optimization and Application of W. C. Y. Lee Microcell Model in 850 MHz Frequency Band." Proceedings of Wireless Networking Symposium, Austin, TX, October 22–24, 2003.

14. de la Vega, David, Lopez, Susanna, Gil, Unai, Matias, Jose M., Guerra, David, Angueira, Pablo, and Ordiales, Juan L. "Evaluation of the Lee Method for the Analysis of Long-Term and Short-Term Variations in the Digital Broadcasting Services in the MW Band." 2008 IEEE International Symposium on Broadband Multimedia Systems and Broadcasting, March 31, 2008–April 2, 2008, pp. 1–8.

15. Matias, Jose Maria, Pena, Ivan, de la Vega, David, Fernandez, Igor, Lopez, Susana, and Angueira, Pablo. "Location Correction Factor for Coverage Planning Tools for DRM (Digital Radio Mondiale) in the 26 MHz Band." IEEE International Symposium on Broad Band Multimedia Systems and Broadcasting, May 13–15, 2009, pp. 1–5.

16. Lee, W. C. Y. "Elements of Cellalar Mobile Radio Systems." *IEEE Transactions on Vehicular Technology* 35 (May 1986): 48–56.

17. Lee, W. C. Y. *Mobile Communications Design Fundamentals*, Ibid, Chap. 1.

18. Lee, W. C. Y. and Yeh, Y. S. "On the Estimation of the Second-Order-Statistics of Long-Normal Fading in Mobile Radio Environment." *IEEE Transactions on Communications* 22 (June 1974): 869–73.

19. Lee, W. C. Y. "Estimate of Local Average Power of a Mobile Radio Signal." *IEEE Transactions on Vehicular Technology* 34 (February 1985): 22–27.

20. Chuang, J. "The Effects of Time Delay Spread on Portable Communications Channels with Digital Modulation." *IEEE Journal on Selected Areas in Communications* 5, no. 5 (June 1987): 879–89.

21. Fung, V., Rappaport, T. S., and Thomas, B. "Bit Error Simulation for $\pi/4$ DQPSK Mobile Radio Communication Using Two-Way and Measurement-Based Impulse Resonse Models." *IEEE Transactions on Vehicular Technology* 21 (February 1972): 27–38.

22. Rappaport, Theodore S. *Wireless Communications*. Englewood Cliffs NJ: Prentice Hall, 1996.

23. Lee, W. C .Y., *Mobile* Communications *Engineering*, Ibid, p. 256.

24. Lee, W. C. Y., *Wireless and Cellular Telecommunications*, Ibid, Chap. 2.

25. Lee, W. C. Y., *Mobile Communications Design Fundamentals*, Ibid, Chapter 10.

26. Lee, W. C. Y., "Smaller Cells for Greater Performance." *IEEE Communications Magazine* (1991): 19–23.

27. The six patents of Lee's microcell system:
    1. U.S. Patent Office No. 4,932,049, 6/5/1990, "Cellular Telephone System"
    2. U.S. Patent Office No. 5,067,147, 11/19/1991, "Microcell System for Cellular Telephone System"
    3. U.S. Patent Office No. 5,193,109, 3/9/1993, "Zoned Microcell with Sector Scanning for Cellular Telephone System"
    4. U.S. Patent Office No. 5,243,598, 9/7/1993, "Microcell System in Digital Cellular"
    5. U.S. Patent Office No. 5,504,936, 4/2/1996, "Microcells for Digital Cellular Telephone Systems"
    6. U.S. Patent Office No. 5,678,186, 10/14/1997, "Digital Microcells for Cellular Networks"

28. Ramo, S., Whinnery. J. R., and Van Duzer, T. *Fields and Waves in Communication Electronics*. New York: John Wiley & Sons, 1965.

29. Stutzman, W. L. *Polarization in Electromagnetic Systems*. Boston: Artech House, 1993.

30. Lee, W. C. Y. *Wireless and Cellular Telecommunications*, Ibid, Sec. 15.3.

31. Lee, W. C. Y. "Estimating Unbiased Average Power of Digital Signal in Presence of High-Level Impulses." *IEEE Transactions on Instrumentation and Measurement* 32 (September 1983): 403–9.

32. Lee, W. C .Y. *Mobile Communications Engineering*, Ibid, Sec. 3.2

33. Saunders, S. *Antennas and Propagation for Wireless Communication Systems*. New York: John Wiley & Sons, 2000.

34. Born, Max, and Wolf, Emil. *Principles of Optics: Electromagnetic Theory of Propagation, Interference and Diffraction of Light*. Cambridge: Cambridge University Press, 1999.

35. Anderson, L. J. and Trolese, L. G. "Simplified Method for Computing Knife-Edge Diffraction with Shadow Region." *IEEE Transactions on Antennas and Propagation* (July 1958): 281–86.

36. Lee, W. C. Y., *Mobile* Communications *Engineering*, Ibid, Sec. 4.2.

37. Parsons, J. D. *The Mobile Radio Propagation Channel*. 2nd ed. New York: John Wiley & Sons, 2000.

38. Millington, G., Hewitt, R., and Immirzi, F. S. "Double Knife-edge Diffraction in Field Strength Predictions." *Proceedings of the IEEE* 109c (1962): 419–29.

39. Deygout, J. "Multiple Knife-Edge Diffraction of Microwaves." *IEEE Transactions on Antennas and Propagation* 14, no. 4 (1996): 480–89.

40. Wilkerson, R. E. "Approximate to the Double Knife-Edge Attenuation Coefficient." *Radio Science* 1, no. 12 (1996): 1439–43.

41. Bullington, K." Radio Propagation at Frequencies above 30 megacycles." *Proceedings of the IEEE* 35 (1947): 1122–36.

42. Epstein, J. and Peterson, D. W. "An Experimental Study of Wave Propagation at 840 MC." *Proceedings of the IRE* 41, no. 5 (1953): 595–661.

43. Stutzman, W. L. *Polarization in Electromagnetic Systems*. Boston: Artech House, 1993.

44. Steele, R. and Hanzo, L. *Mobile Radio Communications*. New York: Wiley-IEEE Press, 1999.

45. Atlas of Radio Wave Propagation Curves for Frequencies Between 30 and 10,000 Mc/s, Radio Research Lab, Ministry of Postal Services, Tokyo, Japan, 1957 pp. 172–9.

46. Giovaneli, C. L. "An Analysis of Simplified Solutions for Multiple Knife-Edge Diffraction." *IEEE Transactions* AP32, no. 3 (March 1984): 297–301.

47. Lee, W. C. Y. *Mobile Communications Design Fundamentals*, Ibid. Sec. 4.7.

48. Lee, W. C. Y. "Effects on Correlation between Two Mobile Radio Base-Station Antenna." *IEEE Transactions on Communications*, 21 (November 1973): 1214–24.

49. Lee, W. C. Y. *Wireless and Cellular Telecommunications*, Ibid. Sec. 8.5.2.

Additional References

Bertoni, Henry L. *Radio Propagation for Modern Wireless Systems*. Englewood Cliffs, NJ: Prentice Hall, 2000.

Fujimoto, J. K., and James, J. R., eds. *Mobile Antenna System Handbook.* Boston: Artech House, 1994.

Jakes, W. C., Jr. *Microwave Mobile Communications.* New York: John Wiley & Sons, 1974.

Levis, Curt A., Johnson, Joel T., Teixeira, Fernando L., "Radiowave Propagation." New Jersey: Wiley & Sons, 2010.

Parsons, J. D. *The Mobile Radio Propagation Channel.* 2nd ed. New York: John Wiley & Sons, 2000.

Rappaport, Theodore S. *Wireless Communications: Principles and Practice*, 2nd ed. Englewood Cliffs, NJ: Prentice Hall, 1996.

Saunders, S. *Antennas and Propagation for Wireless Communication Systems.* New York: John Wiley & Sons, 2000.

Seybold, John S. *Introduction to RF Propagation.* New York: John Wiley & Sons, 2005.

Steele, R., and Hanzo, L. *Mobile Radio Communications.* New York: Wiley-IEEE Press, 1999.

Stuber, Gordon L. *Principles of Mobile Communication.* Norwell, MA: Kluwer Academic Publishers, 1996.

CHAPTER 2

Macrocell Prediction Models—Part 1: Area-to-Area Models

2.1 Free Space Loss

Free space path loss occurs when a transmitted signal propagates over a line-of-sight path through a free space with no obstacles nearby to cause reflection or diffraction. The signal strength of the signal decreases with increasing distance. It excludes factors such as the gain of the antennas used at the transmitter and receiver and any loss associated with imperfections in hardware.

Free space loss can be calculated, as the signal strength decreases at a rate inversely proportional to the distance traveled and proportional to the wavelength of the signal, as shown from the Friis transmission formula:

$$\frac{P_r}{P_t} = g_b g_m \left(\frac{\lambda}{4\pi d}\right)^2 \tag{2.1.1}$$

where g_b and g_m are gains of the terminal antennas, d is the distance between the antennas, and λ is the wavelength. This can be visualized as arising from the spherical spreading of power over the surface of a sphere of radius d centered at the transmitting antenna. Since power is spread over the surface area of the sphere, which increases as d^2, the available power at a receiver antenna of fixed aperture decreases in proportion to d^2.

Equation (2.1.1) can be rearranged to express it as a propagation loss in free space:

$$L_F = \frac{P_t G_b G_m}{P_r} = \left(\frac{4\pi d}{\lambda}\right)^2 = \left(\frac{4\pi d f}{c}\right)^2 \tag{2.1.2}$$

This expression defines L_F, the free space loss in which the square law depends on both frequency and distance.

Expressing the free space loss in decibels, with frequency in megahertz and distance d in kilometers or in miles, we obtain

$$L_F(\text{in dB}) = 32.4 + 20\log d + 20\log f_{\text{MHz}} \quad d \text{ in kilometers} \tag{2.1.3}$$

$$= 36.57 + 20\log d + 20\log f_{\text{MHz}} \quad d \text{ in miles}$$

Thus, the free space loss increases by 6 dB for each doubling in either frequency or distance (or 20 dB per decade of either).

The free space value of the loss is used as a basic reference, and the loss experienced in excess of this free space loss (in decibels) is referred to as the excess loss, L_{ex}. Thus,

$$L = L_F + L_{ex} \tag{2.1.4}$$

The free space loss is usually considered a practical minimum to the path loss for a given distance.

2.2 Plane Earth Model

The ground plays a key role in the propagation of a radio wave. It acts as a partial reflector and a partial absorber, and it affects the radio energy along the radio distance. The principal effect of plane earth propagation is shown in Eq. (2.2.1).

$$E = E_0[\underbrace{1}_{a)} + \underbrace{Re^{j\Delta}}_{b)} + \underbrace{(1-R)Ae^{j\Delta}}_{c)} + \underbrace{\ldots}_{d)}] \tag{2.2.1}$$

where $a)$ is the direct wave, $b)$ is the reflected wave, $c)$ is the surface wave, and $d)$ is the induction field and secondary effects of the ground. R is the reflection coefficient of the ground and is approximately equal to -1 when the incident angle and reflected angle θ is less than 1 radian. The symbol Δ is the phase difference between the transmitted wave and the reflected wave. A is a constant loss due to the reception of surface wave.

The sum of three principal waves composes the ground wave. The direct wave and the reflected wave are easy to understands, as they correspond to our common experience with visible light. The surface wave is the principal component of the total ground wave at frequencies less than 3 MHz but usually can be neglected at frequencies above 300 MHz.

The ratio of the received power to the radiated power for transmission over plane earth was derived in Chap. 1, Eq. (1.9.3.1.8), as follows:

$$\frac{P_2}{P_1} = \left[\left(\frac{\lambda}{4\pi d}\right)^2 G_b G_m\right]\left(\frac{4\pi h_b h_m}{\lambda d}\right)^2 = \left(\frac{h_b h_m}{d^2}\right)^2 G_b G_m \tag{2.2.2}$$

where h_b and h_m are the antenna heights of base and mobile and G_b and G_m are the antenna gains of base and mobile.

The plane earth model has been derived theoretically based on the mobile radio environment. Both the base station antenna and the mobile antenna are much smaller compared to the distance between the two antennas. The model is oversimplified, as it does not include terrain profile, vegetation, and buildings.

The plane earth model is a fundamental model used to explain the path loss exponent in the mobile radio environment. The plane earth loss increases far more rapidly than the free space loss and is independent of carrier frequency. The loss increases by 12 dB per octal, or by 40 dB per decade. The plane earth model is not an accurate tool to be used to predict the path loss in real-world propagation.

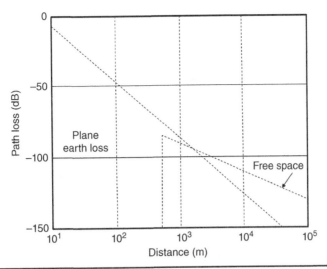

FIGURE **2.2.1**   Free space versus plan earth losses at 900 MHz.

2.3   Young Model

Young[1] reported an important series of measurements in New York at frequencies between 150 and 3700 MHz. Young's experiments were taken in New York City in 1952, and his field trials in which the signal from a base station was received at a vehicle moving in the city streets cover frequencies of 150 (250 W), 450 (125 W), 900 (200 W), and 3700 MHz, where the transmitter height was about 450 ft above the ground.[2,3] His experimental results confirmed that the path loss was much greater than predicted by the plane earth propagation equation. It was clear that the path loss increased with frequency, and there was clear evidence of a strong correlation between path losses at 150, 450, and 900 MHz. The measured data at 3700 MHz were not large enough to justify a similar conclusion, as the data were limited by transmitter power and receiver sensitivity.

The curve presented in Fig. 2.3.1 displays an inverse fourth-power law behavior. His finding was the first one disclosed in mobile radio propagation and has been widely quoted. The model for Young's data is

$$L_{50} = G_b G_m \left(\frac{h_b h_m}{d^2} \right)^2 \beta \tag{2.3.1}$$

where β is called the *clutter factor*. In this case, the clutter factor β represents losses due to buildings rather than terrain features. Figure 2.3.1 shows that at 150 MHz in New York, β is approximately 25 dB. From his experimental results, Young also plotted the path loss not exceeded at 1, 10, 50, 90, and 99 percent of locations within his test area, and these are also shown in Fig. 2.3.1. The data in Fig. 2.3.1 suggest that a lognormal fit to the variation in mean signal level is reasonable.

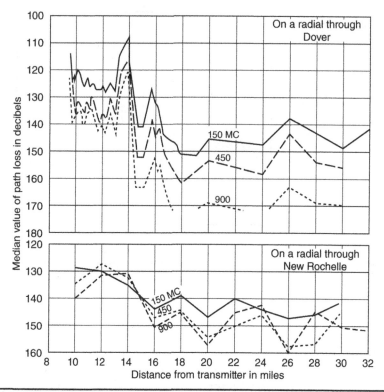

Figure 2.3.1 Median radio frequency path losses along selected routes. (*a*) On a radial through Dover. (*b*) On a radial through New Rochelle.

The median curves of loss have been replotted for three frequencies in Fig. 2.3.2, which shows the median curves of loss for different frequencies. It appears that performance at the various frequencies seems to differ by a constant number of decibels while exhibiting the same trend with distance.

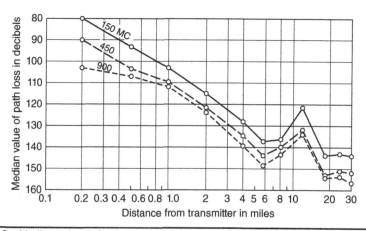

Figure 2.3.2 Median values of measured path losses.

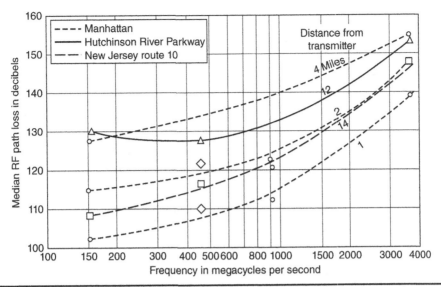

Figure 2.3.3 Radio frequency path losses at locations for the 3700-MC measurement.

As the test setups for transmitter power and receiver sensitivity for 3700 MC were limited, only locations with relatively low path loss can be tested. Figure 2.3.3 compares the results of these locations.

The losses at distances of 10 miles and more are 6 to 10 dB less than might have been predicted from the trend at smaller distances, where the measurements were made in city areas. This is probably because, away from the city, there is a considerable difference in the character of the surroundings, such as the height and number of buildings in the suburbs compared to the city. The measurements at longer ranges were in the suburbs of New York, whereas those nearer the transmitter were in urban Bronx and Manhattan. Except very close to the transmitter, the performance exhibited the same trend with distance at various frequencies and differed by a constant number of decibels.

In summary, Young's results in 1952 made following points:

1. Propagation losses were proportional to the fourth power of the range between transmitter and receiver.

2. The variation around the mean signal strength in a given area was lognormally distributed.

3. Loss increasing with increasing of frequencies.

4. The use of a gain antenna can appreciably lower the transmitter power.

5. The human-made environment has an impact on propagation.

2.4 Bullington Monograms

Bullington[4] made a series of monograms at frequencies above 30 MHz for factors that affect radio propagation, such as frequency, distance, antenna heights, curvature of the earth, atmospheric conditions, and the presence of hills and buildings. Received power can be estimated by means of three or four of these charts. The theory of propagation

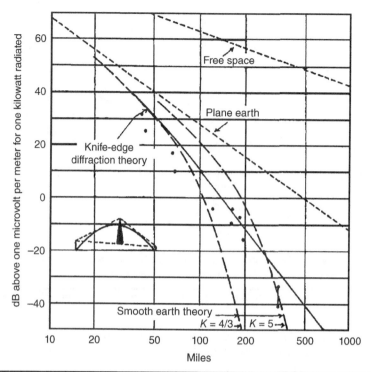

FIGURE 2.4.1   Measured and theoretical field intensities in the 40- to 50-MHz range, adjusted for a radiated power of 1 KW and antenna heights of 500 and 30 ft.[4]

over a smooth spherical earth is used as a baseline, and empirical methods are used to estimate the effects of hills and buildings and the phenomena of atmospheric refraction (bending away from straight-line propagation), atmospheric ducts (tropospheric propagation), and atmospheric absorption. The principal purpose is to provide simplified charts for predicting radio propagation under average weather conditions.

The Bullington model is deducted from free space, a perfectly flat earth, the effect of the curvature of the earth, atmospheric conditions, and irregularities on the earth. But this model is not quite applicable to the mobile radio environment.

Free space and plane earth models were discussed in previous sections. Figure 2.4.1 shows the baseline of the theoretical versus the measured data.

2.4.1   Fading, Refraction from Tropospheric Transmission, and Diffraction[4]

Changing atmospheric conditions cause variations in signal level with time. The severity of the fading also occurs when increasing in either the frequency or the path length. Fading is usually categorized into two general types: (1) inverse bending due to weather and (2) multipath effects due to human-made structures. The path of a radio wave is not a straight line, and bending may either increase or decrease. The effective path clearance and inverse bending may have the effect of transforming a line-of-sight path into an obstructed one. Multipath effects were described in Sec. 1.6.

The dielectric constant of the atmosphere normally decreases gradually with increasing altitude. As a result, the velocity of transmission increases with the height above the ground. As the change in dielectric constant is linear with height, radio waves continue to travel in a straight line over an earth whose radius is now

$$ka = \frac{a}{1 + \dfrac{a d\varepsilon}{2dh}} \tag{2.4.1}$$

where a is true radius of earth and $d\varepsilon/dh$ is the rate of change of dielectric constant with height.

Radio waves are also transmitted around the earth by the phenomenon of diffraction. At grazing incidence, the expected loss over a knife edge is 6 dB while over a smooth, spherical earth, described in Sec. 1.9.2.2. The phenomenon can also been shown in Fig. 2.4.1.1.

2.4.2 Effects of Buildings and Trees

Shadow losses result from buildings, trees, and hills. The data from New York City indicate that the median signal strength (50 percent) is about 25 dB below the corresponding plane earth value. The corresponding values for the 10 and 90 percent points are about 15 and 35 dB, respectively. Typical values of attenuation through a brick wall are from 2 to 5 dB at 30 MHz, and 10 to 40 dB at 3000 MHz, depending on whether the wall is dry or wet.

Over irregular terrain and under the shadow condition, the shadow loss based on knife-edge diffraction theory (see Sec. 1.6) is to be added to the transmission loss, as shown in Fig. 2.4.2.1.

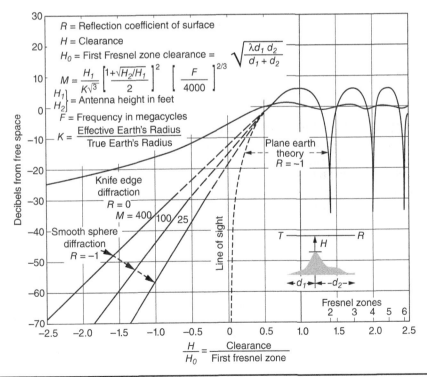

FIGURE 2.4.1.1 Transmission loss versus clearance.

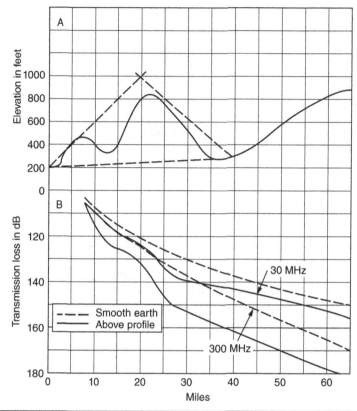

Figure 2.4.2.1   Transmission loss over irregular terrain at 30 and 300 MHz; half-wave dipoles of 250 and 30 ft.[5]

2.5   Egli Model—One of the Clutter Factor Models

The Egli model consists of empirical formulas with a "terrain factor" to be applied to the theoretical plane earth field strength.[6] The terrain factor has a median value of 27.5 dB for 900 MHz. The variation of field strength with antenna heights and distance is based on the plane earth model. Median transmission loss for a 1.5-m-high mobile antenna over a distance d is equal to

$$L = 139.1 - 20 \log h_b + 40 \log d \quad \text{(dB)} \tag{2.5.1}$$

The Egli model is a systematic interpretation of measurements covering a wide range of frequencies (90 to 1000 MHz). It is a greatly simplified model based on the assumption of "gently rolling terrain with average hill heights of approximately 50 ft." Therefore, no terrain elevation data between the transmitting and receiving facilities are needed. Instead, the free space propagation loss formula is used to adjust the height of the transmitting and receiving antennas aboveground. The Egli model is based on measured propagation path losses and then reduced to a mathematical model.

Measurements taken in urban and suburban areas found that the path loss exponent is close to 4, so some models to use the plane earth loss formula plus (in decibels)

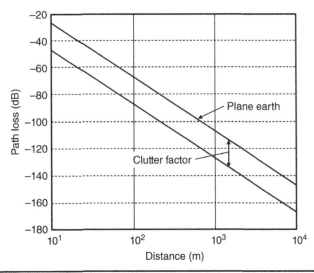

Figure 2.5.1 Clutter factor model.

an extra loss component called the clutter factor, as shown in Fig. 2.5.1. The various models differ based on different frequencies and environments.

The Egli model[6] is an example of a clutter factor model. Based on a large number of measurements taken around American cities over irregular terrain at frequencies of 90 to 1000 MHz, Egli observed, besides that the signal path loss follows an inverse fourth-power law with range, an excess loss over and above that loss. This excess loss depended on frequency and the nature of the terrain. Therefore, a multiplicative factor β is introduced to account for this:

$$L_{50} = G_b G_m \left(\frac{h_b h_m}{d^2}\right)^2 \beta \qquad (2.5.2)$$

where G_b is the gain of the base antenna, G_m is the gain of the mobile antenna, h_b is the height of the base antenna, h_m is the height of the mobile antenna, d is the propagation distance, and $\beta = (40/f)^2$, where f is in MHz. The subscripts b and m refer to base and mobile, respectively, and β is the factor introduced to account for the excess loss and is given by

$$\beta = \left(\frac{40}{f}\right)^2 \quad f \text{ in MHz} \qquad (2.5.3)$$

In Eq. (2.5.3), 40 MHz is the reference frequency at which the median path loss reduces to the plane earth value regardless the irregularity of the terrain.

The value of β is a function of terrain irregularity, the value obtained from Eq. (2.5.3) being a median value. By assuming the terrain undulation height to be a lognormal distribution about its median value, a family of curves is given in Fig. 2.5.2. The departure of β from its median value at 40 MHz is a function of terrain factor (dB) and the frequency of transmission. This model does not explicitly take diffraction losses into account.

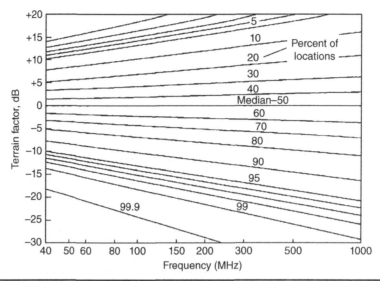

FIGURE 2.5.2 Mobile service field strength terrain factor.[6]

Egli's overall results were presented in a monograph form, but the CCIR[7] has given an approximation to these results as follows:

$$L = 40 \log R + 20 \log f_c - 20 \log h_b + L_m \qquad (2.5.4)$$

where

$$L_m = \begin{cases} 76.3 - 10 \log h_m & \text{for } h_m < 10 \\ 76.3 - 20 \log h_m & \text{for } h_m \geq 10 \end{cases} \qquad (2.5.5)$$

This approximation involves a small discontinuity at $h_m = 10$ m. This equation introduces an additional f_c^{-2} received power dependence, which likely represents the results of real measurements. For very large antenna heights, Eq. (2.5.2) may provide the loss value less than the free space value, in which case the free space value should be used. The Egli model is an empirical model to match the measured data.[6]

For detailed planning, software packages are available that use DTED (Digitized Terrain Elevation Data) to assist this model.

2.6 The JRC Method

The Joint Radio Committee (JRC) of the Nationalized Power Industries has adopted a method[8] that has been in widespread use for many years, particularly in the United Kingdom. It is a terrain-based technique and uses a classical propagation prediction approach similar to that used by Longley–Rice, which will be described in Sec. 3.4. This model[9,10] uses a computer-based topographic database providing height reference points at 0.5-km intervals. The topographical database is a two-dimensional array. Each array element corresponds to a point on a service area map. Using a quantized map, such as a

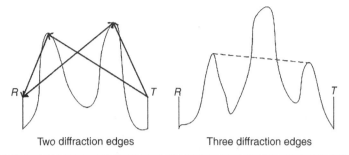

Two diffraction edges   Three diffraction edges

FIGURE 2.6.1 Two or more knife edges.

digital elevation map, in the service area, the computer program uses topographic data to plot a ground path profile between the transmitter and a chosen receiver location. The terrain heights and positions of obstructions are indicated. The computer then calculates both the loss L_F due to free space and the loss L_p due to plane earth and selects the higher one as expressed here:

$$L = \max(L_F, L_p) \tag{2.6.1}$$

In the case that no line-of-sight path exists or that Fresnel zone clearance is inadequate, the computer calculates the diffraction loss L_D along the path. The total loss is

$$L = \max (L_F, L_p) + L_D \tag{2.6.2}$$

The calculation of diffraction loss was described in Sec. 1.9.2.2 for a single diffraction edge and multiple diffraction edges.

The principle embodied in the JRC method[8] is still widely used, even though in its original form it generally tends to underestimate path losses.

This model uses a digital elevation map and performs site-specific propagation coverage. The calculation of radial terrain profiles and the signal strength along the radial is very similar to the Lee model, which will be described in Sec. 3.1.8. It has the limitation of being unable to account for losses due to trees and buildings.

2.7   Terrain-Integrated Rough-Earth Model

2.7.1   Description of TIREM

The Terrain-Integrated Rough-Earth model (TIREM)[11] is a computer software library used by the U.S. government. It consists of hundreds of modeling and simulation tools and calculates basic median propagation loss (path loss) of radio waves over irregular earth terrain. The calculation method was developed in the early 1960s and evolved into a TIREM software version distributed by the Defense Information Systems Agency Joint Spectrum Center for Department of Defense users.

TIREM covers the range of radio frequencies from 1 MHz to 20 GHz over terrain elevations that are specified by a set of discrete points along a great-circle path between the transmitting antenna and receiving antenna. The digital terrain elevation data can provide earth terrain information. TIREM takes into account (1) the transmitting medium (surface refractivity and humidity), (2) antenna properties (height, frequency, and polarization), and (3) geometrical environment (relative permittivity, conductivity, and terrain elevations). The calculation of path loss also includes the effects of reflection, surface wave propagation, diffraction, tropospheric scatter propagation, and atmospheric absorption but not ducting phenomena, fading, ionosphere propagation, and absorption due to rain or foliage.

TIREM computes radio frequency propagation loss over irregular terrain and seawater for ground-based and airborne transmitters and receivers. TIREM is capable of predicting the signal coverage for land mobile radios, sensor acquisition ranges, and television and radio broadcast stations.

Figure 2.7.1.1 depicts a radio path over a curved earth involving both rough earth terrain and seawater. TIREM examines the terrain type (including seawater) and elevation profile and selects the optimum model(s) for calculating propagation loss along the path. Based on the geometry of the profile, all appropriate modes of TIREM are used to calculate the path loss either within line of sight (LOS) or beyond line of sight (BLOS). Figure 2.7.1.2 illustrates both LOS and BLOS zones in a propagation path.

Under the NLOS condition, TIREM's modeling of multiple-knife-edge diffraction is illustrated in Fig. 2.7.1.2. Figure 2.7.1.3 shows a larger image of the multiple-knife-edge diffraction illustration. The calculation of diffraction loss was shown in Sec. 1.9.2.

2.7.2 Summary of Land Propagation Formulas

The path loss predicted by TIREM depends on whether the path is LOS, BLOS, or troposcatter. The total terrain-dependent propagation loss for LOS and BLOS path is

$$L_{TIREM} = \begin{cases} L_{FS} + A_{LOS} + A_{ABSORB} & \text{LOS} \\ \min(L_{FS} + A_{DIF} + A_{ABSORB} + L_{TRO} + A_{ABSORB}) & \text{BLOS} \end{cases} \quad (2.7.2.1)$$

where A_{LOS} is the excess loss for line of sight as determined by the frequency, A_{DIF} is the excess loss for the diffracting region, and A_{BSORB} is the molecular absorption loss.

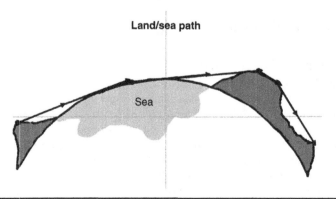

Land/sea path

Sea

Figure 2.7.1.1   TIREM propagation paths.

Profile evaluation

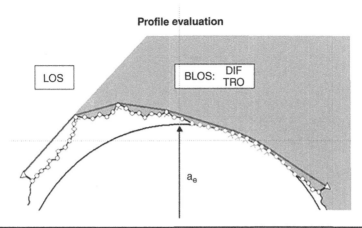

FIGURE **2.7.1.2**   TIREM propagation profile evaluation.

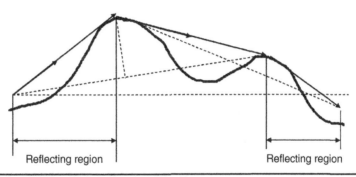

FIGURE **2.7.1.3**   TIREM multiple-knife edge diffraction.

The LOS, knife-edge-diffracted and spherical-earth-diffracted propagation, and tro-poscatter losses are calculated separately. The excess propagation losses as obtained from either the normalized fields for LOS or diffraction shown in Eq. (2.7.2.1) can also listed, as in Table 2.7.2.1.

The formulas listed in Table 2.7.2.1 are semiempirical. They involve certain assumptions and approximations. The major empirical points are summarized here.

- The terrain-reflected energy is assumed to cover a region surrounding the terrain point, which is clear from the radius of the first Fresnel zone. For low antennas, this region is virtually covered by transverse terrain, which is not considered in TIREM.

- The mean scattering coefficient is obtained from measured data.

- The loss due to the surface wave within LOS is derived from a flat earth.

- In the diffracted region, each peak is assumed to be a ridge and is treated as an ideal knife edge regardless of the appearance of the peak.

Region	Excess Propagation Loss	Limits	
Line-of-sight	$A_{LOS} = \begin{cases} \min(A_{SELOSS},\ A_{REF}) \\ \min[\alpha A_{SELOSS} + (1-\alpha)A_{REF}, A_{REF}] \\ A_{REF} \end{cases}$	$f_{MHz} < 16$ $16 \le f_{MHz} \le 20$ $f_{MHz} > 20$	
	$A_{REF} = -20 \log (|E_{SFW}/E_0|)$ A_{SELOSS} from Table 3-1	– –	–
Diffracting	$A_{DIF,HF} = A_{SELOSS}$ $A_{DIF,\alpha} = \alpha A_{DIF,HF} + (1 - \alpha) A_{DIF,VHF}$ $A_{DIF,VHF} = \begin{cases} \min(A_{SELOSS},\ A_{KNIFE}) \\ A_{KNIFE} \end{cases}$	$f_{MHz} \le 16$ $16 < f_{MHz} \le 20$ $K \ge 3$ and $\bar{A}_{kE} \le 7$ dB $f_{MHz} > 20$ $K < 3$ or $A_{kE} > 7$ dB $f_{MHz} > 20$	
	A_{SELOSS} from Table 3-1 $\alpha = -\frac{1}{4} f_{MHz} + 5$ $A_{KNIFE} = -20 \log (|E_{DIF}/E_0|)$ $\bar{A}_{kE} = \dfrac{1}{k} \sum A_{F,k}$	– – – –	–
L_{TIREM}:	Total terrain-dependent propagation loss for LOS and BLOS path		
L_{FS}:	Free space loss		
A_{LOS}:	Excess loss for LOS as determined by the frequency		
A_{ABSORB}:	Molecular absorption loss		
A_{DIF}:	Excess loss for the diffraction region		
L_{TRO}:	Troposcatter loss		
A_{SELOSS}:	Spherical-earth loss due to the contribution from surface wave and diffracted wave, depending on the relative permittivity and conductivity of the terrain		
A_{REF}:	Loss relative to free space due to the reflected ray		
α:	$= 5 - f_{MHz}/4$		
E_{spW}:	Intensity of space wave		
E_0:	Intensity of direct wave		
$A_{DIF,HF}$:	Diffraction loss (estimated by the spherical earth) for frequency less than 16 MHz		
$A_{DIF,\alpha}$:	Loss for the intermediate frequency region $16 < f_{MHz} \le 20$		
$A_{DIF,VHF}$:	Diffraction loss for 20 MHz and above		
A_{KNIFE}:	Knife-edge diffraction loss above free space		
E_{DIF}:	K knife-edge field intensity		
$\bar{A}_{kE}$:	Average knife-edge loss value, if more than two knife edges		
K:	Number of knife edges		
F_k:	Magnitude of the Fresnel integral for each knife edge		
$A_{F,k}$:	$= -20 \log F_k$		

TABLE 2.7.2.1   Summary of Terrain-Dependent Propagation Formulas for LOS or Diffraction

2.8   Carey Model

The Carey propagation model[12] is based on Part 22 of the US FCC Rules and Regulations. The model is essentially a simplified statistical method of estimating field strength and coverage based only on the effective radiated power (ERP) and height above average terrain (HAAT) at the base station. Since the terrain information is averaged, the model takes into account specific individual localized obstructions or shadowing. The range of the terrain for the model to use is between 3 and 16 km from the transmitter site. The terrain obstructions outside of this range are ignored. Any terrain obstructions within 3 km or beyond 16 km that block the line of sight over the radio path would get the same calculation as the LOS condition. The main use for this model is for license applications such as cellular radio licensing applications or other submissions to the FCC that specifically require the use of the methods described in Part 22 of the FCC Rules or other administrative requirements, such as certain frequency coordination procedures.

Carey curves[12] give the $F(L, T)$ field strength versus distance noted that $F(L, T)$ denotes field strength exceed at L percent of locations during T percent of the time.

$F(50, 50)$ and $F(50, 10)$ field strength versus distance for propagation under average terrain conditions with: (1) mobile antenna height of 1.8 m, (2) base station antenna heights ranging from 30 to 1500 m above average terrain, and (3) distances up to 130 km for the $F(50, 50)$ curves and up to 240 km for the $F(50, 10)$ curves.[12] The Carey curves for 450 to 1000 MHz were based on a CCIR recommendation covering the entire band. Median transmission loss for this model can be expressed as

$$L = 110.7 - 19.1 \log h + 55 \log D \quad \text{(dB)}, \quad \text{for } 8 \leq D < 48 \tag{2.8.1}$$

$$L = 91.8 - 18 \log h + 66 \log D \quad \text{(dB)}, \quad \text{for } 48 \leq D < 96 \tag{2.8.2}$$

The curves are derived from CCIR curves for television broadcasting. The latter were adjusted downward by 9 dB to account for the 1.8-m height of mobile station antennas.

The Carey formula is used to calculate the FCC Cellular Geographic Service Area formula. This formula approximates this distance to the 32-dBμ contour predicted by Carey:

$$d = 1.05 * H^{0.34} * P^{0.17}$$

where d is the distance in miles from the cell site antenna to the reliable service area boundary, H is the antenna height in feet above average terrain, and P is the ERP in watts.

The service area usually is defined by connecting the specific coverage distance on the 32-dBμV/m point along different angle of radius. It is called Carey curve or Carey contour. Conversion between dBμ and dBm in power delivery[13] is calculated as

$$\text{dBm} = \text{dBμV/m} - 20 \log(f) - 77.21 \tag{2.8.3}$$

where f is the frequency in MHz. Equation (2.8.3) is based on the antenna gain of 3 dBd (above dipole). The 32-dBmV/m contour was an eased version from that of the 39-dBμV/m contour for cellular license application.

The Carey contour was based on the two-way system (for link balance) or paging/broadcast (forward link only) in early measurements by Carey. Actually, Carey measured only a number of frequencies, and others (nonspecific to his actual measurement) were interpolated. In general, the higher the value in dBμV/m, the stronger the signal received, as shown in Eq. (2.8.3).

2.9 CCIR Model

2.9.1 Description of the Model

The CCIR (Comité Consultatif International des Radio-Communication, now ITU-R) model takes topography into account using only statistical parameters, in contrast to most other methods. These parameters are the effective antenna height of the transmitter and a value Δh (as shown in Fig. 2.9.1.1) defining the degree of terrain irregularity. This model predicts the field strength for a given terrain over an extrapolated frequency range of 100 to 3000 MHz, and its prediction method was conceived for land mobile radio service with vertical polarization. A basic median attenuation is introduced to quasi-smooth terrain in urban areas. Some correction factors are applied for different transmitter and receiver antenna heights and for different terrain and environmental clutter. The correction factors correspond to different effective antenna heights and terrain irregularities as well as to isolated mountains, sloped terrain, and open, quasi-open, or suburban areas and to different receiving antenna heights.

CCIR curves give field strengths at 900-MHz frequencies for 50 percent of locations and 50 percent of the time in urban areas for mobile antenna heights of 1.5 m and base antenna heights between 30 and 1000 m, as shown in Fig. 2.9.1.2.[15]

To adjust for mobile antenna heights of 3 m instead of 1.5 m, a height-gain factor of 3 dB is suggested. Standard deviations are given as a function of distance and terrain irregularity, assuming that propagation variations with location and time are characterized in decibels by the Gaussian distribution. Treatment of the median transmission loss for frequencies near 900 MHz is based on the work of Okumura et al.[14] stated in Sec. 2.12.

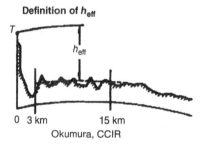

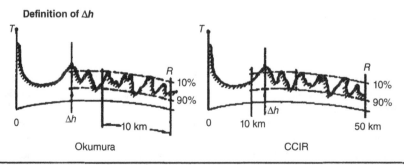

FIGURE 2.9.1.1 Definition of the effective transmitter antenna height h_{eff} and the parameter of the terrain irregularity Δh according to CCIR and Okumura et al.[14]

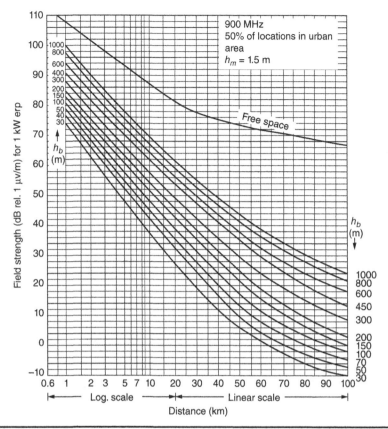

FIGURE 2.9.1.2   CCIR field strength prediction curves for urban areas at 900 MHz: 50 percent of the time, 50 percent of locations, 1 kW ERP, $h_m = 1.5$ m.[15]

An empirical formula for the combined effects of free space path loss and terrain-induced path loss is given by

$$L_{CCIR}(\text{dB}) = 69.55 + 26.16 \log_{10} f_{\text{MHz}}$$
$$- 13.82 \log_{10} h_1 - a(h_2) + (44.9 - 6.55 \log_{10} h_1) \log_{10} d_{\text{km}} - B \qquad (2.9.1.1)$$

where h_1 and h_2 are base station and mobile antenna heights in meters, respectively; d_{km} is the link distance in kilometers; f_{MHz} is the center frequency in megahertz; and

$$a(h_2) = (1.1 \log_{10} f_{\text{MHz}} - 0.7)h_2 - (1.56 \log_{10} f_{\text{MHz}} - 0.8) \qquad (2.9.2)$$

The field strength prediction curves shown in Fig. 2.9.1.2 are based on a considerable amount of experimental data collected in many countries. They are applicable over the kind of rolling, hilly terrain found in many parts of Europe and North America for which the terrain irregularity parameter Δh is typically 50 m. In any specific situation, it is necessary to look up another curve, that gives a correction related to the value of Δh. The values of field strength measured in a small area will be lognormally distributed around the predicted median value. Standard deviations is expressed as functions of distance and terrain irregularity.

CCIR curves are based on a single parameter Δh, which is inadequate to define the required correction factor with sufficient accuracy. Furthermore, terrain variations in the immediate vicinity of the mobile are not explicitly taken into account. A more accurate method is therefore required, as shown in the following section.

2.10   Blomquist–Ladell and Edwards–Durkin Models

The Blomquist–Ladell[16] model is used for calculating transmission loss in the VHF and UHF regions. It is a deterministic model giving the long-term median of the basic transmission loss. The model considers the same type of losses as the JRC method (see Sec. 2.6) but combines them in a different way in an attempt to provide a smooth transition between points where the prediction is based on free space loss L_F and plane earth propagation loss L_p. The ground dielectric constant and the terrain profile, including vegetation, are properly accountable.

The Blomquist–Ladell model considers two limiting propagation cases: the spherical earth can be regarded as smooth, and it can be represented by a number of knife edges. Blomquist and Ladell introduce two empirical propagation models. The first model calculates the propagation factor as

$$F_{loss} = \min\,(F_B, F_{diffraction}) \qquad (dB) \qquad (2.10.1)$$

where F_{loss} is a total loss which is taking smaller number among two factors:

F_B(dB) is the propagation factor based on the smooth, spherical earth field.

$F_{diffraction}$ (dB) is the propagation factor based on the obstacle diffraction field.

The other model is the square root of the sum of the squares of F_B and $F_{diffraction}$, that is,

$$F_{loss} = \sqrt{F_B^2 + F_{diffraction}^2} \qquad (dB) \qquad (2.10.2)$$

Blomquist and Ladell consider the same types of losses as the Edwards–Durkin model[9] but combine them differently. The total propagation loss is given by an expression of the following type:

$$L_T(dB) = L_F + [(L_p' - L_F)^2 + L_D^2]^{1/2} \qquad (2.10.3)$$

where $F_B = L_p' - L_F$ (F_B is given in Eq. [2.10.12])

L_F = free space propagation loss (dB)

L_D = diffraction losses due to irregular terrain (dB) (multiple-knife-edge loss after Epstein and Petersons [Sec. 1.9.2.2.2])

L_B = propagation losses caused by buildings (dB)

L_T = total propagation loss (dB)

L_p or L_p' = plane earth propagation loss (dB) the former is from the Edwards–Durkin model and the latter from the Blomquist–Ladell model

F_B = a correction term for the earth curvature effect of the troposphere and for any factor other than diffraction over irregular terrain

H_b and H_m = effective antenna heights (m) of base and mobile stations, respectively

R = distance between base and mobile station (km)

Edwards and Durkin use an empirical L_p from Bullington:

$$L_p = K_2 - 20 \log\,(H_b) - 20 \log\,(H_m) + 40 \log(R)\ (dB) \qquad (2.10.4)$$

where K_2 is 115.1 dB for transmission between two half-wave dipoles and is 118.7 dB in the case of isotropic antennas. It is shown that for most cases,

$$L_T = L_p + L_D \qquad (2.10.5)$$

One way to perform the calculations is to solve the propagation problem first for a smooth spherical earth and then correct the solution with factors that account for different conditions of wave propagation phenomena.

The plane earth propagation loss (dB) L_p' from the Blomquist–Ladell model is

$$L_p' = F_B + L_F \qquad (2.10.6)$$

where F_B is shown in Eq. (2.10.12).

The total loss L_T can be expressed as

$$L_T(\text{dB}) = L_F + [(L_p' - L_F)^2 + L_D^2]^{1/2} + L_V + L_U + L_A \qquad (2.10.7)$$

where L_V is the vegetation loss, L_U is the urban loss, and L_A is the loss from atmosphere.

The square-root model is frequently used. Over highly obstructed paths, $L_D \gg (L_p' - L_F)$, the total loss of Eq. (2.10.3) can be approximated by

$$L_T(\text{dB}) = L_F + L_D \qquad (2.10.8)$$

For unobstructed paths, L_D approaches 0, and the total loss from Eq. (2.10.3) becomes

$$L_T(\text{dB}) = L_p' \qquad (2.10.9)$$

Besides diffraction losses being equal to 0, this result is similar to the one of Edwards and Durkin, and it would in fact be identical if the same formula were used to estimate L_p (Edwards and Durkin) and L_p' (Blomquist and Ladell).

The essential part of the Blomquist–Ladell[9] model is an empirically developed formula. The total loss is a weighted sum of free space loss, plane earth loss, and multiple-knife-edge loss as

$$L_T = \begin{cases} L_F + \left(F_B^2 + L_D^2\right)^{\frac{1}{2}} & F_B \le 0 \\ L_F + \left(F_B^2 - L_D^2\right)^{\frac{1}{2}} & F_B > 0, \text{ but } \le |L_D| \\ L_F - \left(F_B^2 - L_D^2\right)^{\frac{1}{2}} & F_B > 0, \text{ but } > |L_D| \end{cases} \qquad (2.10.10)$$

where

$$L_F = 32.45 + 20\log_{10}(f_c) + 20\log_{10}(R) \qquad (\text{dB}) \qquad (2.10.11)$$

When an isotropic antennas is used, the propagation factor F_B is

$$F_B = 10\log_{10} \left| \left(\frac{4\pi h_b^2}{\lambda d} + \frac{\lambda c_b^2}{\tau d(c_b - 1)} \right) \left(\frac{4\pi h_m^2}{\lambda d} + \frac{\lambda c_m^2}{\tau d(c_m - 1)} \right) \right| + Y \qquad (2.10.12)$$

$$Y = \begin{cases} -2.8X & X < .53 \\ 6.7 + 10\log_{10}X - 10.2X & .53 \le X < 2 \end{cases} \qquad (2.10.13)$$

$$X = \left(\frac{2\pi}{\lambda} \right)^{\frac{1}{3}} (k\partial)^{-\frac{2}{3}} \quad d \qquad (2.10.14)$$

where L_T = total loss (dB); L_F = free space loss (dB);

$\quad\quad F_B$ = propagation factor (dB); L_D = diffraction loss (dB);

$\quad\quad Y$ = correction factor (dB); f_c = frequency (HHz);

$\quad\quad R$ = distance between antenna (km); h_b, h_m = base and mobile station effective antenna heights (m);

$\quad\quad \lambda$ = wavelength (m); d = distance between antenna (m) = 1000 R;

$\quad\quad \partial$ = earth radius (6.371×10^6m);

$\quad\quad K$ = earth radius factor ($\frac{4}{3}$ for standard ratio atmosphere); and

$\quad\quad c_b, c_m$ = dielectric constant at the base and mobile stations (normally taken as 10 over dry terrain).

2.11   Ibrahim–Parsons Model[17,18]

2.11.1   Findings from the Empirical Data

This empirical model is applied to urban environments. It was based on the field trials in London between an elevated base station and a mobile moving in the city streets at frequencies of 168, 445, and 896 MHz. Measured data were collected with base station antennas at a height of 46 m above local ground.[16] Samples were taken every 2.8 cm.

List of principal symbols:

H = relative mobile spot height, m

L_f = free space propagation loss, dB

L_p = plane earth propagation loss, dB

L_D = terrain diffraction loss, dB

L = land usage factor

U = degree of urbanization factor

P_L = predicted median path loss between two isotropic antennas, dB

d = range, m ($d \leq 10,000$ m)

f = transmission frequency, MHz

h_t = transmitter antenna height above local ground level, m

h_r = receiver antenna height above local ground level, m ($h_r \leq 3$ m)

The data were collected in batches, each batch representing a square of 0.5 km coinciding with the National Grid System. This size of test square was judged suitable so that the propagation data become representative in the urban environment. The layout of the test squares in London is shown in Fig. 2.11.1.1. The average route length within each test square was 1.8 km.

Each square is assigned three parameters, H, U, and L, defined as follows:

H (terrain height) is defined as the actual height of a peak, basin, plateau, or valley found in each square or the arithmetic mean of the minimum and maximum heights found in the square if it does not contain any such features.

U (the degree of urbanization factor) is defined as the percentage of building site area within the square that is occupied by buildings having four or more floors. U varied

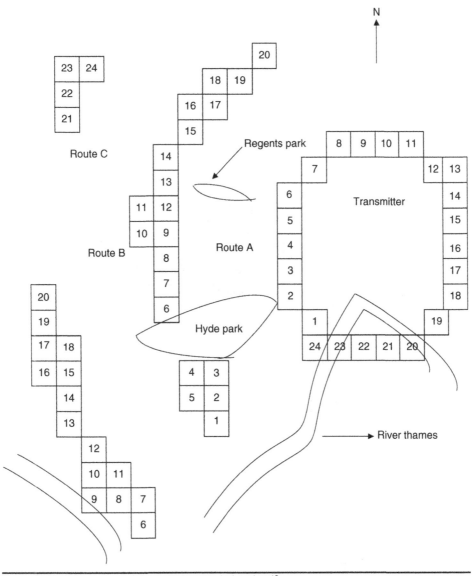

FIGURE 2.11.1.1   Layout of the test squares in London.[18]

between 2 and 95 percent, a very sensitive parameter. The building height distributions in the urban and suburban areas are shown in Figs. 2.11.1.2 and 2.11.1.3, respectively.

L (land usage factor) is defined as the percentage of the test area actually occupied by any buildings.

These parameters were selected empirically as having good correlation with the data.

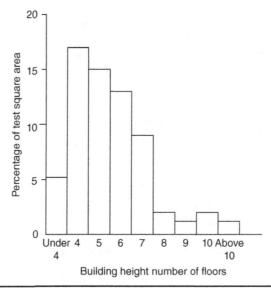

FIGURE 2.11.1.2 Building height distribution in the urban area.[17]

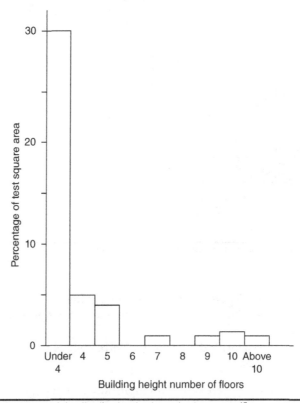

FIGURE 2.11.1.3 Building height distribution in the suburban area.[17]

Frequency (MHz)		Median Path Loss (dB)	RMS Prediction Error
168	Best fit	1.6 + 36.2 log d	5.3
	fourth-power law	– 12.5 + 40 log d	5.3
455	Best fit	15.0 + 43.1 log d	6.18
	fourth-power law	– 4.0 + 40 log d	6.25

TABLE 2.11.1.1   Range Dependence Regression Equations at 168 and 455 MHz

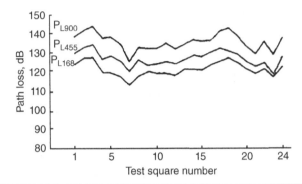

FIGURE 2.11.1.4   Mean path loss to each test square at 2-km range in London.[17]

A total of 64 test squares were selected in three arcs, approximately 2, 5, and 9 km around the base station. The total length of the measurement route was about 115 km.

Two models were proposed. The fully empirical model shows marginally lower prediction errors but relies on a complex formulation that bears no direct relationship to propagation principles. The semiempirical model associated with the Egli clutter factor method is based on the plane earth loss. A clutter factor b is introduced as a function of f_c, L, H, and U.

In general, the median received signal decreased as the mobile moved away from the base station. The range dependence regression equations at 168 and 455 MHz is shown in Table 2.11.1.1.

Figure 2.11.1.4 shows the median path loss to each of the test squares at 2-km range for three frequencies. The correlation coefficient was 0.93 between the measurements at 168 and 455 MHz and 0.97 between the measurements at 455 and 900 MHz.

2.11.2   Two Proposed Models

Two possible approaches were taken. The first one was an empirical expression for the path loss based on multiple regression analysis, and the second was to start from the theoretical plane earth equation and to correlate the excess path loss in terms of "clutter factor." The main difference between the first empirical method and the second semiempirical method is that a fourth-power exponent law is assumed in the second approach.

2.11.2.1 Empirical Model

The following equation was produced as the best fit for the London data:

$$L_{50}(\text{dB}) = -20\log(0.7h_b) - 8\log h_m + \frac{f}{40} + 26\log\frac{f}{40} - 86\log\left(\frac{f+100}{156}\right)$$

$$+ \left[40 + 14.15\log\left(\frac{f+100}{156}\right)\right]\log d + 0.265L - 0.37H + K_1 \qquad (2.11.2.1)$$

where $K_1 = 0.087U - 5.5$ for the highly urbanized area and 0 otherwise. H is the difference in average ground height between the transmitter and receiver, while the distance is limited by $0 \le d \le 10$ km on the OS (Ordnance Survey, UK mapping authority) map.

2.11.2.2 Semiempirical Model

This model is based on the plane earth equation. The median path loss is expressed as the sum of the theoretical plane earth loss and an excess clutter loss termed β. The values of β at 168, 455, and 900 MHz were obtained for each test square.

The model is given as

$$L_T = 40\log r - 20\log(h_m h_b) + \beta \qquad (2.11.2.2)$$

where

$$\beta = 20 + \frac{f_c}{40} + 0.18L - 0.34H + K_2$$

and

$$K_2 = 0\ 0.094U - 5.9$$

K_2 is applicable only in the highly urbanized city center; otherwise, $K_2 = 0$.

The root-mean-square prediction errors produced by the two models are summarized in Table 2.11.2.1.

The estimation of values of parameters L, U, and H of the test squares is under consideration.

Parameter H can be easily extracted from a map.

L and U can sometimes be obtained from other stored information or through the estimation.

	Frequency (MHz)		
	168	455	900
Empirical model	2.1	3.2	4.19
Semi-empirical model	2.0	3.3	5.8

TABLE 2.11.2.1   Root-Mean-Square Prediction Errors in Decibels Produced by the Two models

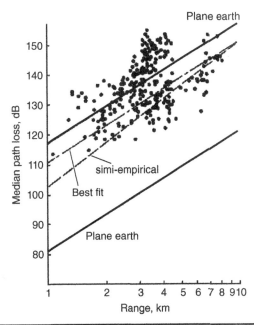

FIGURE 2.11.2.1   Predicted versus measured based on Allsebrook Birmingham's data at 441 MHz.[18]

The value of β in a flat city ($H = 0$) at a frequency of 900 MHz is given as

$$\beta \text{ (dB)} = 42.5 + 0.18P \qquad (2.11.2.3)$$

The symbol P in Eq. (2.11.2.3) lies in the range of 0 to 80 percent, where β is between 42.5 and 57 dB. This agrees well with some independently measured results, shown in Fig. 2.11.2.1, for which $\beta = 49$ dB. The model is used mainly in urban areas.

2.12   Okumara–Hata and the Cost 231 Hata Models

2.12.1   Okumura Method Hata Model

From the Okumara method, the Okumara-Hata model is created. It was developed based on the results of an extensive series of measurements in certain urban and suburban areas around Tokyo, Japan, through the works of Y Okumura et al.[19]

Okumura provides a method for predicting field strength and area coverage for a given service area by using the basic median field strength curve for 900 MHz as a reference for an urban area application. The parameters for the Okumura–Hata model are listed in Table 2.12.1.1.

Predictions are made via a series of graphs and a set of formulas. Quasi-smooth terrain is taken as the reference, and correction factors are added for the other types of terrain.

Frequency (MHz)	Distance (km)	Base Station Effective Antenna Height (m)	Receiver Antenna Height	Correction Factors
200, 453, 922, 1310, 1430, and 1920	1–1000	30–1000	Typical of land mobile applications	Suburban, open and isolated mountain, rolling hills, sloping terrain, and mixed land–sea paths

TABLE 2.12.1.1   Okumura–Hata Model Parameters

The Okumura–Hata model divides the area being investigated into a series of clutter and terrain types, namely, open, suburban, and urban. Definitions of these areas are described here:

- Open area: Open space, such as farmland, rice fields, and open fields
- Suburban area: A village or highway scattered with trees and houses, such as residential areas and small towns
- Urban area: A built-up city or large town with large buildings and houses

The Okumura–Hata model uses the free space loss as baseline and derives the reference curves, as shown in Fig. 2.12.1.1 for a fixed base station antenna height of 200 m and a mobile antenna height of 3 m. Based on the frequencies and distance, the basic median attention (path loss), $A_{mu}(f, d)$, can be found on the curve. The base station antenna height correction factor is shown in Fig. 2.12.1.2, and the mobile antenna height correction factor is shown in Fig. 2.12.1.3.

The baseline predicted value is expressed as

$$Path\ Loss\ \text{(median attenuation) (dB)} = L_f + A_{mu}(f, d) + h_b + h_m \qquad (2.12.1.1)$$

where L_f is the free space loss and A_{mu} is a function of distance of frequency, derived from Fig. 2.12.1.1, and h_b and h_m are the base station and the mobile station antenna height, respectively.

Correction factors have to account for antennas heights other than the reference heights. The basic formulation of the technique can be expressed as

$$Path\ Loss\ \text{(median attenuation) (dB)} = L_f + A_{mu}(f, d) + h_b + h_{te} + h_m + h_{re} \qquad (2.12.1.2)$$

where h_{te} is the base station antenna height gain factor as a function of range and h_{re} is the mobile (receiver) antenna height gain factor as a function of frequency.

The dependency of h_b on propagation loss is shown in Fig. 2.12.1.2. It shows that path loss increases as the antenna height decreases.

As shown in Fig. 2.12.1.3, the correction is presented as relative height gain to a mobile station 3 m in height in an urban area over quasi-smooth terrain.

The dependence of f_c on propagation loss is shown in Fig. 2.12.1.4. It shows that path loss increases as the frequency increases.

More correction factors are also provided in graphical form for rolling hilly terrain, isolated mountain, general sloping terrain, and mixed land-sea path, and can be added or subtracted as appropriate.

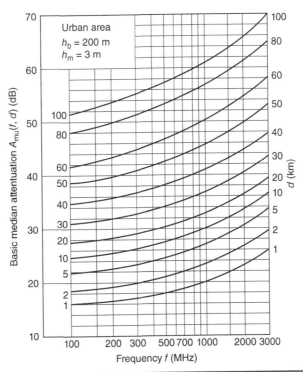

FIGURE 2.12.1.1 Basic median path loss relative to free space in urban areas over quasi-smooth terrain.

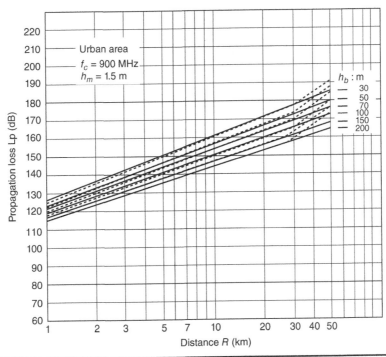

FIGURE 2.12.1.2 Propagation loss in an urban area with different h_b.

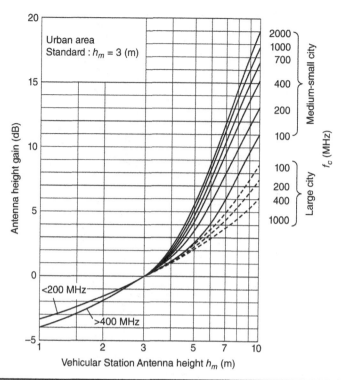

Figure 2.12.1.3 Prediction curves for vehicular antenna height gain in an urban area.[19]

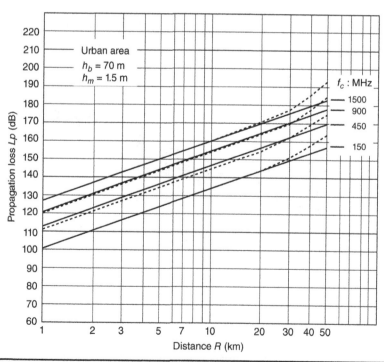

Figure 2.12.1.4 Propagation loss in an urban area with different f_c.

2.12.1.1   Okumura–Hata Model

The Okumura–Hata model, together with related corrections, is probably the most popular model used in planning real systems. Okumura published many empirical curves useful for cellular systems planning based on extensive measurements of urban and suburban radio propagation losses in Tokyo. These empirical curves were condensed to a convenient set of formulas known as the Hata formulas that are widely used in the industry.

2.12.1.2   Hata's Formulation

Hata[15] drew a set of formulas to replace the graphical information given by Okumura. Hata's formulation is confined to certain ranges of input parameters over quasi-smooth terrain.

Hata's formula for Okumura's predictions of median path loss is as follows:

$$L(\text{dB}) = 69.55 + 26.16 \log_{10} f_{\text{MHz}} - 13.82 \log_{10} h_1 - a(h_2)$$
$$+ (44.9 - 6.55 \log_{10} h_1) \log_{10} d_{\text{km}} - K \qquad (2.12.1.3)$$

where h_1 (30 to 200 m) and h_2 (1 to 10 m) are the base station and mobile antenna heights in meters, respectively; d_{km} (1 to 20 km) is the link distance and f_{MHz} (150 to 1500 MHz) is the carrier frequency; $a(h_2)$ is a mobile antenna height-gain correction factor that depends on the environment; and K is the factor used to correct the small-city formula for suburban and open areas.

2.12.1.2.1   Urban Areas   In urban areas, Eq. (1.12.1.3) can be deduced as follows:

$$L(\text{dB}) = 69.55 + 26.16 \log_{10} f_{\text{MHz}} - 13.82 \log h_1 - a(h_2)$$
$$+ (44.9 - 6.55 \log h_1) \log d \qquad (2.12.1.4)$$

where $K = 0$, 150 MHz $\leq f_c \leq$ 1500 MHz, 30 m $\leq h_1 \leq$ 200 m, and 1 km $\leq d \leq$ 20 km.
For a small or medium-size city,

$$a(h_2) = (1.1 \log f_{\text{MHz}} - 0.7)h_2 - (1.56 \log f_{\text{MHz}} - 0.8) \qquad (2.12.1.5)$$

For a large city,

$$a(h_2) = \begin{cases} 8.29 (\log 1.54 h_2)2 - 1.1 & f \leq 300 \text{ MHz} \\ 3.2 (\log 11.75 h_2)2 - 4.97 & f \geq 300 \text{ MHz} \end{cases} \qquad (2.12.1.6)$$

2.12.1.2.2   Suburban Areas

$$L \text{ (dB)} = L \text{ (urban)} - 2 \, [\log(f_c/28)]^2 - 5.4 \qquad (2.12.1.7)$$

where L (urban) is from Eq. (2.12.1.4), $K = 4.78(\log_{10} f_{\text{MHz}})^2 - 18.33 \log_{10} f_{\text{MHz}} + 40.94$, and $a(h_2) = (1.1 \log_{10} f_{\text{MHz}} - 0.7)h_2 - 1.56 \log_{10} f_{\text{MHz}} - 0.8$.

2.12.1.2.3 Open Areas

$$L \text{ (dB)} = L \text{ (urban)} - 4.78 (\log f_c)^2 + 18.33 \log f_c - 40.94 \qquad (2.12.1.8)$$

where L(urban) is from Eq. (2.12.1.4), $K = 2[\log_{10}(f_{MHz}/28)]^2 + 5.4$, and $a(h_2) = (1.1 \log_{10} f_{MHz} - 0.7) h_2 - 1.56 \log_{10} f_{MHz} - 0.8$.

2.12.1.2.4 Hata's Formulas for $f_{MHz} = 800$ MHz   In this case, Eq. (2.12.1.3) at $f_{MHz} = 800$ MHz reduces to the following formula:

$$L_{Hata} = 145.49 + (44.9 - 6.55 \log_{10} h_1) \log_{10} d_{km} - a(h_2) - 13.83 \log_{10} h_1 - K \qquad (2.12.1.9)$$

where

Type of Area	$a(h_2)$	K
Open	$2.52h_2 - 3.77$	28.26
Suburban		9.79
Medium-size to small city		0
Large city	$32(\log_{10} 11.75h_2)^2 - 4.97$	0

Many methods were derived to improve the Hata model, including those of Allsebrook,[18] Delisle et al.,[20] Aurand and Post,[21] and Akeyama et al.[22] The Okumura–Hata model is perhaps the most widely deployed model for macrocells as it is accurate and easy to implement.

2.12.2 Cost 231 Hata Model

The European Co-operative for Scientific and Technical Research (EURO-COST) formed the COST-231 working committee to develop an extended version of the Hata model from 1500 MHz to 2 GHz.

The COST 231 model, sometimes called the Hata model PCS extension, is valid between 1500 and 2000 MHz. The COST 231 median path loss is given by

$$L(\text{dB}) = 46.3 + 33.9 \log(f_{MHz}) - 13.82 \log(h_1) - a(h_2)$$
$$+ [44.9 - 6.55 \log(h_1) \log(d) + C \qquad (2.12.2.1)$$

where f_{MHz} is the frequency in MHz (1500 to 2 Ghz), h_1 is the base station height in meters, h_2 is the mobile station height in meters, $a(h_2)$ is the mobile antenna height correction factor (see Eq. 1.12.1.5 for suburban and Eq. 2.12.1.6 for urban), d is the link distance in kilometer (1 to 20 km), $C = 0$ dB for medium-size cities or suburban centers with medium tree density, and $C = 3$ dB for metropolitan centers.

The application of this model is restricted to macrocells where the base station antenna is above the rooftops of adjacent buildings. It is not applicable to microcells where the antenna height is low.

2.12.2.1 Extension of the Hata Model to Longer Distances

An empirical formula for modifying the Hata model from Eq. (2.12.1.4) to extend distances 20 to 100 km was developed by ITU-R and is given by

$$L_{ITU} \text{ (dB)} = 69.55 + 26.16 \log_{10} f_{MHZ} - 13.82 \log_{10} h_1 - a(h_2)$$
$$+ (44.9 - 6.55 \log_{10} h_1)(\log_{10} d_{km})^b - K \qquad (2.12.2.2)$$

where

$$b = \begin{cases} 1, & d_{km} < 20 \text{ km} \\ 1 + (0.14 + 0.000187 f_{MHz} + 0.00107 h_1')(\log_{10}(d_{km}/20))^{0.8} & d_{km} \geq 20 \text{ km} \end{cases}$$

and h_1' in the equation of b is

$$h_1' = \frac{h_1}{1 + 7 \times 10^6 h_1^2}.$$

2.12.2.2 Hata–Davidson Model

The Hata–Davidson model[23] is derived from the Hata model, which was in turn based on the Okumura' curves, as described previously.

The main factors included in Hata–Davidson model are the area type (urban, suburban, quasi-open, and open) as well as corrections for the receiver antenna height. The model also includes frequency and distance corrections to extend the distance range to 300 km.

The Hata–Davidson model computes the basic median field strength based on the transmit antenna height measured on the HAAT rather than the topography of the entire path. HAAT is calculated on the undulation of terrain from a 3 to 16 km segment over the radio path.

The Telecommunications Industry Association recommends in its publication TSB-88A that the Hata–Davidson model is the modification of Hata model, which consists of the addition of correction terms:

$$L_{HD} = L_{hata} + A(h_1, d_{km}) - S_1(d_{km}) - S_2(h_1, d_{km}) - S_3(f_{MHz}) - S_4(f_{MHz}, d_{km}) \qquad (2.12.2.3)$$

where $A(h_1, d_{km})$ and $S_1(d_{km})$ are the distance correction factors extending the range to 300 km. They are listed below:

Distance	$A(h_1, d_{km})$	$S_1(d_{km})$
$d_{km} < 20$	0	0
$20 \leq d_{km} < 64.38$	$0.62137(d_{km} - 20)[0.5 + 0.15\log_{10}(h_1/121.92)]$	0
$64.38 \leq d_{km} < 300$	$0.62137(d_{km} - 20)[0.5 + 0.15\log_{10}(h_1/121.92)]$	$0.174(d_{km} - 64.38)$

S_2 is a base station antenna height correction factor extending the range of h_1 values to 2500:

$$S_2(h_1, d_{km}) = 0.00784 |\log_{10}(9.98/d_{km})|(h_1 - 300) \quad \text{for } h_1 > 300 \text{ m}$$

S_3 and S_4 are frequency correction factors extending the frequency to 1500 MHz:

$$S_3(f_{MHz}) = f_{MHz}/250 \log_{10}(1500/f_{MHz})$$

$$S_4(f_{MHz}, d_{km}) = [0.112 \log_{10}(1500/f_{MHz})](d_{km} - 64.38) \quad \text{for } d_{km} > 64.38 \text{ km}$$

2.13 Walfisch–Bertoni Model

The Walfisch–Bertoni model is a physical model that addresses propagation in urban environments. Figure 2.13.1 illustrates the geometry used in the Walfisch–Bertoni model.[24,25] As shown in Fig. 2.13.1, there are four paths. The most important path is the primary path to the mobile that is diffracted over the tops of the buildings in the vicinity of the mobile,[26,27] as shown by path 1. Another path over the rooftop to the mobile from reflection is shown as path 2. Other possible propagation paths exist due to building penetration (path 3) and multiple reflections and diffractions (path 4). The last two are generally negligible.

To determine the field diffracted down to street level, it is necessary to establish the field incident on the rooftop of the building right before the mobile. For a large number of buildings between the base station and the mobile, the field at the rooftop, E_{Roof}, can be obtained from

$$E_{\text{Roof}} \approx 0.1 \left(\frac{\alpha \sqrt{\frac{b}{\lambda}}}{0.03} \right)^{0.9} \tag{2.13.1}$$

where the glancing angle α, and the separation between two adjacent buildings b are shown in Fig. 2.13.1.

The total path loss is the sum of two losses: the free space path loss between the antennas, and the additional loss due to the final diffraction down to the mobile on the street level.

The free space path loss formula is shown in Eq. (2.13.2):

$$L_{FS} \text{ (dB)} = 32.4 + 20 \log f_{\text{MHz}} + 20 \log d_{\text{km}} \tag{2.13.2}$$

Due to the final diffraction down to the mobile on the street level, the amplitude of the field at the mobile E_r is obtained by multiplying the rooftop field E_{Roof} by a factor:

$$E_r = E_{\text{Roof}} \frac{\sqrt{\lambda}}{2\pi} \left[\left(\frac{b}{2} \right)^2 + (h - h_m)^2 \right]^{-\frac{1}{4}} \left[-\frac{1}{\gamma - \alpha} + \frac{1}{2\pi + \gamma - \alpha} \right] \tag{2.13.3}$$

where h is the height of the buildings and h_m is the height of the mobile antenna. The angles γ and α are measured in radians with

$$\gamma = \tan^{-1} [2(h - h_m)/d] \tag{2.13.4}$$

For a level terrain, α (radians) is given by

$$\alpha = \frac{h_b - h}{d} - \frac{d}{2r_e} \tag{2.13.5}$$

where r_e is the effective earth radius, $r_e = 8.5 \times 10^3$ km.

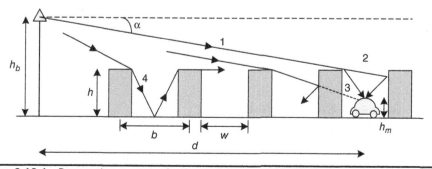

FIGURE 2.13.1   Propagation geometry for the model proposed by Walfisch and Bertoni.[24]

Equation (2.13.3) is further simplified by neglecting $1/(2\pi + \gamma - \alpha)$ as compared to $1/(\gamma - \alpha)$ and assuming that a is small compared to γ.

Using Eqs. (2.13.2), (2.13.3), (2.13.4), and (2.13.5), the excess loss is given by

$$L_{ex}(\text{dB}) = 57.1 + A + \log f_c + 18 \log d - 18 \log(h_b - h) - 18 \log\left(1 - \frac{d^2}{17(h_b - h)}\right) \qquad (2.13.6)$$

The last term in Eq. (2.13.6) accounts for the earth curvature. It can often be neglected. The building geometry A can be expressed as

$$A = 5 \log\left[\left(\frac{b}{2}\right)^2 + (h - h_m)^2\right] - 9 \log b + 20 \log\left\{\tan^{-1}\left[\frac{2(h - h_m)}{b}\right]\right\} \qquad (2.13.7)$$

When using isotropic antennas, the total path loss is obtained by adding L_{ex} to the free space path loss L_{FS}.

2.14   Ikegami Model[27]

The Ikegami model tries to predict the field strengths at each point in an urban coverage area using a ray-theoretical approach. Ikegami analyzed the controlling propagation factors based on the geometrical optics assuming a simple two-ray model.

Ikegami has made extensive drive tests in Kyoto City. He analyzed building heights, shapes, and positions with detailed city maps and traced ray paths between the transmitter and receiver using only single reflections from the walls of the buildings. Diffraction is calculated using a single-edge approximation at the building nearest the mobile unit. The building-wall reflection loss is assumed to be fixed as a constant value. The two rays, reflected and diffracted, are received by the mobile unit as follows:

$$E_m = E_{FS} - 10 \log f_{\text{MHz}} - 10 \log(\sin\phi) - 20 \log(h_b - h_m)$$
$$+ 10 \log W + 10 \log\left(1 + \frac{3}{L_r^2}\right) + 5.8 \qquad (2.14.1)$$

where E_m = local mean field strength, E_{FS} = free space field strength, L_r = reflection loss, set to a fixed value at −6 dB, ϕ = angle between the street and the direct line from base to mobile, h_b = building height in meters, h_m = receiver height in meters, W = street width in meters, and f_c = frequency in MHz.

The model represents the situation illustrated in Fig. 2.14.1. Further, the elevation angle θ of the base station from the top of the knife edge shown in the figure is negligible in comparison to the diffraction angle down to the mobile level. Therefore, there is no dependence on base station height. This model has been compared with the measurements at 210, 400, and 600 MHz as well as with transmitting antenna heights of 600, 500, and 120 m above mean ground level.

The Ikegami model assumes that all urban buildings in the area have the same height and that the base station antenna height does not affect propagation in the urban area. The model predicts that mean field strength is approximately determined by diffracted and reflected rays, which are subject only to geometrical optics.

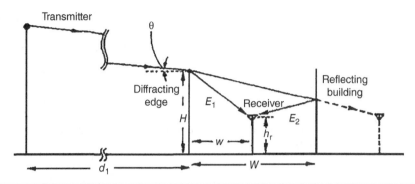

FIGURE 2.14.1   Ikegami model illustrated.[27]

2.15   Walfisch–Ikegami Model[28]

The Walfisch–Ikegami model is a semi deterministic model for medium-size to large cells in built-up areas. This model was adopted by COST (Cooperation of Scientific and Technical Research) Project 231 in Europe and is recommended by ITU for cellular and PCS applications.

The Walfisch–Ikegami model is applied in the in the range of 800 to 2000 MHz and path distances in the range of 0.02 to 5 km. The model provides two propagation situations: LOS and NLOS. In a LOS situation, there is no obstruction in the direct path between the transmitter and the receiver. This model assumes that the base station antenna height is $h_b \geq 30$ m such that the path has a high degree of Fresnel zone clearance.

For LOS paths, the path loss equation for the Walfisch–Ikegami model is

$$L_{WI}\,(\text{dB}) = 42.6 + 26 \log d_{\text{km}} + 20 \log f_{\text{MHz}} \quad (d \geq 20\ \text{m}) \tag{2.15.1}$$

while the free space path loss is

$$L_{FS}\,(\text{dB}) = 32.4 + 20 \log f_{\text{MHz}} + 20 \log d_{\text{km}} \tag{2.15.2}$$

When the distance $d_{\text{km}} \to 20$,   the loss $L_{WI} \to L_{FS}$.

For NLOS path situations, four factors are included:

- Height of buildings
- Width of roads
- Building separation
- Road orientation with respect to the LOS path

The Walfisch–Ikegami model for the NLOS path gives the path loss using the following parameters:

h_b = base antenna height over street level in meters (4 to 50 m)

h_m = mobile station antenna height in meters (1 to 3 m)

h_B = nominal height of building roofs in meters

$\Delta h_b = h_b - h_B$ = height of base antenna above rooftops in meters

$\Delta h_m = h_B - h_m$ = height of mobile antenna below rooftops in meters
b = building separation in meters (20 to 50 m recommended if no data)
W = width of street ($b/2$ recommended if no data)
ϕ = angle of incident wave with respect to street (use 90° if no data)

Using the parameters listed above, for NLOS propagation paths, the Walfisch–Ikegami model gives the following expression for the path loss in decibels:

$$L_{NLOS} = \begin{cases} L_{FS} + L_{rts} + L_{mds}, & L_{rts} + L_{mds} \geq 0 \\ L_{FS} & L_{rts} + L_{mds} < 0 \end{cases} \qquad (2.15.3)$$

where L_{FS} = free space loss = $32.45 + 20\log_{10} d_{km} + 20\log_{10} f_{MHz}$, L_{rts} = roof-to-street diffraction and scatter loss (see Eq. (2.15.4)), and L_{msd} = multiscreen diffraction loss (see Eq. (2.15.6)).

The formula given for L_{rts} involves an orientation loss, L(ϕ). Angle ϕ is the street orientation angle, as shown in Fig. 2.15.2.

$$L_{rts} = -16.9 - 10\log_{10} W + 10\log_{10} f_{MHz} + 20\log_{10} \Delta h_m + L(\phi) \qquad (2.15.4)$$

where f_{MHz} is the frequency in MHz and w_m is the distance between the building and the mobile on either side of the street, typically $w_m = W/2$, and the final term of Eq. (2.15.4) is the orientation loss for street orientation at an angle ϕ to the radio path:

$$L(\phi) = \begin{cases} -10 + 0.354\phi, & 0 \leq \phi \leq 35° \\ 2.5 + 0.075(\phi - 35°), & 35° \leq \phi \leq 35° \\ 4.0 + 0.114(\phi - 55°), & 55° \leq \phi \leq 90° \end{cases} \qquad (2.15.5)$$

The formula given for the multi screen diffraction loss term L_{msd} is

$$L_{msd} = L_{bsh} + k_a + k_d \log_{10} d_{km} + k_{MHz} \log_{10} f_{MHz} - 9\log_{10} b \qquad (2.15.6)$$

In this expression, L_{bsh} is shadowing gain, which occurs when the base station antenna is higher than the rooftops:

$$L_{bsh} = \begin{cases} -18\log_{10}(1 + \Delta h_b), & \Delta h_b > 0 \\ 0, & \Delta h_b \leq 0 \end{cases} \qquad (2.15.7)$$

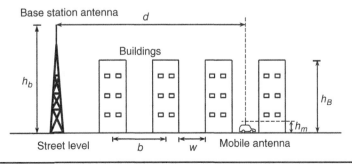

Figure 2.15.1   The geometry of the Walfisch–Ikegami model.

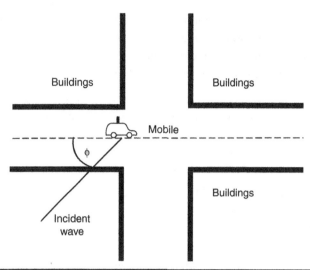

FɪɢᴜʀE **2.15.2** Defining the street orientation angle.

L_{msd} decreases for wider building separation (b). The quantities k_a, k_d, and k_{MHz} determine the dependence of loss on the distance (d_{km}) and the frequency (f_{MHz}). The term in the formula for the multiscreen diffraction loss is given by

$$
k_a = \begin{cases}
54, & \Delta h_b > 0 \\
54 + 0.8|\Delta h_b|, & \Delta h_b \leq 0 \text{ and } d_{km} \geq 0.5 \\
54 + 0.8|\Delta h_b|(d_{km}/0.5), & \Delta h_b \leq 0 \text{ and } d_{km} < 0.5
\end{cases} \tag{2.15.8}
$$

This relation in Eq. (2.15.8) shows that there is a 54-dB diffraction loss if the base station antenna is above the rooftops ($\Delta h_b > 0$), but more than a 54-dB loss if it is below the rooftops.

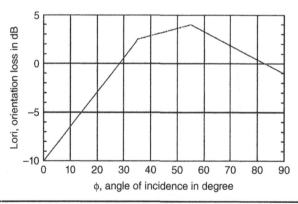

FɪɢᴜʀE **2.15.3** Orientation loss versus angle of incidence.

The loss will be just a little more than 54 dB if the link distance is less than 500 m. The distance factor k_d in Eq. (2.15.6) for L_{msd} is given by

$$k_d = \begin{cases} 18, & \Delta h_b > 0 \\ 18 + 15(|\Delta h_b|/h_B) & \Delta h_b \leq 0 \end{cases} \qquad (2.15.9)$$

L_{msd} increases with distance at 18 dB per decade if the base antenna is above the rooftops ($\Delta h_b > 0$). But if the antenna is below the rooftops, the k_d is higher than 18 db per decade.

The frequency factor k_{MHz} in the formula for the multiscreen diffraction loss is given by

$$k_{MHz} = -4 + \begin{cases} 0.7\left(\dfrac{f_{MHz}}{925} - 1\right) & \text{medium-size city and suburban areas} \\ 1.5\left(\dfrac{f_{MHz}}{925} - 1\right) & \text{metropolitan (urban) areas} \end{cases} \qquad (2.15.10)$$

L_{FS} and L_{rts} together give an increase of 30 dB per decade of frequency. For a typical cellular frequency of 850 MHz, the value of k_{Mhz} is about -4 dB for either situation, as shown in Eq. (2.15.10), so the total dependence on frequency for the 800-MHz cellular band is about 26 dB per decade.

If data are unavailable, the following default values are recommended:

$h = 3\text{m} \times (\text{number of floors}) + \text{roof height}$

Roof height, $h_B = \begin{cases} 3\text{ m} & \text{for pitched roofs} \\ 0\text{ m} & \text{for flat roofs} \end{cases}$

Building separation $b = 20$ to 50 m
Width of the street $w = b/2$
Angle of incident wave $\phi = 90°$

The COST model is applicable to the following range of parameters:

$f_{MHz} = 800$ to 2000 MHz
$h_b = 4$ to 50 m
$h_m = 1$ to 3 m
$d_{km} = 0.02$ to 5 km

The Walfisch–Ikegami model matches measurements quite well when the base station antenna is above rooftop height, producing mean errors of about 3 dB with standard deviations in the range 4 to 8 dB.

2.16   Flat-Edge Model

The flat-edge model was created by Saunders et. al.[29,30] assuming that all buildings are of equal height and spacing. The geometry is shown in Fig. 2.16.1, illustrating the following parameters:

r_1 = distance from the base station to the first building in meters
α = elevation angle of the base station antenna from the top of the final building in radiant when = 0, the base station

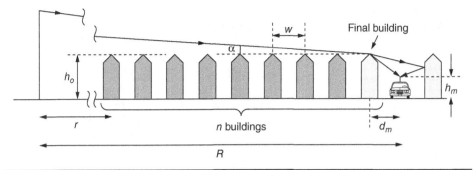

FIGURE **2.16.1**   Definition of geometry for the flat-edge model.

w = building separation

h_0 = building height

d_m = distance from the mobile to the last building in meters

The value of w used should be an effective way to account for the longer paths between the buildings for oblique incidence. The excess path loss is then expressed as

$$L_E = L_{n-1}(t)L_{ke} \tag{2.16.1}$$

where L_{ke} is the single-edge diffraction loss over the final building and $L_n - 1$ is the loss for multiple diffraction over the remaining $(n-1)$ buildings. The result is that $L_n - 1$ is a function of a parameter t only, where t is given by

$$t = -\alpha\sqrt{\frac{\pi w}{\lambda}} \tag{2.16.2}$$

It is given by the following formula:

$$L_n(t) = \frac{1}{n}\sum_{m-0}^{n-1} L_m(t)F_s(-jt\sqrt{n-m}) \quad \text{for } n \geq 1 \quad L_0(t) = 1 \tag{2.16.3}$$

where

$$F_s(jx) = \frac{e^{-jx^2}}{\sqrt{2j}}\left\{\left[S\left(x\sqrt{\frac{2}{\pi}}\right) + \frac{1}{2}\right] + j\left[C\left(x\sqrt{\frac{2}{\pi}}\right) + \frac{1}{2}\right]\right\} \tag{2.16.4}$$

and S and C are Fresnel integrals.

The flat-edge model may be calculated either directly from Eq. (2.16.3).

Walfisch[28] has introduced the concept of the "settled field": if the number of building n is large enough for buildings to occupy the whole of the first Fresnel zone from the line-of-sight ray, then the attenuation function can be given by

$$A_n(\alpha) = \exp\left(\frac{jkw\alpha^2}{2}\right)L_n\left[\alpha\sqrt{\frac{\pi w}{\lambda}}\right]$$

$$= \exp\left(\frac{jkw\alpha^2}{2}\right)L_n[t] \tag{2.16.5}$$

Equation (2.16.5) is applied to the distance r to the base station is very large, $r \gg n\,w$.

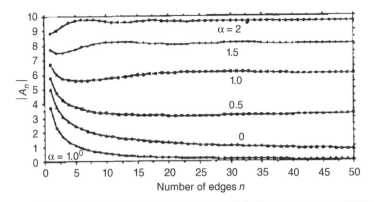

FIGURE 2.16.2   Variation of attenuation with number of edges - $w = 150\,\lambda$.

The field $A_n(\alpha)$ becomes independent of further increases in n, as can be seen from Fig. 2.16.2. From the results of numerical simulations Walfisch proposed the approximate function

$$|A| = 0.1 \left[\frac{\alpha}{0.03} \sqrt{\left(\frac{w}{\lambda}\right)} \right]^{0.9} \qquad \text{for } n \gg 1 \qquad (2.16.6)$$

where $|A|$ is the attenuation relative to the free space field strength. This function suggests a propagation loss dependence of approximately 38 dB/decade of range, which is in a close agreement with many reported measurements. This function is $|A|$ is only an approximate, however, and is not a function of n.

The overall path loss from the flat-edge model is a summation of three losses: the free space loss, the final building diffraction loss, and the reflections from the buildings across the street using the Ikegami model, shown as follows:

$$L_T = L_{FS} + L_E + L_n(t) \qquad (2.16.7)$$

where L_{FS} is the free space loss, L_E is given in Eq. (2.16.1), and $L_n(t)$ can be found from Eq. (2.16.3).

This formulation is extremely simple to compute and applies for any values of angle α, even when the base station antenna height is below the rooftop level. The number of buildings can be increased to extremely high values without difficulty.

2.17   ITU Model

It is often difficult to evaluate the best model for a given application from the many described in this chapter. Recommendations produced by the ITU are a good reference source in this situation, as they summarize in simple form some recommended procedures, as shown in Table 2.17.1.

The ITU's recommendations may not always represent the most accurate model for a given case, but they have the benefit of being widely accepted and used for coordination and comparison purposes.

We'll focus on ITU-R P.1546 and ITU-R P.530 in this section. ITU-R P.1141 is more focused on microcells and will be discussed in Chap. 4.

ITU-R P.1411	Propagation data and prediction methods for the planning of short-range outdoor radio communication systems and radio local area networks in the frequency range 300 MHz to 100 GHz
ITU-R P.1546	Method for point-to-area predictions for terrestrial services in the frequency range 30 MHz to 3000 MHz
ITU-R P.530	Propagation data and prediction methods required for the design of terrestrial line-of-sight systems
ITU-R P.341	UHF frequency range, basic transmission loss
ITU-R P.676	Gaseous attenuation
ITU-R P.526	Knife-edge diffraction
ITU-R P.833	Propagation through trees specific attenuation in vegetation
ITU-R P.679	Building entry for terrestrial systems
ITU-R P.1057	Definitions of probability distributions
ITU-R P.1238	Indoor propagation over the frequency range 900 MHz to 100 GHz
ITU-R P.370 and ITU-R P.529	Propagation for over distances of 1 km and greater, and over the frequency range 30 MHz to 3 GHz
ITU-R P.1407	Multipath propagation and definition of terms
ITU-R P.453	Effects of refraction

TABLE 2.17.1   ITU-R Recommendations Table

2.17.1   ITU-R Recommendation P.1546[31]

This is the recommendation of ITU-R P.1546 for point-to-area prediction of field strength for broadcasting, land mobile, maritime mobile, and certain fixed services, such as point-to-multipoint systems. This model uses a semiempirical method for the reliable prediction of radio propagation at VHF and UHF bands. It covers point-to-area predictions for terrestrial services in the frequency range 30 to 3000 MHz and for the distance range of 1 to 1000 km.

This method can include the effects of terrain, scattering objects of the environment, and other propagation conditions, among various factors and corrections. There are 21 tunable parameters in the point-to-area prediction method of ITU-R P.1546, providing a high degree of freedom for tuning the model.

The propagation curves represent field strength values for 1 kW of ERP at nominal frequencies of 100, 600, and 2000 MHz, respectively, as function of various parameters, including frequency, transmitting antenna height (10 to 1200 m), time variability (percentage of time exceeded 50, 10, and 1 percent), and path type (land, cold sea, and warm sea), the height of the receiving (mobile) antenna being equal to the representative height of ground cover and distance. These curves are based on measurement data relating mainly to mean climatic conditions in temperate regions containing cold and warm seas.

A rigorous interpolation/extrapolation procedure is given to allow prediction for any input values within the specified range. The model is based on series of curves (or tables) originating from measurements and allowing predictions for wide area macrocells and for broadcasting and fixed wireless access applications.

The field strength at each measurement point is calculated for a given percentage of time inside the range from 1 to 50 percent. The field strength will be exceeded for $t\%$ of times at each receiver location given by

$$E(t) = E_t(\text{median}) + Qi(t/100)\,\sigma_T \text{ dB}(\mu V/m) \tag{2.17.1.1}$$

where $E_t(\text{median})$ is the median field strength with respect to the time at the receiver location, $Qi(x)$ is the inverse complementary cumulative normal distribution as a function of probability, and σ_T is the standard deviation of normal distribution of the field strength at the receiver location. ITU-R P1546 is intended to provide the statistics of reception conditions over a given area, not at any particular point. The field strength value at $q\%$ of location within an area represented by a square of 200×200 m is given by

$$E(q) = E_L(\text{median}) + Qi(q/100)\,\sigma_L \text{ dB}(\mu V/m) \tag{2.17.1.2}$$

where $E_L(\text{median})$ and σ_L are the median and standard deviation of field strength over the defined area, respectively and q is the percentage of location varying between 1 and 99.

When the terrain information is available, the transmitting (base) antenna height h_1 should be obtained as follows:

Condition 1: For land paths shorter than 15 km,

$$h_1 = h_b$$

where h_b is the height of the antenna above terrain height averaged between $0.2\,d$ and d in kilometers and d is the distance between the transmitter and the receiver.

Condition 2 : For land paths of 15 km or longer,

$$h_1 = h_{eff}$$

where h_{eff} is defined as the transmitter height in meters over the average level of the ground between distances of 3 and 15 km from the transmitting (base) antenna in the direction of the receiving (mobile) antenna.

The following formulas are used according to the recommendation for field strength prediction:

$$l_d = \log_{10}(d) \tag{2.17.1.3}$$

$$k = \frac{\log_{10}\left(\dfrac{h_1}{9.375}\right)}{\log_{10} 2} \tag{2.17.1.4}$$

$$E_1 = \left(a_0 k^2 + a_1 k + a_2\right)l_d + (0.1995 k^2 + 1.8671 K + a_3) \tag{2.17.1.5}$$

$$E_{ref1} = b_0[\exp(-b_4 10^{l_d^{b5}}) - 1] + b_1 \cdot \exp\left[-\left(\frac{l_d - b_2}{b_3}\right)^2\right] \tag{2.17.1.6}$$

$$E_{ref2} = -b_6 l_d + b_7 \tag{2.17.1.7}$$

$$E_{ref} = E_{ref1} + E_{ref2} \tag{2.17.1.8}$$

$$E_{off} = c_5 k^{c_6} + \frac{c_0}{2} k \left\{ 1 - \tan h \left[c_1 \left(l_d - \left(c_2 + \frac{c_3^k}{c_4} \right) \right) \right] \right\} \tag{2.17.1.9}$$

$$E_2 = E_{ref} + E_{off} \tag{2.17.1.10}$$

$$p_b = d_0 + d_1 \sqrt{k} \tag{2.17.1.11}$$

$$E_u = \min(E_1, E_2) - p_b \log_{10} \left(1 + 10^{\frac{|E_1 - E_2|}{p_b}} \right) \tag{2.17.1.12}$$

$$E_{fs} = 106.9 - 20 l_d \tag{2.17.1.13}$$

$$E_b = \min(E_u, E_{fs}) - 8 \log_{10} \left(1 + 10^{\frac{|E_u - E_{fs}|}{8}} \right) \tag{2.17.1.14}$$

$$\text{Corrections} = C_{\text{ERP}} + C_{h_2} + C_{\text{urban}} + C_{\text{TCA}} + C_{h_1 < 0} \tag{2.17.1.15}$$

$$E_c = E_b + \text{corrections} \quad \text{dB} \left(\mu V \middle/ m \right) \tag{2.17.1.16}$$

In the above equations, d and h_1 are in kilometers and meters, respectively; E_{fs} is the free space field strength; and E_b is the propagating field strength without considering the corrections. Both E_{fs} and E_b are received from 1 kW ERP. All coefficients $a_0, a_1, \ldots, a_7$, $b_0, b_1, \ldots, b_7, c_0, c_1, \ldots, c_6, d_0$ are given for nominal frequencies and time percentage in the recommendation. These coefficients are defined as the optimization parameters in the optimization algorithm. $C_{\text{ERP}}, C_{h_2}, C_{\text{urban}}, C_{\text{TCA}}$, and $C_{h_1} < 0$ are the corrections for effective radiated power, receiving (mobile) antenna height, short urban/suburban paths, terrain clearance angle, and negative values of h_1, respectively. The related formulas for calculation of $C_{h_2}, C_{\text{urban}}, C_{\text{TCA}}$ and $C_{h_1} < 0$ can be found in.[31] The correction C_{ERP} must be added to E_b if the ERP of the transmitter antenna is not equal to the nominal value of 1 kW:

$$C_{\text{ERP}} = 10 \log_{10} \left(\frac{\text{ERP}}{1000} \right) \tag{2.17.1.17}$$

The above formulas are used according to the recommendation P.1546 for point-to-area field strength prediction.

2.17.2 Recommendation ITU-R P.530-9[32]

This ITU terrain model is based on diffraction theory and provides a relatively quick means of determining a median path loss.[32] Diffraction loss will depend on the type of terrain and the vegetation. For a given path ray clearance, the diffraction loss will vary from a minimum value for a single-knife-edge obstruction to a maximum for smooth spherical earth. Methods for calculating diffraction loss for the single-knife-edge case and for the smooth spherical earth case as well as for paths with irregular terrain are discussed in Recommendation ITU-R P.526—Knife-Edge Diffraction. These two cases are shown in Fig. 2.17.2.1.

The diffraction loss over average terrain can be approximated for losses by the formula

$$A_d = -20 h / F_1 + 10 \quad \text{dB} \tag{2.17.2.1}$$

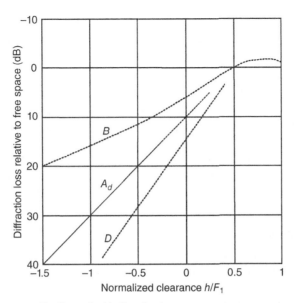

B: theoretical knife-edge loss curve
D: theoretical smooth spherical Earth loss curve,
       at 6.5 GHz and k_e= 4/3
A_d: empirical diffraction loss based on equation
       (2) for intermediate terrain
h: amount by which the radio path clears the Earth's
       for intermediate terrain
F_1: radius of the first Fresnel zone

Figure 2.17.2.1   Diffraction loss for obstructed LOS radio paths.[32]

where *h* is the height difference (m) between most significant path blockage and the path trajectory. The height difference *h* is negative if the top of the obstruction of interest is above the virtual line of sight.

F_1 is the radius of the first Fresnel ellipsoid given by

$$F_1 = 17.3 \sqrt{\frac{d_1 d_2}{fd}} \quad \text{m} \tag{2.17.2.2}$$

where f = frequency (GHz), d = path length (km), and d_1 and d_2 = distances (km) from the two terminals to the path obstruction.

A curve, referred to as A_d, based on Eq. (2.17.2.1), is also shown in Fig. 2.17.2.1. This curve, strictly valid for losses larger than 15 dB, has been extrapolated up to 6-dB losses.

Figure 2.17.2.1 shows three plots of the expected diffraction loss due to terrain roughness versus the normalized terrain clearance. Curve *B* is the theoretical knife-edge diffraction curve. Curve *D* is the theoretical smooth-earth loss at 6.5 GHz using a 4/3 earth radius. Curve A_d is the ITU terrain loss model over intermediate terrain. Each of these curves represents the excess terrain loss beyond the free space loss.

The ratio h/F_1 is the normalized terrain clearance. When the terrain blocks the line of sight, $h/F_1 < 0$. This model is generally considered valid for losses about 15 dB, but it is acceptable to extrapolate to as little as 6 dB of loss. The other two curves shown represent extremes of clear terrain and very rough terrain, so they provide insight into the variability that can be expected for any given value of normalized clearance.

2.18   On-Body Model[33]

The on-body model is a new concept. When the terminal handset is used, the body size, shape, and posture, as well as the orientation and distance of the antenna from the body of the user will become a part of the system. It has been called the body-area network (BAN).[33] This BAN will affect the reception of the mobile signal and need to be included in the prediction of the overall path loss of the mobile radio propagation. We include it in this chapter because it is an area network. Also, it affects the performance of all cellular networks, from macrocell to in building (picocell). There are two models.

2.18.1   Model 1

The received signal may be represented as the combination of a constant LOS signal and a Rayleigh-distributed time-varying component. The spatial subchannel linking the ith receive element and the jth transmit element in an $(N \times M)$ MIMO channel matrix H can be represented as

$$h_{ij}(t) = \sqrt{\frac{p_r}{K+1}} \left[\sqrt{K} e^{j\varphi_{ij}} + Z_{ij}(t) \right] \qquad (2.18.1.1)$$

where K is the Rician factor, φ_{ij} is the phase of the jth transmit element to ith subchannel in the constant component, and z_{ij} (t) is the correlated NLOS component. The phase φ_{ij} is randomly distributed over $[0, 2\pi]$. The received power term p_r can be modeled for a given transmitter-receiver separation as

$$p_r(d) = p_r(d_0) - 10\mathrm{n} \log\left(\frac{d}{\mathrm{d0}}\right) + X_{\mathrm{shad}}(d)$$

where $X_{\mathrm{shad}}(d)$ is the lognormally distributed shadowing term. The first term in Eq. (2.18.1.1) corresponds to the LOS component, and the second term corresponds to NLOS component.

2.18.2   Model 2

In model 1, the receive antenna did not encounter identical statistics owing to the variation in their position, orientation, and the amount of shadowing. In model 2, the index I is attached to the receive power $(p_r)_i$ and to the Rician K-factor in order to indicate that they depend on the receive antenna. It follows that the $h_{ij}(t)$ of the MIMO channel matrix in this model can be represented as

$$h_{ij}(t) = \sqrt{\frac{K_i(p_r)i}{K_i+1}} e^{j\varphi_{ij}} + \sqrt{\frac{(p_r)i}{K_i+1}} z_{ij}(t) \qquad (2.18.2.1)$$

The two models are used to predict the on-body antenna orientation and position. For each on-body channel, the transmitting array was placed at the belt position at the left side of the body, while the receiving array was placed at the right side. The receiving array that was placed on the head is called belt-head channel, and that placed at the chest is called the belt-chest channel. Model 1 has been shown to well represent the belt-chest channel, while model 2 provides a more reliable representation of the belt-head channel and a good estimation of the capacity in such a channel. Readers who are interested in this topic should read the reference.[33]

2.19   Summary

In this chapter, we have selected only those macrocell prediction models that are widely used by the industry. Some of these models are simple and some are complicated. In general, we can categorize on the basis of the different approaches of prediction into three kinds of models: physical, empirical, and statistical. Physical models are not practical to use in macrocell prediction because manipulating the model in coverage areas is complex. Statistical and empirical models are commonly used. For planning purposes, there are three different models: area-to-area, point-to-area, and point-to-point models. When exploring an area for a cellular system, the area-to-area model or point-to-area model is enough to get the information we are looking for. However, if deploying a cellular system in an area, then the point-to-point model is more adequate. Two point-to-point models will be introduced in Chap. 3: the Lee model and the Longley–Rice model.

We also include the on-body model in this chapter to give readers an overall picture of predicting the outcome from a cellular system, not only the radio path from the base station to the receiver but also the received signal affected by the human body.

References

1. Young, W. R. "Comparison of Mobile Radio Transmission at 150, 450, 900 and 3700 MC." *Bell System Technical Journal* 31 (1952): 1068–85.
2. Parsons, J. D. *The Mobile Radio Propagation Channel*. 2nd ed. West Sussex: Wiley, 2000.
3. Blaunstein, N. *Radio Propagation in Cellular Networks*. Norwood, MA: Artech House, 2000.
4. Bullington, Kenneth. "Radio Propagation Variations at VHF and UHF." Proceedings of the Institute of Radio Engineers, 1950.
5. Bullington, Kenneth. "Radio Propagation for Vehicular Communications." *IEEE Transactions on Vehicular Technology* 26, no. 4 (November 1977): 295–308.
6. Egli, J. J. "Radio Propagation above 40MC over Irregular Terrain." *Proceedings of the Institute of Radio Engineers* (1957): 1383–91.
7. CCIR. "Methods and Statistics for Estimating Field Strength Values in the Land Mobile Services Using the Frequency Range 30MHz to 1GHz." CCIR XV Plenary Assembly, Geneva, Report 567, Vol. 5, 1983.
8. Palmer, F. H. "The CRC VHF/UHF Propagation Prediction Program: Description and Comparison with Field Measurements." *NATO-AGARD Conference Procedings* 238 (1978): 49/1–49/15 (including JRC method).
9. Edwards, R. and Durkin, J. "Computer Prediction of Service Area for VHF Mobile Radio Networks." *Proceding of the IEE* 116, no. 9 (1969): 1493–500.
10. Dadson, C. E. "Radio Network and Radio Link Surveys Derived by Computer from a Terrain Data Base." NATO-AGARD Conference Publication CPP-269, 1979.
11. The Terrain-Integrated Rough-Earth Model TIREM, http://handle.dtic.mil/100.2/ADA296913 (accessed June 17, 2008).
12. Carey, R. "Technical Factors Affecting the Assignment of Frequencies in the Domestic Public Land Mobile Radio Service." Report R-6406. Washington DC: Federal Communications Commission, 1964.

13. Lee, W. C. Y. *Mobile Communications Design Fundamentals*. 2nd ed. New York: John Wiley & Sons, 1993.

14. Grosskopf, Rainer, "Comparison of Different Methods for the Prediction of the Field Strength in the VHF Range," *IEEE Transactions on Antennas and Propagation*, 35, no. 7 (July 1987): 852-59.

15. Hata, M. "Empirical Formula for Propagation Loss in Land Mobile Radio Services." *IEEE Transactions on Vehicular Technology* 29 (1980): 317–25.

16. Blomquist, A., and Ladell, L. "Prediction and Calculation of Transmission Loss in Different Types of Terrain." NATO AGARD Conference Publication CP 144, Research Institute of National Defense, Stockholm, 1974, 32/1–32/17.

17. Ibrahim, M. F., and Parsons, J. D. "Signal Strength Prediction in Built-Up Areas, Part I." *Proceedings of the IEEE* 130, no. 5, Pt. F (1983): 377–84.

18. Allsebrook, K., and Parsons, J. D. "Mobile Radio Propagation in British Cities at Frequencies in the VHF and UHF Bands." *IEEE Transactions on Vehicular Technology* 26, no. 4 (1977): 95–102.

19. Okumura, Y., Ohmori, E., Kawano, T., and Fukuda, K. "Field Strength and Its Variability in the VHF and UHF Land Mobile Radio Service." *Rev. Elec. Commu. Lab*. 16, no. 9/10 (1968): 825–73.

20. Delisle, G. Y., Lefevre, J. P., Lecours, M., and Chouin, J. Y. "Propagation Loss Prediction: A Comparative Study with Application to the Mobile Radio Channel." *IEEE Transactions on Vehicular Technology* 34, no. 2 (1985): 86–95.

21. Aurand, J. F. "A Comparison of Prediction Models for 800 MHz Mobile Radio Propagation." IEEE Transactions on Vehicular Technology 34, no. 4 (1985): 149–53.

22. Akeyama, A., Nagatsu, T., and Ebine, Y. "Mobile Radio Propagation Characteristics and Radio Zone Design Method in Local Cities." *Rev. Elec. Comm. Lab*. 30 (1982): 308–17.

23. The Telecommunications Industry Association (TIA) recommends its publication TSB-88A. Also, the Hata–Davidson model appeared in "A Report on Technology Independent Methodology for the Modeling, Simulation and Empirical Verification of Wireless Communications System Performance in Noise and Interference Limited Systems Operating on Frequencies between 30 and 1500 MHz." TIA TR8 Working Group, IEEE Vehicular Technology Society Propagation Committee, May 1997.

24. Walfisch, J., and Bertoni , H. L. "A Theoretical Model of UHF Propagation in Urban Environments." *IEEE Transactions on Antennas and Propagation* 36, no. 12 (October 1988): 1788–96.

25. Maciel, L. R., Bertoni, H. L., and Xia, H. H. "Unified Approach to Prediction of Propagation over Buildings for All Ranges of Base Station Antenna Height." *IEEE Transactions on Vehicular Technology* 42, no. 1 (February 1993): 41–45.

26. Ikegami, F., and Yoshida, S. "Analysis of Multipath Propagation Structure in Urban Mobile Radio Environments." *IEEE Transactions on Antennas and Propagation* (1980): 531–37.

27. Ikegami, F., Yoshida, S., Tacheuchi, T., and Umehira, M. "Propagation Factors Controlling Mean Field Strength on Urban Streets." *IEEE Transactions on Antennas and Propagation* 32, no. 8 (1984): 822–29.

28. COST 231 Final Report, Digital Mobile Radio: COST 231. "View on the evolution towards 3rd Generation Systems." Brussels: Commission of the European Communities and COST Telecommunications, 1999.

29. Saunders, S. R., and Bonar, F. R. "Explicit Multiple Building Diffraction Attenuation Function for Mobile Radio Wave Propagation." *Electronics Letters* 27, no. 14 (1991): 1276–77.

30. Saunders, S. R., and Bonar, F. R. "Prediction of Mobile Radio Wave Propagation over Buildings of Irregular Heights and Spacings." *IEEE Transactions on Antennas and Propagation* 42, no. 2 (1994): 137–44.

31. International Telecommunication Union, ITU-R Recommendation P.1546. "Method for Point-to-Area Predictions for Terrestrial Services in the Frequency Range 30 MHz to 3 000 MHz." Geneva: ITU, 2012.

32. ITU-R Recommendations. "Propagation Data and Prediction Methods Required for the Design of Terrestrial Line-of-Sight Systems." Geneva: ITU-R, 2001.

33. Ghalida, G., Khan, I., Hall, P., and Hanzo, L. "MIMO Stochastic Model and Capacity Evaluation of On-Body Channels." *IEEE Transaction on Antenna and Propagation* 60 (June 2012): 2980–86.

CHAPTER 3

Macrocell Prediction Models—Part 2: Point-to-Point Models

3.1 The Lee Model[1-5]

The calculation of macrocell coverage is based on signal propagation prediction techniques. The Lee "point-to-point" model predicts each local mean of the received signal strength at the mobile terminal based on the radio path between the base station and the mobile but plots all the predicted local means along a mobile path in a mobile cellular environment. The point-to-point model is a cost-saving tool used for planning a system in a large coverage area. Otherwise, selecting the proper sites of a base station based on the measured data in a large area is very costly and labor intensive.

The predicted signal strengths also can be calculated on a sector-by-sector basis. It is used as an area-to-area model or point-to-area model. A recommendation is to use a sector of 60° or 120°. In each sector, we may predict the local means along every 0.25° or 0.5° radial increment within a sector and also use a small distance (radial length) increment along each radial line. The prediction of local means for each sector is stored in a data file. These files can be used to support most analyses, such as to generate a neighboring cell list, frequency planning, traffic demand distribution, hardware dimensioning, and so on.

The Lee model uses a reference frequency of 850 MHz; however, the model has been validated with the measured data within the frequency range from 150 to 2400 MHz.

Measurements taken from over 500 drive tests conducted in Europe, Asia, and the United States have been used to test the Lee model's implementation. These data cover many different cities and many different characteristics of morphologies, such as dense urban, urban, suburban, and rural.

The Lee model is also integrated with measurement data and can be customized in any particular area or to support the use of path loss values that are derived from drive test data as the basis for any analysis supported. This can be handled alone or in conjunction with predicted data. Empirical data can also be used to "tune" path loss slopes and intercepts, using a technique known as measurement integration, to improve prediction accuracy for the Lee model.

Based on the measurement results, the Lee model is enhanced to combine line-of-sight (LOS) gain and non–line-of-sight (NLOS) loss together for more realistically predicting the signal strength. This is covered in Sec. 3.3 of enhanced Lee model.

This chapter focuses on theoretical and technical aspects of the algorithm of Lee macrocell modeling.

3.1.1   Implementation of the Lee Macrocell Model

3.1.1.1   Models of Implementation

Three modes of the Lee macrocell model are analyzed here. The basic mode of the Lee model is the Lee single breakpoint model; the other two modes are the multiple breakpoint model and the automorphology model.

The other two modes of macrocell model are derived from the Lee single breakpoint model. Therefore, the theoretical discussion in this chapter focuses primarily on the Lee single breakpoint model, with additional explanations of how the other two modes differ from it.

In implementation, the easiest way to select a mode among the three modes is based on the granularity in a sector-by-sector basis. Thus, a coverage plot may consist of different sectors using different modes, based on the appropriate modeling objectives for dealing with different specific propagation conditions in a given sector.

The accuracy of using each mode depends on a number of factors, as discussed below. Users have the opportunity to customize a certain mode among the three modes in the Lee macrocell model by feeding the empirical data into the modeling parameters to increase prediction accuracy and to meet the specific.

3.1.1.2   Accuracy of Model Variations

The multiple breakpoint model is currently considered to provide the most accurate prediction and is the one recommended to be used. This model was designed specifically to improve prediction accuracy for "near in" distances (within 1 mile from the base station).

Imported empirical data can be used to increase the accuracy of both the single breakpoint and the multiple breakpoint models through a technique known as measurement integration. Many radial zones can be identified due to the user's specified requirements from running this process; the more zones identified, the more accurate the prediction results provided.

The degree of accuracy of signal strength predictions is based on the automorphology model and depends on a careful selection of a path loss slope value for each land-use class based on an appropriate propagation condition. But measurement integration technique cannot be used with this mode.

All three modes of the Lee macrocell model may be customized to the extent that certain model parameters and constants can be specified in a particular mode.

- *Model parameters*, which include the radial length increment dx (sampling interval), gain factor, loss factor, and a distance from reflection point and a mobile signal smoothing. Values for these model parameters may be used for the choice of modes, path loss, intercept point value, and so on associated with each different environment type.

- *System constants*, which include parameters related to the use of water enhancement, earth curvature, and terrain averaging.

- *Center frequency*, which can be used to further optimized the prediction.
- *Path loss slope* values per land-use class, for use with the automorphology model.

Sector parameters can be specified for each sector, which include environment type, the mode of the model to be used, path loss slope(s)/intercept(s), radial length, and number of radials.

When using the single breakpoint model or the automorphology model, you may specify a single slope and a 1-mile intercept. When you choose to use the multiple breakpoint model, you may specify additional slopes and intercepts.

The choice of environment type brings with it the associated values for the model parameters, such as radial length dx, gain factor, loss factor, and distance from reflection point and mobile smoothing. This set of parameters is independent of the choice for all other sector parameters related to macrocell modeling.

All sector parameters may be specified for each sector individually or on a group basis for any group of cells/sectors.

The model is based on the experimentation that can be created through the terrain and signal profiles or by including the slope and intercept calculations through measurement integration.

3.1.2 The Lee Single Breakpoint Model—A Point-to-Point Model

The Lee single breakpoint model forms the basis for all three modes of macrocell modeling. This model is given by the following equation:

$$P_r = P_{r_0} - \gamma \cdot \log\left(\frac{r}{r_0}\right) - A_f + G_{effh}(h_e) - L + \alpha \qquad (3.1.2.1)$$

where P_r = the received power in dBm

P_{r_0} = the received power at the intercept point r_0 in dBm

γ = the slope of path loss in dB per decade

r = the distance between the base station and the mobile in miles or kilometers

r_0 = the distance between the base station and the intercept point in miles or kilometers

h_e = the effective base station antenna height in feet or meters

h_2 = the mobile antenna height in feet or meters

h_1' = actual antenna height at the base station

h_e' = maximum effective antenna height

A_f = the frequency offset adjustment in dB from the default frequency f_0

$$= 20 \log\left(\frac{f}{f_0}\right) \qquad (3.1.2.2)$$

$G_{effh}(h_e)$ = gain from effective antenna heights h_e due to the terrain contour

$$= 20 \log\left(\frac{h_e}{h_1'}\right) \qquad \text{(under a nonshadow condition)} \qquad (3.1.2.3)$$

Max. $G_{effh}(h_e')$ = the maximum effective antenna height gain

$$= 20 \log\left(\frac{h_e'}{h_1'}\right) \qquad (3.1.2.4)$$

$L =$ (the knife-edge diffraction loss) $-$ (the maximum effective antenna height gain)

$$= L_D - 20 \log\left(\frac{h_e'}{h_1'}\right) \geq 0 \quad \text{(under a shadow condition)} \qquad (3.1.2.5)$$

The effective antenna height h_e' is measured the height at the base station from the intersected point of a line that is drawn from the tip of the hill along the slope of the hillside to the base station. The description of h_e' is shown in Sec. 3.1.2.4.

$\alpha =$ the signal adjustment factor in dB, such as additional gains if the actual antenna gains g_b' and g_m' and antenna heights h_1' and h_2' at two terminals are different from the standard conditions

$$= (g_b' - g_b) + (g_m - g_m') + 20 \log\left(\frac{h_1'}{h_1}\right) + 10 \log\left(\frac{h_2'}{h_2}\right)$$

$$= \Delta g_b + \Delta g_m + \Delta g_{h1} + \Delta g_{h2} \qquad (3.1.2.6)$$

The signal strength value given by the Lee single breakpoint model is composed of four components:

1. The area-to-area path loss, used as a baseline for the model, is derived from a propagation slope (γ) and 1-mile (or extrapolate to 1 km if not measured at 1 km) intercept value (P_{r_0}). P_{r_0} may be obtained from the measured data. Because P_{r_0} and γ vary from city to city due to human-made structures and only measured data can give the answer, the effect due to human-made structures is accounted in the model. The area-to-area component includes a frequency-offset adjustment (A_f), which is used to adjust the actual center frequency of your system to the model's reference frequency of 850 MHz (see Table 3.1.2.1.2).

2. The effective antenna height gain G_{effh} is determined by the terrain contour between the base station and the mobile where a specular reflection point is located. This component is significant for the case of a nonobstructed direct signal path, including LOS and NLOS paths. It accounts the terrain-contour effect into the model.

3. Diffraction loss L is predicted using Fresnel–Kirchoff diffraction theory. For multiple knife edges, Lee uses both a modified Epstein–Petersen method[3] and a separated knife-edge check to evaluate diffraction loss. This component is significant for the condition of an obstructed direct signal path or called diffraction path. It is also the effect due to the terrain contour and is also accounted for in the model.

4. An adjustment factor compensates for the difference between a set of default conditions for base station transmit with mobile parameters (which are used as assumptions for calculating the area-to-area path loss component) and the actual values for these parameters in each sector to be predicted.

Once received signal strength values have been predicted for each point along the signal path (radial), the Lee model can use a signal-smoothing process to produce the final prediction.

A point-to-point model constantly calculates along the signal path for path loss and checks for signal obstructions. There are three conditions for the direct path. When the direct path is not obstructed by the terrain, it has two conditions: LOS path and the NLOS path. Also, there is a diffraction path that is obstructed by the terrain:

1. Nonobstructed direct path or LOS path—When a direct path is not obstructed by any objects between the transmitter and the mobile and the reflected path from the ground is weak, the free space path loss formula will be used. If the mobile is very close to the ground, the reflected path from the ground is strong, and the open area case is considered.

2. NLOS path—When the direct path is obstructed from human-made structures and trees, common condition in mobile communications, there is a gain or loss according to the terrain contour along the radial path. The effective antenna height gain is a significant contributor to the received signal. This gain varies according to the terrain contour along the radial path. The effective antenna height gain is represented by $G_{effh}(h_e)$ in Eq. (3.1.2.3)

3. Diffraction path—When the direct path is obstructed by the terrain contour between the transmitter and the mobile, the signal actually incurs a loss due to terrain-related diffraction. This diffraction loss is designated by symbol L in Eq. (3.1.2.5).

The direct path and a shadow region are illustrated in Fig. 3.1.2.1. In the area of transition from the shadow region to the clear region, the nonobstructed direct path appears only in the clear region, and the signal received at the mobile may have two wave components: a direct path and a reflected path. The reflected path occurs if the effective reflective point exits clear from the shadow region and an effective antenna height gain is yielded from the reflected wave.

Both the effective antenna height gain and diffraction loss are also known as "attenuation adjustments." The calculation of these adjustments depends on terrain elevation effects.

3.1.2.1 Area-to-Area Path Loss

Area-to-Area Path Loss is the first component of the model as shown in Eq. (3.1.2.1). It is due to human-made effect on the environment such as structures.

There are a number of possible path loss curves measured in different areas used as references in predicting the area-to-area path loss in the modeling of a macrocell. The Lee single breakpoint model uses the suburban area-to-area path loss curve as its starting reference curve.

The path loss curve is derived from the slope γ and 1-mile intercept values (P_{r_0}), which are used to predict the received signal strength (P_r) at each mobile point along the radio

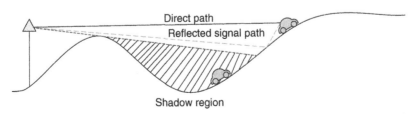

FIGURE 3.1.2.1   Direct path and shadow region.

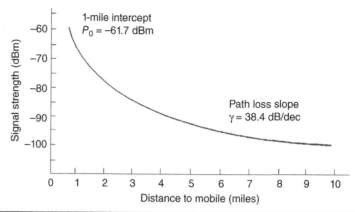

FIGURE **3.1.2.1.1**   Path loss reference curve (suburban).

path from the base station to the mobile. The 1-mile intercept value is based on the given ERP at the base station and the antenna heights of both at the base and at the mobile.

The slope and intercept values, which can be adjusted by the user, are defined as follows:

- The slope indicates signal path loss along the radio path in decibels per decade (dB/dec). The suburban curve uses a default value of 38.4 dB/dec for the path loss slope over land.

- The 1-mile intercept indicates the signal level (dBm) received, under the standard conditions as shown below, at a distance of 1 mile from the transmitter. The reference value for this 1-mile intercept is –61.7 dBm.

The predicted signal strength (P_r) from the area-to-area path loss of the model is given by one component of Eq. (3.1.2.1) as follows:

$$P_{r(\text{area-to-area})} = P_{r_0} - \gamma \log\left(\frac{r}{r_0}\right) - A_f + \alpha \qquad (3.1.2.1)$$

where P_{r_0} = the received power at the intercept point in dBm
γ = the slope of path loss in dB per decade
r = the distance between the base station and the mobile in miles (or kilometers)
r_0 = the distance between the base station and the intercept point in miles (or kilometers)
A_f = the frequency-offset adjustment from 850 MHz in dB

$$= 20 \log\left(\frac{f}{850}\right) \qquad (3.1.2.2)$$

$$\alpha = \Delta g_b + \Delta g_m + \Delta g_{h_1} + \Delta g_{h_2}$$

$$= (g'_b - g_b) + (g'_m - g_m) + 20 \log\left(\frac{h'_1}{h_1}\right) + 10\log\left(\frac{h'_2}{h_2}\right) \qquad (3.1.2.3)$$

where g_b, g_m, h_1, and h_2 are described below with their standard values shown. Those g'_b, g'_m, h'_1, and h'_2 are the values different from the standard ones.

The standard conditions assumed for the path loss prediction include the following base station transmit power and mobile parameters:

P_t = base station transmit power = 10 W
h_1 = base station antenna height = 100 ft (~30.5 m)
h_2 = mobile antenna height = 10 ft (3 m)
g_b = base station antenna gain = 6 dBd (dB over dipole)
g_m = mobile antenna gain = 0 dBd (dB over dipole)

The Lee model uses the standard base station antenna height to determine effective antenna height gain. The model also uses the actual base station antenna height to determine the diffraction loss.

An adjustment factor is also applied to the signal strength prediction to fine-tune for the difference between the standard condition values and the values of the actual parameters.

3.1.2.1.1   Significance of the 1-Mile Intercept   The 1-mile intercept (P_{r_0}) used by the Lee single breakpoint model is an initial parameter at which the signal received, under standard conditions, at a distance of 1 mile (1.609 km) from the base station. There are four main reasons the model uses a 1-mile intercept for predicting propagation for near-in distances, as follows:

1. Within a radius of 1 mile, the antenna beam width is narrow in the vertical plane; this is especially true of high gain omnidirectional antennas. Thus, the signal reception is reduced at a mobile less than 1 mile away because of the large elevation angle. This angle causes the mobile to be in the shadow region outside the main beam. The larger the elevation angle, the weaker the reception level due to the antenna's vertical pattern.

2. There are fewer roads within a radius of 1 mile around the base station, and the data are insufficient to create a statistical curve. Also, the road orientation—both in-line and perpendicular—close to the base station can cause a significant difference (from 10 to 20 dB) in signal reception levels on those roads.

3. The nearby surroundings of the base station can bias the reception level—either up or down—when the mobile is within the 1-mile radius. When the mobile is more than 1 mile away from the base station, the effect due to the nearby surroundings of the base station becomes negligible.

4. For a land-to-mobile propagation, the antenna height at the base station strongly affects mobile reception in the near field; therefore, the mobile reception at 1 mile away has to refer to a given base station antenna height.

For distances of less than 1 mile, the Lee macrocell model projects the path loss curve predicted by the single breakpoint model extending backward from 1 mile to the base station.

3.1.2.1.2   Slope and Intercept Reference Values   Slope and intercept values for a specific city can be obtained from the mean value of measured data. There are slopes and intercept values for the major cities in a list. For future predictions in similar areas or cities as in the list, we may copy the slopes and intercept values without further measurement data. The available values from Table 3.1.2.1.1 shown below can be used.

Environment	(1-Mile Intercept)	(Path Loss Slope)
Free space	−45 dBm	20dB/dec
Open area	−49 dBm	43.5dB/dec
Suburban	−61.7 dBm	38.4dB/dec
Newark	−64 dBm	43.1dB/dec
Philadelphia	−70 dBm	36.8dB/dec
Tokyo, Japan	−84 dBm	30.5dB/dec
New York City	−77 dBm	48.0dB/dec

TABLE 3.1.2.1.1   Slope and Intercept Reference Values

The measured data shown in Table 3.1.2.1.1 can be illustrated by plotting signal loss against distance on a logarithmic scale, as shown in the area-to-area path loss curves in Fig. 3.1.2.1.2. Some cities have human-made structures similar to those of cities shown in the chart. They can use the value of that similar city. Therefore, the area-to-area prediction model is based on the human-made structures in different environment, such as open areas, suburban and urban areas, and cities. In the area-to-area model, the terrain contour is not considered, but it will be considered by the point-to-point prediction model as shown in the following section.

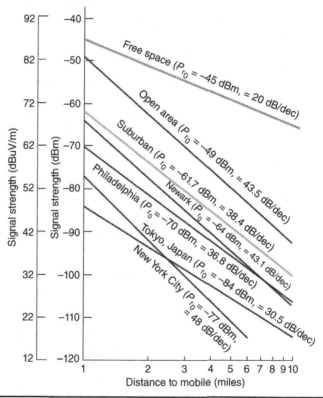

FIGURE 3.1.2.1.2   Area-to-area path loss reference curves.

3.1.2.2 Frequency-Offset Adjustment

Frequency offset adjustment is the second component of the model shown in Eq. (3.1.2.1). The reference frequency for the lee single breakpoint model is 850 MHz, and the model is valid within the frequency range of 150 to 2400 MHz.

When the center frequency used in the prediction is different from the reference frequency, the single breakpoint model calculates a frequency-offset adjustment (A_f) in Eq. (3.1.2.1), which is given in dB. The algorithm for this adjustment depends on the following two factors:

1. The frequency range that includes the specified center frequency, such as 150 to 450 MHz, 451 to 850 MHz, and 851 to 2400 MHz, are defined for the adjustment.

2. Environment selection can be specified per sector. The Lee model can support many environment types, such as dense urban, urban, commercial suburban, residential suburban, cluttered rural, open rural, deciduous forest, water, and evergreen forest.

For purposes of the frequency-offset adjustment, the Lee model categorizes the environment types into two classes, as follows:

1. Urban, which includes dense urban and urban

2. Nonurban, which includes commercial suburban, residential suburban, cluttered rural, open rural, deciduous forest, and evergreen forest

Table 3.1.2.2.1 gives the frequency adjustment algorithms for each of the three frequency ranges and within each of the two environment cases.

Urban	
$150 \text{ MHz} \leq f \leq 450 \text{ MHz}$	$A_r = -30 \log\left(\dfrac{450 \text{ MHz}}{850 \text{ MHz}}\right) + \left(-20 \log\left(\dfrac{(f)\text{MHz}}{450 \text{ MHz}}\right)\right)$
$451 \text{ MHz} \leq f \leq 850 \text{ MHz}$	$A_r = -30 \log\left(\dfrac{f}{850 \text{ MHz}}\right)$
$851 \text{ MHz} \leq f \leq 2400 \text{ MHz}$	$A_r = -30 \log\left(\dfrac{f}{850 \text{ MHz}}\right)$
Nonurban	
$150 \text{ MHz} \leq f \leq 450 \text{ MHz}$	$A_r = -20 \log\left(\dfrac{f}{850 \text{ MHz}}\right)$
$451 \text{ MHz} \leq f \leq 850 \text{ MHz}$	$A_r = -20 \log + \left(\dfrac{f}{850 \text{ MHz}}\right)$
$851 \text{ MHz} \leq f \leq 2400 \text{ MHz}$	$A_r = -30 \log + \left(\dfrac{f}{850 \text{ MHz}}\right)$

TABLE 3.1.2.2.1 Frequency-Offset Adjustment Algorithms

3.1.2.3 Effective Antenna Height Gain

Effective antenna height gain is the third component of the model as shown in Eq. (3.1.2.1). It is due to the terrain-contour effect. When the direct path from the base station to the mobile is not obstructed by terrain, the received signal consists of direct and reflected waves. In this situation, the adjustment of received signal strength due to a change in antenna height at the base station does not depend solely on the actual antenna height above the local ground level. Rather, it depends on the effective antenna height (h_e) as determined by the terrain contour between the base station and the mobile.

Once the effective antenna height is found, we can use it with the standard condition antenna height (h_1) to calculate the effective antenna height gain (G_{effh}) from the Lee single breakpoint model, shown in Eq. (3.1.2.3).

3.1.2.3.1 Determining Effective Antenna Height
Effective antenna height is determined by deriving from a specular reflection point, which can be found using the following parameters:

- Base station antenna height, h_1
- Mobile antenna height, h_2
- Distance from the base station to the mobile, d
- Terrain elevations between the base station and the mobile

Note that both the base station antenna height and mobile antenna heights are considered. Both antennas are vertically lined up with y-axis, not perpendicular to the ground slope, as shown in the figure. This is because, for illustration, the scales of antenna heights and the scale of the ground are not the same. If the scales of x-axis and y-axis are the same, then both antennas are almost perpendicular to the ground.

Effective antenna height is determined using the following steps:

1. Find the specular reflection point.
   a. Connect the negative image of the transmitter antenna to the positive image of the mobile antenna; the intercept point at the ground level is considered a reflection point (R_1).
   b. Connect the negative image of the mobile antenna to the positive image of the transmitter antenna; the intercept point at the ground level is also considered a reflection point (R_2).

   The Lee model uses the reflection point found closest to the mobile as the specular reflection point.

2. Extend a ground plane—tangent to the average elevation along the terrain contour at the specular reflection point—from the specular reflection point back to the location of the base station (transmitter) antenna.

3. Measure the effective antenna height (h_e) from the intersection of the extended ground plane and the y-axis on which the vertical mast of the base station antenna is located.

Four different cases are considered for calculating effective antenna height gain.

3.1.2.3.2 Calculating Effective Antenna Height Gain   The effective antenna height gain calculation considers the terrain contour and the relation of effective antenna height (h_e) to the standard condition antenna height (h_1) of 100 feet (~30.5 m). The Lee model considers four different cases: (1) terrain sloping is upward when $h_e > h_1$, (2) over a flat terrain when $h_e > h_1$, (3) terrain sloping is downward when $h_e < h_1$, and (4) over a flat terrain when $h_e < h_1$. These cases are illustrated in Figs. 3.1.2.3.1 and 3.1.2.3.2.

The formula of the effective antenna height gain from the Lee model is shown in Eq. (3.1.2.3) when the actual antenna height is the same as the standard antenna height ($h_1' = h_1$) as follows:

$$G_{effh} = 20\log\left(\frac{h_e}{h_1}\right) \qquad (3.1.2.3.1)$$

1. For these two cases (for a terrain sloping upward and for a flat terrain when $h_e > h_1$), Eq. (3.1.2.3.1) results in a positive gain ($G_{effh} > 0$ dB).

2. For these two cases (for a terrain sloping downward and for a flat terrain when $h_e < h_1$), Eq. (3.1.2.3.1) results in a negative gain ($G_{effh} < 0$ dB). If $h_e < h_1/10$, then h_e is forced to cap at $h_1/10$.

Figure 3.1.2.3.3 illustrates the path loss prediction that is obtained based on the area-to-area path loss curve, adding or subtracting the effective antenna height gain at each local point due to the influence of the local terrain contour; as a result, the overall point-to-point signal strength prediction is plotted for each local point along the mobile path. Note that the variation in prediction can be significant, as shown at points D through G in the figure.

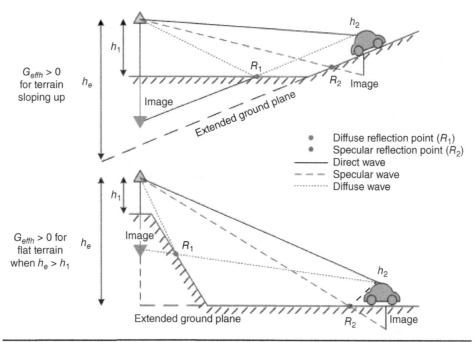

FIGURE 3.1.2.3.1   Effective antenna height gain (G_{effh})—positive gain.

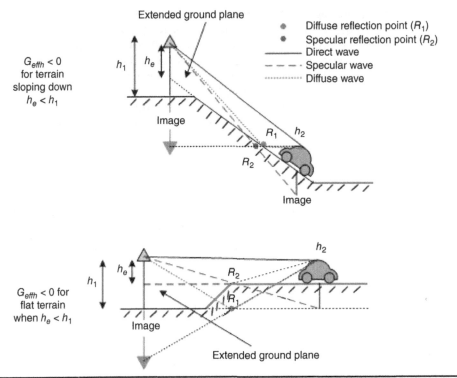

FIGURE 3.1.2.3.2 Effective antenna height gain (G_{effh})—negative gain.

3.1.2.4 Diffraction Loss from Diffraction Path

3.1.2.4.1 Single-Knife-Edge Case The diffraction loss is the fourth component as shown in Eq. (3.1.2.1) that is due to the obstruction from the hills. The diffraction path is where the direct path from the base station to the mobile is obstructed (in shadow) or partially obstructed (near shadow) by one or more knife edges. We introduced diffraction loss briefly in Sec. 1.9.2.2. In this case, the radio signal experiences losses due to the diffraction.

The Lee model uses Fresnel–Kirchoff diffraction theory to predict the diffraction loss component (L) in Eq. (3.1.2.1) for the Lee single breakpoint model. The diffraction loss L consists of two parts. One is the based on the knife-edge diffraction loss L_D, and the other is the correction factor due to the loss from a real obstacle, the shape of which is not a knife edged. The correction factor obtained from the effective antenna height will be described in the next section. The knife-edge diffraction loss L_D is obtained based on a dimensionless parameter v, a diffraction factor, given by Eq. (1.9.2.2.1.6) as

$$v = (-h_p)\sqrt{\left(\frac{2}{\lambda}\right)\left(\frac{1}{r_1} + \frac{1}{r_2}\right)} \qquad (3.1.2.4.1)$$

where λ = wavelength, r_1 = distance from the base station to the knife-edge ($r_1' \approx r_1$), r_2 = distance from the knife edge to the mobile ($r_2' \approx r_2$), and h_p = height of the knife edge, which can be above or below the line that connects the base station and mobile antennas.

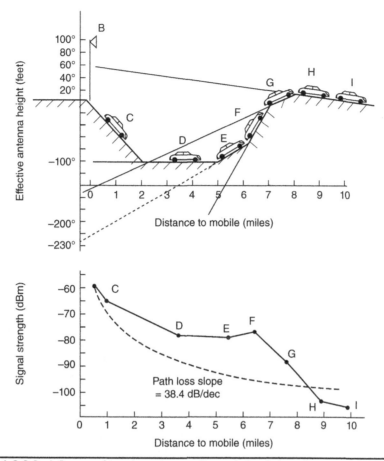

Figure 3.1.2.3.3 Influence of terrain contour on G_{effh} and signal strength.

Once the value of diffraction factor v is obtained, the knife-edge diffraction loss $L_D(v)$ can be found from the curve, which is shown in Fig. 1.9.2.2.1. Table 3.1.2.4.1 gives an approximation of this curve for different values of diffraction factor v.

The method for calculating the height of a knife-edge obstruction was discussed earlier. When a signal is blocked by multiple knife edges, the Lee model evaluates the parameters for all knife edges together and treats each one separately.

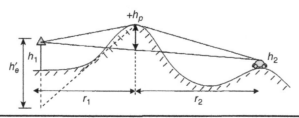

Figure 3.1.2.4.1 Determining for the diffraction loss calculation.

$1 \leq v$	$L = 0$ dB
$0 \leq v < 1$	$L = 20 \log (0.5 + 0.62v)$
$-1 \leq v < 0$	$L = 20 \log (0.5e^{0.95v})$
$-2.4 \leq v < -1$	$L = 20 \log (0.4\sqrt{0.1184 - (0.1v + 0.38)^2})$
$v < -2.4$	$L = 20 \log \left(-\dfrac{0.255}{v}\right)$

TABLE **3.1.2.4.1**   Diffraction loss (L) Values[6]

3.1.2.4.2   A Real Situation: When Diffraction Is from a Real Hill   In past years, engineers have found that the Fresnel diffraction loss formula always gave the calculate values more pessimistic than the measured data. The diffraction loss formula is based on the knife-edge scenario, but the curvature of the earth is not a knife edge. Also, the bending of the curvature is different from every hilltop. The knife-edge diffraction formula does not include the curvature factor. Therefore, in this section, a method[7] to predict the diffraction loss L more realistically is presented.

The slope of the hillside can gauge the bending of the curvature of the hilltop. In the Lee prediction model, the slope of the hilltop is used to calculate the effective antenna gain G_{effh} when the mobile is in a nonshadow area. In the area of transition to or from shadow, the diffraction loss is also affected by effective antenna height gain (G_{effh}). Therefore, use of the G_{effh} to adjust the diffraction loss from the formula is the right approach.

Logically, the G_{effh} of a mobile on top of different hills are different due to the hillside slope. The more steep is the hillside, the more gain is the G_{effh}. From the knife-edge diffraction loss formula, the loss is obtained based on the diffraction parameter v, which is a function of r_1, r_2, h_p, and λ only, as shown the formula in Eq. (3.1.2.4.1). *This equation does not involve the parameter of curvature.* Therefore, we have added a correction factor to the knife-edge diffraction loss.

Now the prediction tool takes the maximum G_{effh} into the calculation of diffraction loss as

$$L = L_D(v) - \max G_{effh} \qquad (3.1.2.4.2)$$

where max G_{effh} is the effective antenna height gain, G_{effh}, calculated from an effective antenna height h'_e, which is measured the height at the base station from the intersected point of a line that is drawing from the top of the hill along the slope of the hillside to the base station.[8] In Fig. 3.1.2.4.1, h'_e is depicted in the figure. Now comparing the calculation from Eq. (3.1.2.4.2) with the measurement data, we have found a great match.

Actually, the rigorous derivation should include the curvature parameter in the diffraction formula. At present, no one has done this yet. Thus, our approach is to calculate the gain from max G_{effh} and use it to reduce the theoretical diffraction loss. It can be interpreted as equivalently lowering the actual height of h_p in the diffraction formula so that the loss becomes less. This is an indirect method of solving the diffraction loss. However, this approach can be applied only to the slope angle of a small hill, say, less than 10°. Another methods for predicting the diffraction loss at a high hill case are described in Sec. 3.2.4.

3.1.2.4.3   Multiple-Knife-Edge Case with Single-Knife-Edge Check   When a signal path is blocked by multiple knife edges, the Lee model compares diffraction losses calculated by adding all knife edges together and from each one separately. The greatest loss value is then used for the diffraction loss component of the signal strength prediction.

The multiple-knife-edge evaluation involves these steps:

1. For each knife edge, the Lee model calculates the obstruction height (h_p) using the Epstein–Petersen method (see Sec. 1.9.2.2.2). The Lee models consider two different scenarios—when there are two knife edges and when there are three or more knife edges.

2. The parameter v is evaluated for each single knife edge, assuming no other knife edge, and the individual single-knife-edge diffraction loss is computed from the appropriate formula in Table 3.1.2.4.1.

3. The individual values L_i of two or three knife edges are calculated as shown in Fig. 1.9.2.2.2.2, and the individual values are summed for all knife edges as a total diffraction loss (L_t) using the formulas in Table 3.1.2.4.1.

4. The "single-knife-edge check" is then used to compare the greatest diffraction loss caused by any individual knife edge with the total diffraction loss. The Lee model uses the greatest of these values for the diffraction loss component of the signal strength prediction. This ensures that the total loss caused by the multiple-knife-edge calculation is at least as great as the loss caused by any one knife edge considered individually.

3.1.2.4.4   Transition to and from the Shadow Region   In the area of transition to and from the shadow, region, the received signal strength at a knife edge is affected through a transition region; the Lee model considers the case by using the positive effective antenna height gains G_{effh} as a reference.

The transition region is that when h_p is negative, the value v is between 0 and –1. As h_p becomes positive, the effective antenna height gain disappears and G_{effh} is used in the transition region from the nonobstructed direct path to shadow, as shown in Fig. 3.1.2.4.2. When $h_p = 0$ at the knife edge ($v = 0$), the diffraction loss is 6 dB. The diffraction loss becomes 0 dB when $v \geq 1$.

As h_p becomes negative, the signal path is partially obstructed by the terrain contour, and G_{effh} is used in the transition region. The signal path is out of the shadow and becomes the nonobstructed direct path, as shown in Fig. 3.1.2.4.3. This gain is phased in as v goes from 0 to 1.

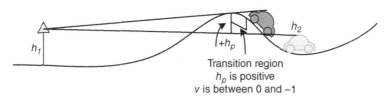

Transition region
h_p is positive
v is between 0 and –1

Figure 3.1.2.4.2   Transition to shadow.

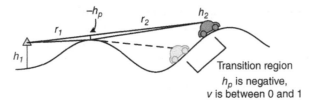

Figure 3.1.2.4.3   Transition from shadow.

3.1.2.5   Adjustment Factor

The adjustment factor is the fifth component of the model as shown in Eq. (3.1.2.1). The initial signal strength prediction at each point along the radial assumes a set of standard conditions in making certain parameters at a base station transmit and at a mobile unit, as follows:

P_t = base station output power = 10 W
h_1 = base station antenna height = 100 ft (~30.5 m)
h_2 = mobile antenna height = 10 ft (~3.0 m)
g_b = base station antenna gain = 6 dBd
g_m = mobile antenna gain = 0 dBd

The fifth component of the Lee model in Eq. (3.1.2.1) also applies an adjustment factor α for the signal strength prediction to compensate for the difference between the standard conditions and the actual values of these parameters, which are defined as follows:

P'_t = actual base station output power in watts
h'_1 = actual base station antenna height in feet
h'_2 = actual mobile antenna height (a default at 5 ft or 1.5 ms)
g'_b = actual base station antenna gain in dBd
g'_m = mobile antenna gain in dBd (a default at 0 dBd)

Note that the adjustment factor calculated here does not include an adjustment for the standard base station antenna height. The standard base station antenna height is used in determining the effective antenna height gain, whereas the actual base station antenna height is used in determining diffraction loss.

The adjustment factor α is given by the following equation:

$$\alpha = 10 \log \frac{P'_t}{P_t} + 10 \log \frac{h'_2}{h_2} + (g'_b - g_b) + (g'_m - g_m) \qquad (3.1.2.5.1)$$

Because the actual mobile antenna height (h'_2) is always fixed at a half of the standard condition mobile antenna height (h_2), the adjustment for mobile antenna height is $10 \log (0.5) = -3$ dB. Also, because the mobile antenna gain is fixed a 0 dBd for both the standard and the actual conditions, no adjustment is needed. Thus, the adjustment factor needs only to adjust the actual base station output power and actual base station antenna gain and includes the −3dB mobile antenna height adjustment. The adjustment factor α in Eq. (3.1.2.5.1) can be simplified as follows:

$$\alpha = 10 \log\left(\frac{P_t'}{P_t}\right) + (-3 \text{ dB}) + (g_b' - g_b) \qquad (3.1.2.5.2)$$

An example calculation is given next.

3.1.2.5.1 Example of Adjustment In this example, the unadjusted signal strength of −61 dBm is predicted at a point in the field under the standard conditions. However, the actual base station output power and antenna gain are as follows:

$$P_t' = 30 \text{ W} \qquad g_b' = 5 \text{ dBd}$$

The adjustment factor in Eq. (3.1.2.5.2) is calculated as follows:

$$\alpha = 10 \log\left(\frac{30 \text{ W}}{10 \text{ W}}\right) + (-3 \text{ dB}) + (6 \text{ dB} - 5 \text{ dB})$$

$$\alpha = 10 \log(3) + (-3 \text{ dB}) + (1 \text{ dB})$$

$$\alpha = (4.8 \text{ dB}) + (-3 \text{ dB}) + (1 \text{ dB}) = 2.8 \text{ dB}$$

The 2.8-dB adjustment factor is applied to the unadjusted prediction, giving an adjusted signal strength of −58.2 dBm at this point. This value is subject to signal smoothing if necessary as it will be described Sec. 3.1.2.6.

3.1.2.5.2 Determining Actual Base Station Antenna Gain The Lee model assumes a base station antenna gain of 6 dBd as one of the standard conditions. However, in any given model, the antenna may have a different standard gain. Also, the actual gain at any point is a function of the radiation pattern observed from both vertical and horizontal angles that in turn depend on the angle of mechanical or electrical downtilt. We will describe this issue in Sec. 3.1.7.1.1.

Besides adjusting the antenna gain difference from the standard antenna gain, the Lee model calculates the actual different base station antenna gains from the observed angle at the mobile after the antenna is either electrically or mechanically downtilted. When the electrical downtilt of an antenna is applied, the actual antenna gain from the observed angle at the mobile can be obtained from the free space antenna pattern at the angle.

However, when the mechanical downtilt is applied, the antenna pattern is no more the same as a free space pattern but is in a different shape. Therefore, the calculation of the antenna gain from the mechanical downtilted antenna is different, as shown in Sec. 3.1.7.

3.1.2.6 Signal-Smoothing Process

3.1.2.6.1 From the Raw Predicted Values The Lee single breakpoint model predicts a signal strength value for each local point on every signal path (radial). These points are spaced evenly along the radial at an equal distance. Once raw signal strength values have been predicted from the Lee model, a signal-smoothing process is used to determine the final predicted value (P_r) for each point. This process enhances the model and

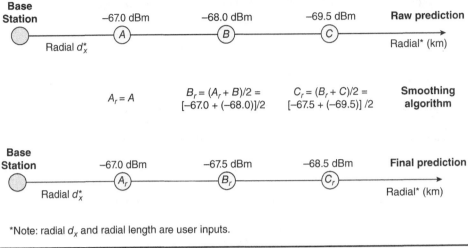

*Note: radial d_x and radial length are user inputs.

FIGURE 3.1.2.5.1   Signal-smoothing example.

is called the enhanced Lee model. This process, illustrated in Fig. 3.1.2.5.1, calculates a "running average" at each point as follows:

1. The predicted signal at the initial point (A in the diagram) has not been averaged, thus let $A_r = A$ for this point.

2. Beginning with the second point (B in the diagram), the final predicted signal strength is determined by adding the raw signal strength at this point with the final signal strength at the previous point and dividing it by two. Thus, $B_r = (A_r + B)/2$ and so on.

3.1.2.6.2   From the Measured Data   We have to use the running average to get the local means from the measured data. The local mean has been determined by averaging 50 samples of a piece of data over a distance of 40 wavelengths at 800 MHz. If the carrier frequency is at 400 MHz, the averaging of a distance of 20 wavelengths is adequate. It has been determined by Lee[9] and described in Sec. 1.6.3.1.

3.1.3   Variations of the Lee Model

There are several variations of the Lee single breakpoint model. The variations differ from the basic model only in predicting the path loss component (exclusive of the frequency-offset adjustment) in certain areas, as described below.

The basic method for determining path loss for the single breakpoint model was described in detail earlier. For distances of less than 1 mile, the Lee model projects the path loss curve predicted by the single breakpoint model extending backward from 1 mile to the base station, as illustrated in Fig. 3.1.3.1.1.

3.1.3.1   Lee Multiple Breakpoint Model

The multiple breakpoint model was developed to improve the accuracy of prediction for distances within 1 mile of the base station. Among the currently provided model

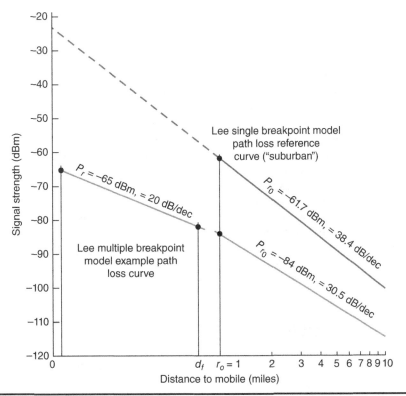

FIGURE 3.1.3.1.1   Lee single and multiple breakpoint models—sample path loss curves.

variations, this model is considered the most accurate, particularly when measurement integration is used to further "tune" the slope and intercept values.

Two intercepts are defined—one is the free space slope from the transmitter to the near-in distance (d_f) and the other is the environment slope connected from the near-in distance to the 1-mile (r_o) point. The user may specify two slope values—one using the free space slope from a point (one wavelength from the base station) to distance d_f and the other using the extended environment slope backward from the 1-mile intercept (r_o) beginning at d_f. On some occasions, the model needs to interpolate the path loss slope between these two points d_f and r_o, as illustrated in Fig. 3.1.3.1.1.

An optional 10-mile intercept may also be specified, with a different slope value for distances beyond 10 miles from the base station. Therefore, values for multiple slopes and intercepts may be specified by the individual users under different circumstances.

Once the path loss is determined for each point along the signal path (radial), the frequency-offset adjustment and other component adjustments of the single breakpoint model will be added to the signal strength prediction.

When the imported empirical data collected from the field are used with the multiple breakpoint model, the model can have as many as 12 radial zones defined in the measurement integration process. These intercepts and path loss slopes can be automatically generated for each of the defined zones.

3.1.3.2 Automorphology Model

The automorphology model was developed to provide users with an additional option for modeling with multiple path loss slopes due to the absence of measured data. With this model, morphology (for the cover-of-land conditions, such as forest or water, and the use of land, such as human-made structures) data are used to determine the path loss slopes beyond the 1-mile intercept. The morphology attribute types can be grouped appropriately and correlated to path loss slope values.

The effect of the automorphology prediction is that, for distances beyond 1 mile, the path loss slope changes as the morphology class changes.

The accuracy of signal strength predictions from the automorphology model depends on a careful selection of a path loss slope value for each land-use class. Users may revise these values at any time to help increase prediction accuracy by customizing the specific sectors in the database.

Although the measurement integration cannot be used directly with the automorphology model, the results obtained from measurement integration using the single breakpoint or multiple breakpoint models can be used in the automorphology model to help determine values for the 1-mile intercept and the path loss slope within 1 mile of the base station.

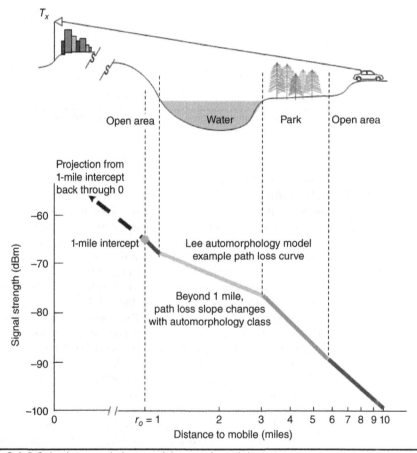

Figure 3.1.3.2.1   Automorphology model—sample path loss curve.

3.1.4   Effects of Terrain Elevation on the Signal Strength Prediction

The calculation of two of the Lee model components—that is, the effective antenna height gain and the diffraction loss—is affected by the local terrain elevation. Terrain elevation data are extracted from the terrain files and used to create a terrain contour profile of each radial, which support s the signal strength prediction of a sector with the terrain contour profile of the radial array.

The model can also consider the effects of terrain averaging and the effective earth curvature when determining terrain elevations in use. Terrain elevation data can also be used for support when an optional water enhancement affects the signal strength calculation in the Lee model variations.

The Lee model can use different terrain elevation data, terrain averaging, and effective earth curvature in dealing with a complicated mobile environment.

3.1.4.1   Terrain Elevation Data

Terrain elevation data[10,11] are stored and divided into a grid of data points spaced at 3-arc-second intervals. The average of all the data points in one grid is stored in that grid. When a final file of all the grids is completed, the Lee model extracts the average terrain elevations from the files to calculate the signal strength prediction.

3.1.4.1.1   Extracting Terrain Elevations   First, the area of interest will be determined by the radial distance of the terrain map (specified by the individual users) and the base station location.

When the corners of the area of interest only have partial 3-arc-second grid points, the area of interest has to expand each corner to align with the entire grid. This "aligned area of interest" is illustrated in Fig. 3.1.4.1.1.

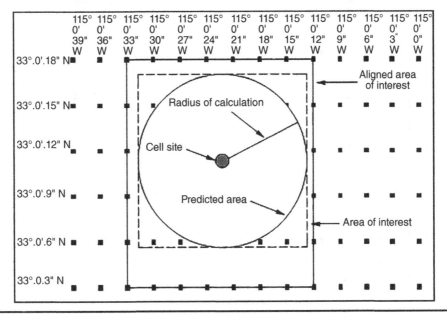

Figure 3.1.4.1.1   Prediction area and aligned area of interest.

When the aligned area of interest has been determined, the Lee model creates a list of terrain contour (elevation profile) files covering this area. The elevation data needed for the signal strength prediction are extracted from this list of files and formed in radial arrays before being stored in memory.

3.1.4.1.2   Terrain Radials and Radial Arrays   A terrain radial represents a signal path from the base station to the mobile, and a radial array represents the points along the radial at which the signal strength is predicted.

The number of radial arrays extracted from the terrain elevation data depends on the number of radials specified by the user. The number of points contained in each radial array is determined by the radial distance of the prediction divided by the radial increment unit (equidistant spacing of points along the radial).

In creating the terrain radial, the Lee model uses three methods for choosing the elevation values for radial points, as follows:

1. When a local point is located at a 3-arc-second grid, the elevation of that grid is used.

2. When a local point is located between two 3-arc-second grids, the Lee model takes the average elevation from the two grids along either a latitude or a longitude line.

3. When a local point is located at the corner of four 3-arc-second grids, such as shown in Fig. 3.1.4.1.2, the Lee model calculates the weighted average of the four closest 3-arc-second grids surrounding the point.

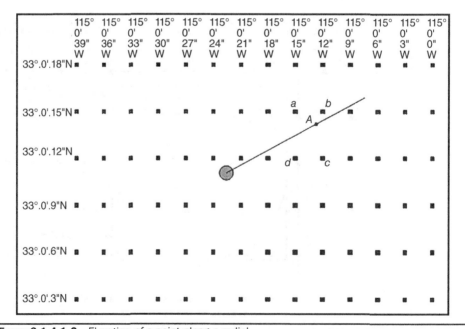

FIGURE 3.1.4.1.2   Elevation of a point along a radial.

3.1.4.2 Terrain Averaging

When a terrain-averaging program is turned on, the Lee model uses a simple running average by averaging the elevations from three points, the current local point, the preceding point, and the succeeding point to find a new elevation at the current local point.

For example, when the averaging scope is set to 5 points, the new elevation for any given point on the radial is obtained by averaging the current elevation with the 2 previous points and 2 succeeding points along the radial.

For those points where only fewer than 3 or 5 points can be averaged (such as the last point on a radial), the extrapolate elevations are used for those points.

3.1.4.3 Effective Earth Curvature[12,13]

When the distance is greater than 10 miles, we may consider the effect of earth curvature. Terrain elevation values can be adjusted with an offset value to compensate for the horizon effect caused by the earth's curvature. This offset represents a reduction in elevation at the point of the mobile, which increases with distance from the base station.

When the earth curvature offset is used, a parameter known as the K factor can be specified to represent the earth's radius. This parameter is used to adjust the earth curvature offset to compensate for the slight curvature of the signal path.

The earth curvature offset can be turned on or off based on a specified value for the K factor, as described below.

3.1.4.3.1 Earth Curvature Offset The earth curvature offset value is the difference between a straight line drawn from the site elevation point to a point along the radial as compared with the curve of the earth through these same points, as illustrated in Fig. 3.1.4.3.1.

A formula for the earth curvature offset can be derived as follows:

1. Consider a mobile elevation point is at a distance (D) in meters from the site elevation and assume the earth's radius (R) to be 6,370,997 m.

2. The angular distance from the site to a point is given by

$$\text{angle} = \text{arc}\sin\frac{D}{R} \tag{3.1.4.3.1}$$

3. The dip below the horizon is given by

$$\text{dip} = (1.0 - \cos(\text{angle})) \times R \tag{3.1.4.3.2}$$

3.1.4.3.2 *K* Factor When the earth curvature option in the prediction model is turned on, a parameter known as the K factor can be specified to adjust the calculated earth curvature offset. The K factor represents a proportion of the actual radius of the earth. The Lee model uses a default value of 1.33:

$$\text{Offset} = 2D^2/3K \tag{3.1.4.3.5}$$

Any value of K factor larger than 1 assumes that the radius of the earth is larger than the actual one, thus resulting in a smaller earth curvature offset. Because the signal path is somewhat curved rather than straight, the signal horizon is lower than would be represented by drawing a straight line from the transmitter to the mobile. Thus, the earth curvature offset should be less than that calculated by using the actual radius of the earth, and use of K factor values larger than 1 is recommended.

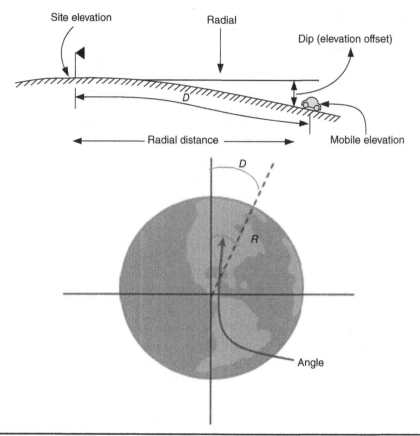

Figure 3.1.4.3.1   Effective earth curvature.

3.1.5   Effects of Morphology on the Signal Strength Prediction

The morphology (different land covered and different land used) data affect the calculation of path loss and can be stated as follows:

1. The effects of morphology for many different types of land covered and land used from the basis of the model for the path loss prediction are described in the automorphology model. With this model, the morphology data are used to adjust the path loss slope for each grid along the radial beyond the 1-mile intercept.

2. When the optional water enhancement is used, the path loss calculation for the effects of water-reflected waves is adjusted. The morphology data can be used to determine which grid contains water. This enhancement can be applied to all of the Lee macrocell models.

The morphology data are stored in the morphology directory as a set of files with a certain extension. Each file can cover a 1 × 1 degree area and is divided into a number of grids with an average elevation value from the data points spaced at 3-arc-second intervals.

In the automorphology model, each attribute is correlated to a path loss slope value. The slope values specified from different types of land covered and land used are used in the automorphology model for those grids beyond the 1-mile intercept.

In the optional water enhancement, the slope value assigned to any type of water in the morphology class is irrelevant. The fact that a bin contains water (or not) is the only consideration for this option.

3.1.5.1 Network Engineering Process

There are still some issues with most of the currently used network planning tools. One of the problems is that no flexible model can handle the different types of morphology that affect propagation, coverage, interference, handoff, and capacity. The morphology model needs to be enhanced from many different perspectives. A simple 2D model comprised of 2D morphology data is not good enough today. The third dimension (height) is playing a more important role in the planning arena of network engineering. Trees can make LOS mobile become non-LOS. Tall, dense forest can cause a difference in decibels up to two digits from the predicted path loss, impacting the accuracy of the morphology model. Another drawback is that some field data are sensitive to the time of the year, which has not been integrated with the network engineering process.

This section addresses innovative ways to deal with the impact of morphology on network engineering. There are many different morphologies, and each has a unique impact on network engineering and planning. For example, different types of trees will have different impacts on propagation. Also, they behave differently if they are weather sensitive. In this section, a 4D (time of the year, and height, width, and length of the morphology) database associated with algorithms is proposed to deal with the impact on system engineering.[14] The key point of this enhancement is that it includes a means of gaining feedback and of integrating measured data from the field so that the 4D morphology model can be tuned (full flexibility) to handle different situations better.

The innovations include; first, the fundamental concept of dealing with morphology in an integrated network with entire system propagation (the specific path loss slope from the morphology will be derived and applied to the existing model); second, the algorithm of deriving these morphology slopes; third, the calculation of shadow loss; and fourth, the identification and integration of the impact on time variance and associated engineering algorithm.

The minutes of use of wireless transmission for calculating the prediction model need to be substantially increased for the enhancement of the morphology. Fortunately, with today's technology, mobile operators in network engineering can afford the cost of the minutes of use with more accurate processes and cheaper data available.

3.1.5.2 The Algorithm of Categorizing Different Morphologies

A generic propagation model is comprised of a slope and a 1-mile (or 1-km) intercept as shown in Fig. 3.1.5.2.1. It applies the slope and intercept to each radial in calculating the signal strength along the radial.

The morphology area can be classified into three different categories, such as type 1-water, type 2-foliage, and type 3-tunnel, as shown in Fig. 3.1.5.2.2. The detailed description of the signal propagation over the water will be shown in Sec. 3.1.6.

Figure 3.1.5.2.2 shows the morphology model that deals with the in-morphology area proposed by this innovation. It is integrated with the traditional slope and intercept of the model.

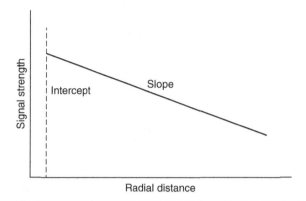

FIGURE **3.1.5.2.1** Generic propagation model.

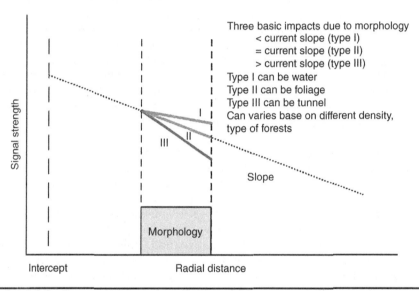

FIGURE **3.1.5.2.2** Proposed integrated morphology model.

The first and most difficult part is to categorize different morphologies. Once the morphology type is defined, the next step is to calculate the cause from the impact of the morphology. The measured data are extracted from the drive test and then normalized to calculate the slope and the intercept for the identified morphology area, as shown in Fig. 3.1.5.2.3.

The measure data need to be compared with the morphology database. To derive the correct slope for different morphologies, the morphology data need to be updated as follows:

1. Take out the normal propagation factor (the slope and intercept) from the area of interest.

2. Identify the radial distance from the measured point to the origin of the morphology along the radial path in the area of interest.

3. Do step 2 until points in the same morphology area of interest are processed.

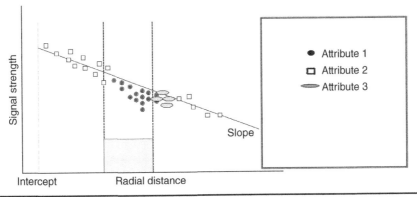

Figure 3.1.5.2.3 Integrated results for different morphologies/attributes.

4. Use the best-fit algorithm for these points to yield a slope in each different morphology area.

5. Store the multiple slopes in the morphology database.

6. Compare with the previous slope in the same area of interest and set the time trigger if the slope is different.

A pseudocode is used to implement this algorithm for storing the measurement data, as shown below:

Preparation

    Allocate all radial paths

    Allocate morphology slope array

    Allocate all data points along a specific radial path

Do

    Allocate a radial array

    Get the radial distance from cell site

    If the data point is not within the boundary of morphology area

        Store the measured signal strength in the radial array

    If the data point is within the boundary of morphology area

        Allocate morphology radial array

        Get the morphology radial distance from the morphology origin point

        Store the measured signal strength in the morphology radial array

Done

The slope and intercept for the radial arrays are calculated by the best-fit algorithm Then store the slope and intercept of each radial path. The algorithm for the process of morphology/attribute data is shown in Fig. 3.1.5.2.4

A different pseudocode, used to implement the measurement integration and to derive the slope in an area of interest for the morphology/attribute, is shown below:

Preparation

    Allocate all morphology data point along the morphology radial path

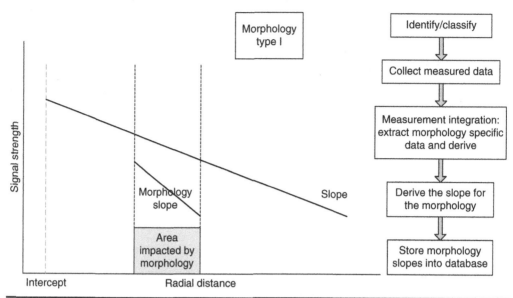

FIGURE 3.1.5.2.4 The algorithm for morphology/attribute data processing.

Do

    Allocate morphology for the radial-normalized signal

    Morphology-radial-normalized signal = (measured-morphology-signal) − (predicted-nonmorphology signal)

    Store morphology-radial-normalized signal in the morphology database

Done

The slope from the morphology-radial-normalized-signal is calculated by using the best-fit algorithm

The slope is stored for this morphology and used for the morphology-slope array along this radial. This morphology slope is shown in Fig. 3.1.5.2.4.

In determining the transition slopes, two "for" loops are used to calculate the two slopes, one before and one after the morphology area. The only key factor is to decide on what distance from the morphology area is needed for calculating the transit slopes. Although the distance does not have much impact on the final results, it might have some impact on the calculation time.

Another added flexibility of the model is to support finding the path loss slopes before and after the morphology area. In many instances (except the water situation in the morphology), the slope after the morphology area will be the main issue. The diagram in Fig. 3.1.5.2.5 shows the scenario of both slopes before and after the morphology area for the purpose of demonstration. The same algorithm that is used to derive the in-morphology slope can be applied in this scenario also.

3.1.5.2.1 Through a Tunnel Figure 3.1.5.2.5 shows the difference due to the enhancement from the beginning to the end of a radial path with the morphology. In the case of passing through a tunnel, the morphology of the tunnel is very important to consider.

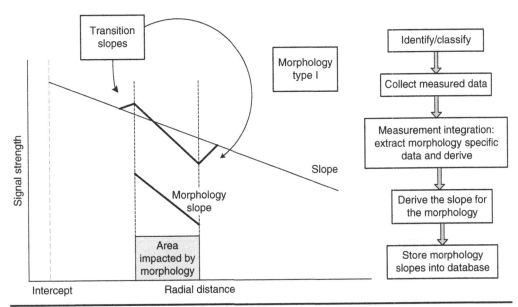

Figure 3.1.5.2.5   Morphology/attribute flexibility.

Otherwise, the mobile signal will be very weak toward the end of the tunnel (morphology = tunnel). Immediately after existing the tunnel, a spike of signal strength will be seen. This creates the issues of interference, handoff, and capacity.

The calculation of generic blockage loss is described in the algorithm diagram shown in Fig. 3.1.5.2.5. It provides the foundation for further enhancement with morphology.

3.1.5.2.2   Blocked by Forest   Something else to consider is the morphology of the blockage along the radio path. As shown in Fig. 3.1.5.2.6, a simple example of blockage is a forest.

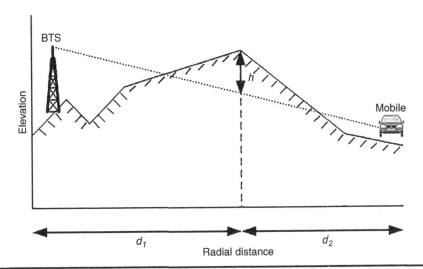

Figure 3.1.5.2.6   Handling of a special case.

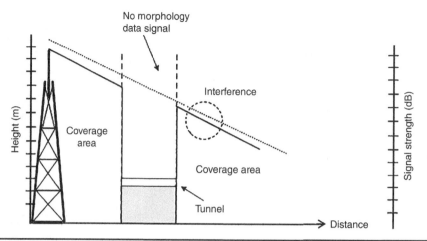

Figure 3.1.5.2.7   The importance of integrated attribute/morphology.

Some trees are taller than 30 ft and will have some impact on signal propagation, especially in situations where signals pass through a boundary from LOS to non-LOS. A sudden change of signal strength can easily be from 10 to 20 dB.

Figure 3.1.5.2.7 shows the whole scenario of a radial path from the beginning at the BTS to the end at the mobile with morphology. Immediately after exiting a tunnel, a spike of signal strength is expected.

3.1.5.2.3 Knife-Edge Diffraction   As shown in Fig. 3.1.5.2.8, the edge h, which is generally used to calculate the shadow loss is replaced by H, which equals the sum of h and the human-made structure height a. Also, the formula of calculating the diffraction loss with morphology will be different.

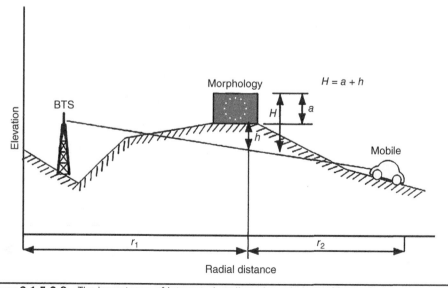

Figure 3.1.5.2.8   The importance of integrated attribute/morphology with terrain.

The chart of the diffraction loss L versus the diffraction parameter V is shown in Fig. 1.9.2.2.1.2, but the diffraction parameter v is replaced by V for calculating the integrated diffraction loss, which needs to include the factor of morphology/attributes:

- Use of the diffraction loss curve (diffraction loss (L) versus normal diffraction parameter v) for finding the terrain diffraction loss. The height of edge h_p is used. Therefore, $H = h_p$.

- Use of the diffraction loss curve for finding the morphology diffraction loss. The morphology loss $L(V\_\text{morphology}) = L(V)$, where the new diffraction parameter V can be found from the diffraction loss curve for $H = h_p + a$. The knife-edge height H is increased by the morphology of additional height, as shown in Fig. 3.1.5.2.8.

  The figure of morphology diffraction loss curve is the same as the figure of terrain diffraction loss curve, as shown in Fig. 1.9.2.2.1.2.

3.1.5.2.4 The Fourth Dimension: Time The fourth dimension of the Lee morphology/attribute model includes the time variable. In certain forests, such as pine forests, leaf length glows to a quarter or half of the wavelength of the signal carrier in a certain season, and those leaves will impact the propagation loss of the signal. Also, in most forests, leaves will be falling during the wintertime. As shown in Fig. 3.1.5.2.9, this fourth dimension changes the characteristic of propagation loss. The Lee model has integrated this as well.

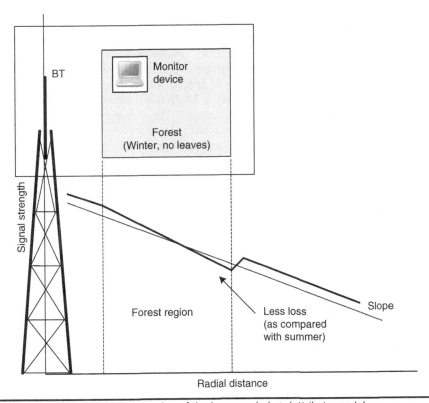

FIGURE 3.1.5.2.9   The fourth dimension of the Lee morphology/attribute model.

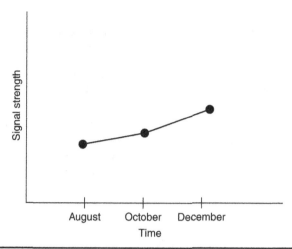

FIGURE **3.1.5.2.10**   The propagation characteristics based on seasons.

As we have mentioned, the fourth dimension, time, is another important morphology. For example, leaves on trees change sizes based on season. Some forests will affect the signal from different propagation characteristics as time passes. Sometimes the changes from these effects can be swung up to double digits in decibels. This innovation includes a feedback loop to incorporate the changes of time. It is implemented in two different steps. First, a sensor is placed to monitor the system's behavior and to provide the feedback loop. Once the sensor triggers the loop, the system characteristics will be tuned to the new status based on the criteria. For example, in dense foliage areas in Atlanta, the propagation loss can swing 10 dB from summer to a cold winter. The seasonal change spans about six months or more. The sensor placed in the field can provide feedback on the measured data every month or every other month. The prediction model can be tuned according to the changes. Once the part of propagation path loss is tuned, as shown in Fig. 3.1.5.2.10, the associated frequency plan, coverage plan, and capacity plan for the designing system can be also tuned accordingly. Therefore, the system will always have a parameter of time as one of the inputs and adjust it dynamically.

The slope of the path loss occurring each month (or at any time frame) is different since the morphology might be sensitive to weather. For example, leaves are in full bloom in the summertime and gradually fall to become thinner layers. When winter comes, there are no tree leaves, and the coverage area will be different, except for pine trees the leaves, of which, called pine needles, will never fall in the wintertime. The system has a program to deal with this situation.

First based on the peculiarity of the area under test to update the propagation model, then the coverage, the best servers (base stations), the interference issues, the handoffs and the system capacity together with dimensioning other network elements (BSC, MSC, and so on) will be predicted from the model.

In the summertime, as we have mentioned, the whole forest can be densely covered with leaves and the impact on propagation can be significant. Especially when the dimensions of the pine tree needles are very close to the quarter, half, or one wavelength of radio wave, major signal loss will be encountered due to the absorption of the propagation

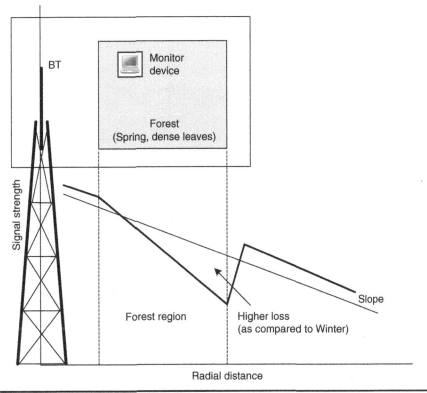

FIGURE 3.1.5.2.11   Dynamics of the Lee model.

energy by the leaves. The coverage can quickly shrink, and associated network engineering issues arise, as shown in Fig. 3.1.5.2.11.

As weather changes, the leaves on trees also change. Through autumn and winter, the forest becomes thinner and thinner. The forest can be better penetrated, and the associated network engineering also needs to be adjusted.

As we can see, the morphology factor has become more important in engineering today. It is important to have an intelligent algorithm that can work with measured data and also be able to link with the network engineering. The Lee model provides an integrated system solution for dealing with the morphology issue. As more morphology data become available, more effective and efficient means of doing network engineering become possible. Thus, it is important for cellular engineering to consider the factors of the morphology and to develop the algorithms and processes for dealing with these factors.

3.1.6   Water Enhancement

An optional water enhancement is available for use with any of the three modes of the Lee macrocell model. Like effective antenna height gain, water enhancement behaves as an additional attenuation factor that is applied to the calculation of the Lee model for determining signal strength.

When water enhancement is turned on, the Lee model checks to see if the mobile is located on land but receives both reflected waves, one from water and one from land,

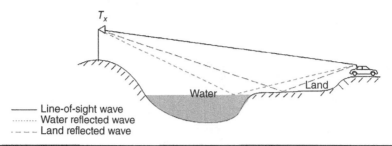

FIGURE 3.1.6.1   Effect of water-reflected waves.

as illustrated in Fig. 3.1.6.1. In this case, the Lee model compensates for the effect of the water-reflected wave by adjusting the path loss at the mobile upward to approach the free space curve.

3.1.6.1 Over-the-Water Conditions[15]

When the mobile is traveling on the other side of the water from the base station antenna as shown in Fig. 3.1.6.1, three waves arrive at the mobile. Since there are no human-made structures over the water, the reflected wave over the water is still considered a speculative reflected wave, although the reflection point of the wave is at a distant location away from the mobile. The received signal strength from three waves can be expressed by extending from two waves, shown in Eq. (1.9.1.3.1), as

$$P_r = P_0 \, (1/(4 \, \pi d/\lambda))^2 \, |\, 1 + a_1 \exp(j \, \Delta\phi_1) + a_2 \exp(j \, \Delta\phi_2)\,|^2 \qquad (3.1.6.1.1)$$

where a_1 and a_2 are the reflection coefficients of water and land, respectively, and $\Delta\phi_1$ and $\Delta\phi_2$ are the phase differences between a direct path and a reflected path from water and from land, respectively.

In a mobile environment, a_1 and a_2 are equal to -1 because the energy of the signal is totally reflected with a phase reversed. Then Eq. (3.1.6.1.1) becomes

$$P_r = P_0(1/(4 \, \pi d/\lambda))^2 \, |\, 1 - (\cos \Delta\phi_1 + \cos \Delta\phi_2) - j \, (\sin \Delta\phi_1 + \sin \Delta\phi_2)\,|^2 \qquad (3.1.6.1.2)$$

$$= P_0 \, (1/(4 \, \pi d/\lambda))^2 \bullet L_r$$

where L_r is the loss factor and

$$L_r = |\, 1 - (\cos \Delta\phi_1 + \cos \Delta\phi_2) - j \, (\sin \Delta\phi_1 + \sin \Delta\phi_2)\,|^2$$

$$= [1 - (\cos \Delta\phi_1 + \cos \Delta\phi_2)]^2 + [\sin \Delta\phi_1 + \sin \Delta\phi_2]^2 \qquad (3.1.6.1.3)$$

since

$$\cos \Delta\phi = 1 - 2 \sin^2 (\Delta\phi/2) \qquad (3.1.6.1.4)$$

Substituting Eq. (3.1.6.1.4) into Eq. (3.1.6.1.3) and simplifying the equation yields

$$L_r = 1 - [-2 + 2 - 2 \sin^2 ((\Delta\phi_1 - \Delta\phi_2)/2)]$$
$$= 1 - 2 \sin^2 ((\Delta\phi_1 - \Delta\phi_2)/2)$$
$$\approx 1 \qquad (3.1.6.1.5)$$

This is because the value of $\sin((\Delta\phi_1 - \Delta\phi_2)/2)$ is very small. From Eq. (3.1.6.1.5), the loss factor L_r becomes one. It means no loss. Therefore, Eq. (3.1.6.1.1) becomes

$$P_r = P_0 \left(1/(4\,\pi d/\lambda)\right)^2 \tag{3.1.6.1.6}$$

It is a free space path loss. We may conclude that the free space path loss will be observed when propagation over the water occurs.

For treating the water enhancement, there are two options for determining where water is located, as follows:

1. Attribute—When the morphology is used as the basis for determining the location of water.

2. Terrain—If the elevation for any point along the terrain radial path is within 0.2 m of the specified elevation for the water base, then the point is considered to be located on water for implementing the water enhancement.

Determining the water location from the morphology of water is a more accurate method. Generally, terrain elevation should be used for this condition only when the morphology files of water files are not available.

3.1.6.2 The Unique Challenges

The location of water presents a unique challenge for radio wave propagation. With inexpensive and easier access to the morphology and terrain data, it is imperative for a prediction model to effectively adapt and manage unique water scenarios. The potential impact on the system performance and resources would be drastic when propagation over the water occurs.

The objective of this section is to specifically address the enhancement added to the Lee macrocell and microcell prediction models in order to handle the unique impact on radio propagation as a result of water surface. It is generally accepted that water enhances radio signals. However, there are many different impacts at varying levels, depending on where a mobile is located relative to the positions of water and the base station. An algorithm has been developed to deal with various scenarios in which water plays a critical role in predicting the effect of propagation loss. With water enhancement, the model can better predict the radio propagation, which has directly impacted cellular system engineering, most notably in coverage, handoffs, interferences, and thus capacity.

This issue focuses on when and in which cases the water enhancement is implemented. A high-level flow diagram and an individual case-by-case analysis is shown in this section. It is assumed that the propagation characteristics of uplink are identical to that of the downlink; therefore, only the characteristics of downlink are illustrated.

There are two generalized cases for the prediction in the proximity of water: case 1, the mobile receiver has a LOS condition, and case 2, the mobile is blocked from the transmitter of the base station. In each case, a number of possible situations can occur in which the water surface enhances the radio waves. When the "basic model" is referred to, only the regular prediction algorithms are used; therefore, the water enhancement is not implemented.

3.1.6.3 The Algorithm Diagram

The Lee macrocell prediction model has treated the radio wave as a theoretical ray path. The reflection point of that ray is obtained by finding a mirror image point of the mobile antenna in the image plane below either the land level or the water level, depending on the whether mobile location is on the land or water. Connect the image point at the mobile to the base station antenna by a line. A water reflection point is located where the line intersects the water level. The reflected ray is then tested by drawing a straight line from the water reflection point(s) to the existing mobile antenna.

The prediction result is raised if the reflected wave from the water level is detected along the ray path unblocked from the reflected point to the mobile. That includes the case in which the mobile is located on water. The flow diagram shown in Fig. 3.1.6.3.1 is the algorithm for testing the effective wave reflection from the water.

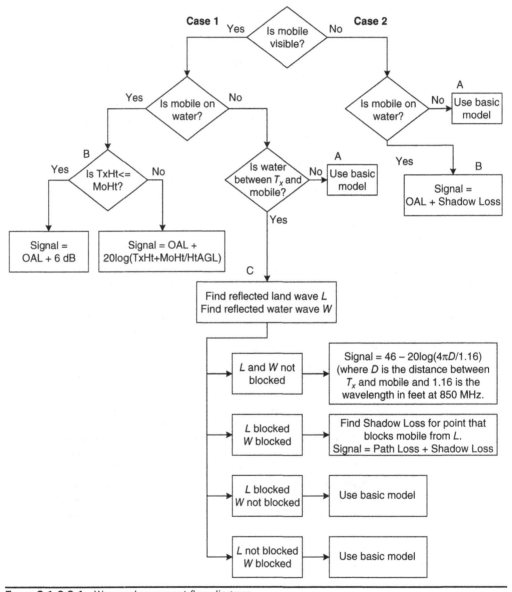

FIGURE 3.1.6.3.1   Wave enhancement flow diagram.

The brief notations shown in Fig. 3.1.6.3.1 stand for following:

TxHt = base station antenna height + above mean sea level (AMSL) [ft]

MoHt = mobile antenna height

HtAGL = base station antenna elevation above ground level [ft]. The antenna height in the standard condition is 100 ft. If the actual height h_1 is not 100 ft, a gain difference 20 log (h_1/100) should be added to the final prediction value.

OAL = open area loss. If the default

FSL = free space loss

T = base station transmitter

L = land

W = water

If the standard condition (see Sec. 3.1.2.1) is applied, the OAL includes a loss slope of 43.5 dB/decade and 1-mile intercept of −49 dBm, and the FSL equation uses the total ERP (antenna output power and gain) to be 46 dBm at the base station.

The shadow loss is obtained from Sec. 3.1.2.4.

Four situations of the mobile locations:

A. Mobile is on land.

   If the reflected wave from the water is not detected along the straight line between the base and the mobile, the signal does not get enhanced. When there are no water-reflected waves, as shown in Fig. 3.1.6.3.2, there are two rays, the LOS and the reflected wave. This case is known as the two-ray model, as shown in Sec. 1.9.1.3.

B. Mobile is on water.

   a. If TxHt = MoHt,
      then Signal = OAL + 6 dB

   If signal > free space loss value
   then signal = free space loss value

   OAL is generally used when a mobile is on water because of the absence of obstacles that can cause one reflected wave from the water. It is similar to an open area effect that can be seen from Fig. 3.1.6.3.3(*a*).

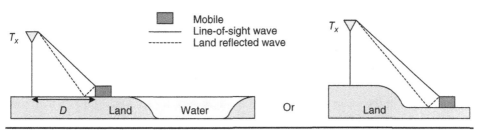

Figure 3.1.6.3.2   Case 1: A. Mobile is not on water, and no water is between the base and the mobile.

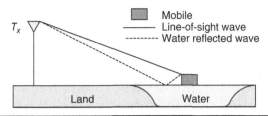

Figure 3.1.6.3.3a   Mobile visible and on water (use base station antenna height above ground level).

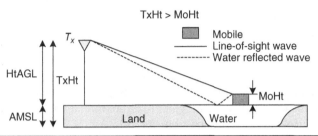

Figure 3.1.6.3.3b   Mobile visible and on water (use base station antenna height above sea level).

b. If TxHt > MoHt

then signal = OAL + 20 log ((TxHt-MoHt)/HtAGL)

     = OAL + effective antenna height gain

If signal > free space loss values

then signal = free space loss value

Under this condition, we use the base station antenna height, TxHt, above the sea level, as shown in Fig. 3.1.6.3.3(b).

C. The mobile is on land and the water reflected wave is detected. This case is shown in Fig. 3.1.6.3.4.

1. If both reflected waves, one from the water and one from the land, are not blocked, then a three-ray model is used. When three rays exist, the propagation loss approaches to the free space loss (see Sec. 3.1.6.1).

2. If both the land- and water-reflected waves are blocked (see Fig. 3.1.6.3.5),
   1. Find the shadow loss from the knife-edge point that blocks the mobile from the land.
   2. Signal = path loss + shadow loss

3. If the path is blocked by the land but not the water (see Fig. 3.1.6.3.6), then the basic model (two-ray model) is used.

4. If the radio path is blocked by the land but not by the water (see Fig. 3.1.6.3.7), then the basic model (two-ray model) is used.

5. If the radio path is blocked from the terrain (see Fig. 3.1.6.3.8), then the knife-edge diffraction loss is applied.
   1. Calculate the shadow loss
   2. Signal = open area loss + shadow loss
   3. No obstruction from buildings

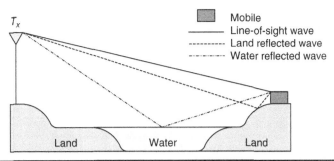

Figure 3.1.6.3.4   Mobile visible, and water between mobile and base and mobile is on land.

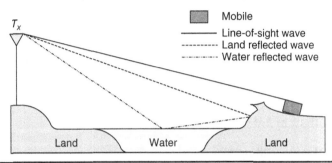

Figure 3.1.6.3.5   Land and water blocked.

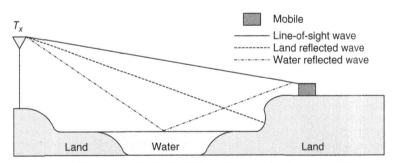

Figure 3.1.6.3.6   Land blocked and water not blocked.

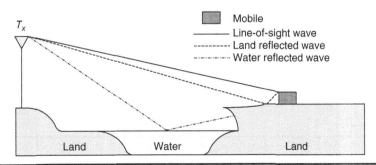

Figure 3.1.6.3.7   Land not blocked and water blocked.

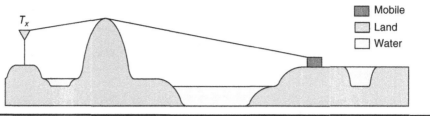

FIGURE **3.1.6.3.8**   Mobile is blocked.

3.1.6.4   Conclusion

The enhancement of the Lee macrocell propagation model for the radio path over the water ensures the inclusion of the result impacting from the water when designing a cellular system. Specifically, when dealing with a CDMA-centric 3G system, it is imperative to be able to predict the behavior of radio waves affected by water. The propagation over the water can introduce an unexpected coverage and create interference problems and thus impact the system capacity and coverage. As a result, the enhancement from the radio path over the water has a significant effect on system resources and performance.

3.1.7   Effect of Antenna Orientation

3.1.7.1   Antenna Tilting[16,17]

There are two types of antenna tilting—mechanical and electrical—used in cellular systems for reducing interference to neighboring cells. Mechanical downtilting of an antenna will change the antenna pattern, but electrical downtilting of an antenna will not. These are depicted in Fig. 3.1.7.1.1.

The horizontal beam shape vary when the mechanical downtilted angle changes.[17] The antenna patterns measured from +90° to –90° of the beam horizontally have different beam shapes, as shown in Fig. 3.1.7.1.1. The horizontal half-power beam width increases as the downtilted angle increases. Also, power reduction depends on the

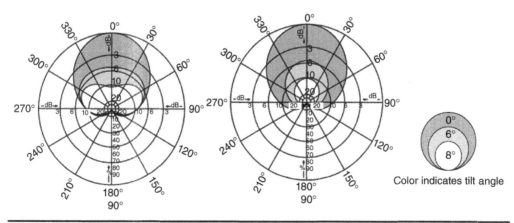

FIGURE **3.1.7.1.1**   Antenna downtilted horizontal pattern—mechanical (left) and electrical (right). Courtesy of Kathrein Inc., Scala Division.

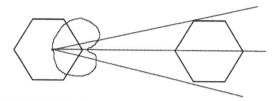

FIGURE 3.1.7.1.2 Reduction of co-channel interference by creating a notch in the pattern.

azimuth angle. The notch will occur at the center (0°) of the beam when the downtilted angle is tilted down to a certain angle. But the electrical downtilting will keep the antenna pattern unchanged over the whole azimuth range, and the horizontal half-power beam width is independent of the downtilted angle.

Mechanical antenna downtilting has been an effective way to reduce interference by confining the signal in its own coverage cell. It can reduce co-channel interference significantly when a notch is developed in the direction of co-channel cell.[16] We will define the condition for developing a notch in the antenna pattern and discuss the impact of downtilting antenna on the system design. We will also introduce the means of achieving the significant and optimal notch. The computer simulation is done to illustrate the notch-developing process in dealing with different antenna patterns.

A major contribution of reducing the interference by the downtilting of the antenna is co-channel interference. One way to reduce co-channel interference is to use a directional antenna at each cell site. In this way, the co-channel interference can be reduced by more than one-half.[3] Also, this interference can be further reduced by downtilting a directional antenna beam pattern. When the antenna pattern is tilted to a certain angle, a notch at the center of horizontal beam pattern is produced. The notch becomes larger when the tilted angle increases. This notch can be effective in reducing the interference in the co-channel cells, as shown in Fig. 3.1.7.1.2.

We will review the idea that downtilting the antenna develops the notch. We also derive the equations for demonstrating the notch phenomenon. The concept of the notch-developing process is discussed. It can be used as a guideline for designing a cellular system.

3.1.7.1.1 Antenna Downtilting Figure 3.1.7.1.1.1 shows that the normal horizontal antenna pattern is on the x-y plane. The maximum beam form is at the x-axis ($\theta_m = 0°$, $\phi = 0°$). When the antenna is downtilted at an angle $\theta_m \neq 0°$, the relationship between the downtilted angle θ_m at the azimuth angle $\phi = 0°$ and the related off-center angle ψ at any azimuth ϕ due to the downtilted angle θ_m can be shown as follows:[16]

$$\sin \frac{\theta_m}{2} = \frac{d}{l} \qquad (3.1.7.1.1.1)$$

$$\frac{\overline{DB}}{\sin \phi} = \frac{l}{\sin(135° - \phi)} \qquad (3.1.7.1.1.2)$$

$$\overline{CD} = l\frac{\sin 45°}{\sin(135 - \phi)} \qquad (3.1.7.1.1.3)$$

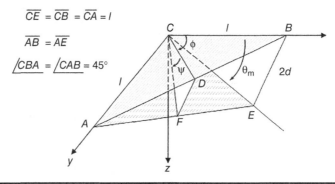

$$\overline{CE} = \overline{CB} = \overline{CA} = l$$

$$\overline{AB} = \overline{AE}$$

$$\underline{/CBA} = \underline{/CAB} = 45°$$

FIGURE 3.1.7.1.1.1 Coordinate of the tilting antenna pattern.

$$\frac{\overline{AD}}{\overline{DF}} = \frac{\overline{AB}}{2d} = \frac{\sqrt{2}l}{2d} \qquad (3.1.7.1.1.4)$$

$$\overline{AD} = \overline{AB} - \overline{DB} \qquad (3.1.7.1.1.5)$$

$$\cos \psi = \frac{2\overline{CD}^2 - \overline{DF}^2}{2\overline{CD}^2} = 1 - \frac{\overline{DF}^2}{2\overline{CD}^2} \qquad (3.1.7.1.1.6)$$

Substituting Eqs. 3.1.7.1.1.1 to 3.1.7.1.1.5 into Eq. 3.1.7.1.1.6, we obtain

$$\cos \psi = 1 - \cos^2\phi \ (1 - \cos\theta_m) \qquad (3.1.7.1.1.7)$$

or

$$\psi = \cos^{-1}[1 - \cos^2\phi \ (1 - \cos\theta_m)] \qquad (3.1.7.1.1.8)$$

where ψ is the off-center angle at an azimuth angle ϕ when the antenna is mechanically downtilted at an angle θ_m.

Equation (3.1.7.1.1.8) is used to determine the off-center angle ψ.

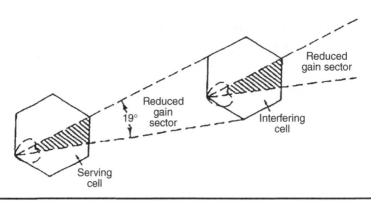

FIGURE 3.1.7.1.1.2 Reduced-gain sector of two co-channel cells.

If the physically tilted angle is $\theta_m = 180°$, then the off-center vertical angle ψ is tilted downward dependent on the azimuth angle ϕ:

$$\phi = \left\{ \begin{array}{c} 0° \\ 45° \\ 90° \end{array} \right. \qquad \psi = \left\{ \begin{array}{c} 18° = \theta \\ 12.7° \\ 0° \end{array} \right.$$

This list tells us that the physically tilted angle ϕ and the angle ψ are not linearly related and that when $\phi = 90°$, then $\psi = 0°$. When the angle θ increases beyond 18°, the notch effect of the pattern in the x-y plane becomes evident, as illustrated in Fig. 3.1.7.1.1.2 and indicated in Fig. 3.1.7.1.1.3.

Because the shape of the antenna pattern at the base station relates directly to the reception level of the signal strength at the mobile unit, the following antenna pattern effect must be analyzed.

When a high-gain directional antenna (the pattern in the horizontal x-y plane is shown in Fig. 3.1.7.1.1.3 and in the vertical x-z plane in Fig. 3.1.7.1.1.4 is physically (mechanically) tilted at an angle θ_m in the x-z plane shown in Fig. 3.1.7.1.1.4, how does the pattern in the x-y plane change? The antenna pattern obtained in the x-y plane after tilting the antenna is shown in Fig. 3.1.7.1.1.3. When the center beam is tilted downward by an angle θ_m, the off-center beam is tilted downward by only an angle ψ, which is smaller than θ_m, as shown in Fig. 3.1.7.1.1.1. The pattern in the x-y plane can be plotted by varying the angle ϕ at any giving downtilted angle θ_m.

3.1.7.1.2 Evaluation of the Notch Condition First, the antenna patterns of the undertested antenna, both vertical and horizontal, are stored in the computer for the simulation program. Then, for any given downtilted antenna angle θ_m, we may use Eq. (3.1.7.1.1.8) to

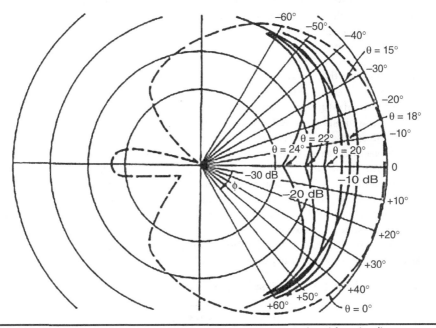

Figure 3.1.7.1.1.3 Notch appearing in tilted antenna pattern (reprinted from Lee[4]).

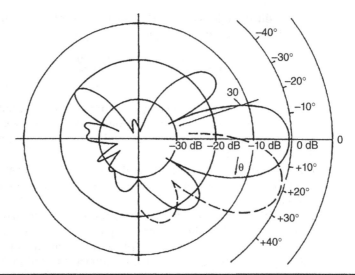

Figure 3.1.7.1.1.4   Vertical antenna pattern of a 120° directional antenna.

find the corrected downtilted antenna angle ψ from the corresponding horizontal direction φ. From the vertical pattern, the gain at angle ψ is higher than the gain at angle θ_m. The additional gain will be added on the horizontal pattern at angle φ, as shown in Fig. 3.1.7.1.1.3. Up to a certain angle θ_m, the notch appears in the pattern as depicted in the figure. The notch appears at a different angle of θ_m, depending on the different antenna patterns. Therefore, a study of the notch condition from the antenna pattern for an under-tested antenna can be done by using the simulation program. However, to obtain a useful notch, other conditions must be considered as well.

   Suppose that we would like to take advantage of this notch effect. From Fig. 3.1.7.1.1.2, we notice that the interfering site could cause interference at those cells within a 19° sector in front of the cell.

   In an ideal situation, such as that shown in Fig. 3.1.7.1.1.2, the antenna pattern of the serving cell must be rotated clockwise 10° by such that the notch can be aimed properly at the interfering cell. The antenna tilting angle θ_m may be between 22° and 24° in order to increase the carrier-to-interference ratio C/I by an additional 7 to 8 dB in the interfering cell, as shown in Fig. 3.1.7.1.1.3. Now we can reduce co-channel interference by an additional 7 to 8 dB because of the notch in the mechanically tilted antenna pattern. Although signal coverage is rather weak in a small shaded area in the serving cell, as shown in Fig. 3.1.7.1.1.2, the use of sufficient transmitting power should correct this situation. Besides, because the antenna is downtilted, the signal strength of the close-in area near the base station becomes strong. Therefore, the mechanically downtilted antenna can serve two purposes.

3.1.7.1.3 Computer Simulation   A simulator is built to simulate the effect of antenna downtilt. It takes the horizontal and vertical gain pattern as an input and generates the 3D composite gain pattern. Then the downtilt is applied to the 3D pattern, and the observed gain pattern is obtained by cutting the 3D pattern with our observing plane.

Three antenna patterns are chosen to demonstrate the effects of mechanical down-tilting of the antenna. The vertical beam main-lobe widths of those antennas are 8°, 30°, and 60°. Their horizontal and vertical beam patterns are listed in Figs. 3.1.7.1.3.1, 3.1.7.1.3.2, and 3.1.7.1.3.3. According to our optimal condition, the first one has a 30° beam width in the horizontal pattern, and its vertical beam is narrow, so we should expect to see a useful notch. However, when either the horizontal beam width is too narrow (see Fig. 3.1.7.1.3.2) or the vertical pattern is too wide (see Fig. 3.1.7.1.3.3), no useful notch can be obtained. So it is impossible to generate a useful notch under these conditions.

Our simulation result shows that all three cases can generate a notch by downtilt, and the notch tilt angle is approximately at the joint angle of their first and second lobes of the vertical pattern. The notch for the first antenna (in Fig. 3.1.7.1.3.1) is shown as a useful one. The other two notches (see Figs. 3.1.7.1.3.2 and 3.1.7.1.3.3) may not be useful because they are too weak and have a large side or back lobe, and that will interfere the co-channel cells behind it.

As our simulator shows, not all patterns can generate useful downtilted patterns. Antenna downtilt might actually create problems for some systems. As shown in Fig. 3.1.7.1.3.2, although a notch is generated, it is not useful for improving system performance. With this simulator, engineers can simulate the effect of downtilt in 3D and decide whether mechanical downtilt is the right approach to improving the system performance.

3.1.7.2 Smart Antenna and Lee's Microcell System

3.1.7.2.1 Smart Antenna A "smart antenna" is viewed as a means of significantly improving spectral efficiency and achieving better quality of service and higher capacities while also achieving considerable savings in base station costs.[18] The smart antenna system uses a small number of antenna panels (usually three). Each of these antenna panels has multiple beams generated by a passive electronic phasing matrix. With the smart antenna, the number of cellular subscribers continues to stay within the antenna coverage. The signals received on the narrow beams of the smart antenna are dynamically connected to the base site radios via an electronic matrix switch. The system capacity is increased in two ways. First, the use of high-gain narrow beams dramatically improves the C/I ratio, thus reducing frequency reuse factor K for the same quality of service. A lower frequency reuse factor means more channels per cell, that is, higher capacity. Second the smart antenna system maintains its channels in one trunk pool. Therefore, it has a higher capacity compared with the sectorial cell, in which three sectors do not share their resources.

3.1.7.2.2 Lee's Microcell System[19] In Lee's patented microcell system, each microcell consists of three zones (noted as zone A, zone B, and zone C), as shown in Fig. 3.1.7.2.1. Each zone has its own small base station just like an access point. This should be the first time the access points were used in the cellular system. A zone switch connects the three zones such that a signal channel can be switched from one zone to another. Thus, within the three zones, a mobile will stay at one signal channel until it moves out from its microcell. Based on the regular cells, the system is $K = 7$, but based on the zones, the system become $K = 3$, as shown in Fig. 3.1.7.2.2. Since the signal channel is confined in one zone at a time, we use $K = 3$ and find that the capacity increases by 2 to 2.5 over that of $K = 7$. The implement of microcell system is very cost effective. The modification of existing cellular equipment

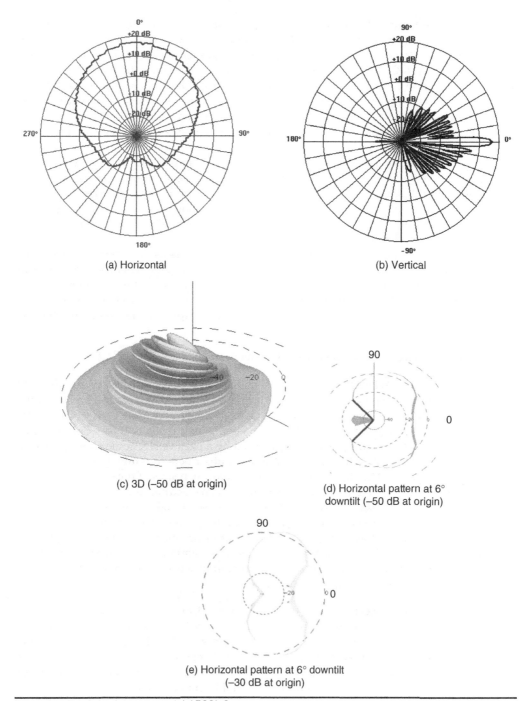

(a) Horizontal

(b) Vertical

(c) 3D (–50 dB at origin)

(d) Horizontal pattern at 6°
downtilt (–50 dB at origin)

(e) Horizontal pattern at 6° downtilt
(–30 dB at origin)

FIGURE 3.1.7.1.3.1   Antenna model 1560L-0.

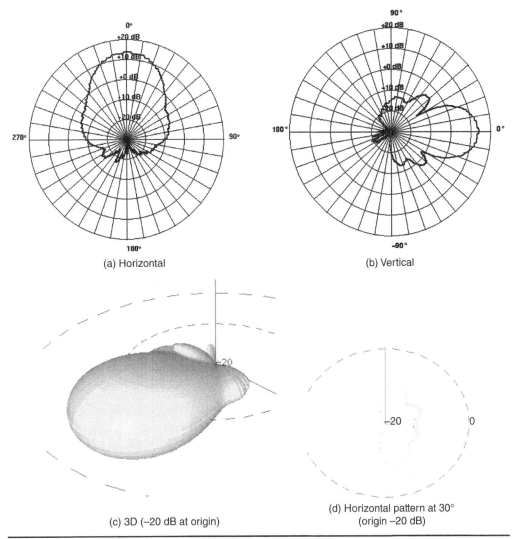

(a) Horizontal

(b) Vertical

(c) 3D (−20 dB at origin)

(d) Horizontal pattern at 30°
(origin −20 dB)

FIGURE **3.1.7.1.3.2**   Antenna model ALP4014-N.

for the microcell system is shown in Fig. 3.1.7.2.3. The microcell system has been deployed in Los Angeles and San Diego since 1991.

Both the smart antenna and Lee's microcell system are believed to be the solutions for achieving a high-capacity system. Both improve both the C/I ratio and trunk efficiency. The computer simulation was implemented to compare the capacities among omni-cell, three-sector cell, Lee's microcell, six-sector cells and a 12-switched beam smart antenna cell. First, the C/I ratios under different path loss exponents Υ and frequency reuse factors K are used to compare these systems. Then the trunking efficiencies among these systems are also compared. As the result shows, the smart antenna has supported a reasonably good C/I ratio and the highest trunking efficiency. This simulator has been also enhanced to calculate the path loss exponent for specific design areas based on the

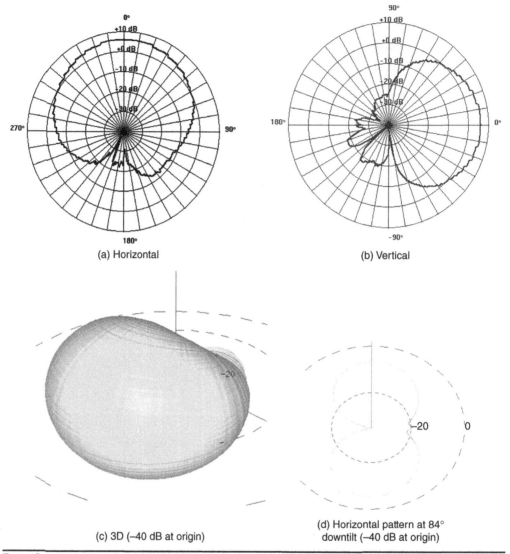

(a) Horizontal

(b) Vertical

(c) 3D (–40 dB at origin)

(d) Horizontal pattern at 84° downtilt (–40 dB at origin)

FIGURE 3.1.7.1.3.3   Antenna model ALP8007-N.

collected measured data. This path loss exponent can then be used as an input to the simulator to simulate the C/I and frequency reuse factor for the area. Measured data were collected in Los Angeles. The path loss exponents were calculated using Lee's prediction model. Based on the calculated path loss exponent and the C/I from the simulator, in a dense urban area, the smart antenna system can support a $K = 3$ reuse scheme.

Since the number of cellular subscribers always continues to grow at a rapid pace in a starting market, service providers are forced to find new methods to increase capacity in their networks. Having sectors in a cell is a means of reducing co-channel interference, and thus increase system capacity. However, it is associated with a trade-off in

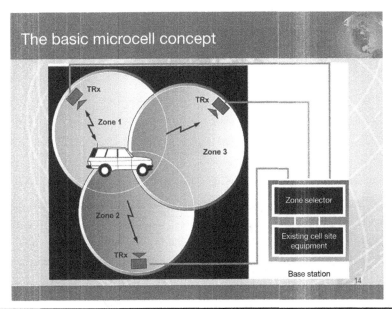

FIGURE 3.1.7.2.1   Lee's microcell system.

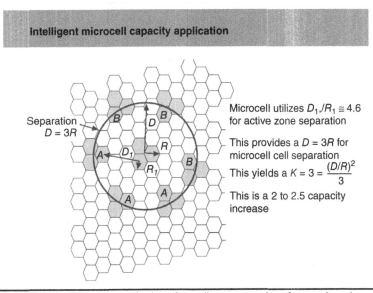

FIGURE 3.1.7.2.2   Capacity increases in the microcell system as interference is reduced.

trunking efficiency. The recent movement in digital modulation technology has helped cope with the increase in subscribers' demands. In the early years, the analog systems were sharing the frequency spectrum with the digital systems. There were more interference coming from the analog system into the digital systems. One economical approach to solve this problem is to use the smart antenna.

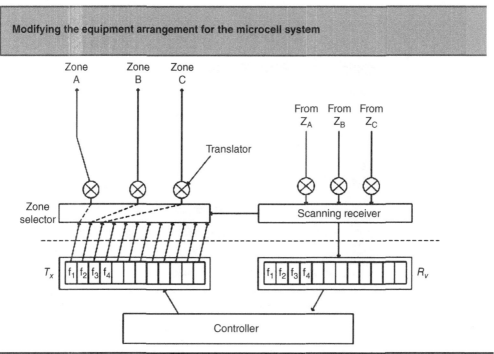

FIGURE **3.1.7.2.3**   Modifying existing cellular equipment for the microcell system.

Computer simulation was implemented to compare the capacity among the following systems: omni-cell, three-sector cell, Lee's microcell, six-sector cell, and switched-beam smart antenna cell. First, the C/I ratios under different path loss exponents and reuse factors are compared within these systems. Then the trunking efficiencies among these systems are also compared.

3.1.7.2.1 Simulation Model   A Monte Carlo simulation model was developed to analyze the cellular system. The simulator supports the Lee,[20] Hata, and flat-fading propagation models. Mobile units can be distributed randomly, uniformly, or linearly in a cell. The ideal (flat) terrain and hexagonal grid (cell) with different sizes are also supported. The multiple sectors within each grid (cell) need to be specified by the users. Usually, three-sector and six-sector cell layouts are the most frequently used. The ideal as well as real antenna patterns can be used in the simulation. Each cell (grid) or sector can have unique cell/sector parameters (ERP, antenna pattern, downtilted angle, percentage of pilot channel power, and so on). Different frequency reuse factors can also be changed interactively ($K = 3, 4, 7, 12$). The power control is simulated with the option of different levels of power steps and errors. The performance of the system can be analyzed using a uniform distribution to get the first order results. After the system is quickly accessed through the uniform distribution of mobile movement, the more complicated and time-consuming analysis can be done based on the outcome from the ideal (i.e., uniform, same-cell parameters) analysis. The flat-fading propagation model supports the path loss with the propagation exponent γ and the lognormal variation of δ dB.

The nominal value for γ is from 2 to 4 and for δ is usually 8 dB. Only the effect of flat fading is considered. The coverage, C/I plots, and statistical charts based on the results from simulation can be generated. Detailed information for each mobile unit and cell site from different layouts and distributions are also supported through all text files.

This simulation is handling an ideal two-tier layout for $K = 3$ and 4; a three-tier layout for $K = 7$; and four-tier layout for $K = 12$. All mobiles have assumed the same ERP if the power control is absent or otherwise have caused the same received power at the base station. The transmission from the mobile antenna is assumed to be omnidirectional. Each cell or sector uses the same antenna for both transmission and reception. The interference from the higher tiers of co-channel cells is compared with the first tier of co-channel cells to show that the interference from the higher tiers can be negligible for the sake of analysis.

Our simulator is run in both forward and reverse links for different path loss exponents (γ = 2, 2.5, 3, 3.5, 4), different sectorial schemes (omni-cell, three-sector cell, Lee's microcell, six-sector cell, and smart antenna cell) and different frequency reuse factors ($K = 3, 4, 7, 12$). Different real antenna patterns are used in different schemes (shown in Figs. 3.1.7.2.1.1 and 3.1.7.2.1.2). In Fig. 3.1.7.2.1.2, the different shapes of left beams and right beams are shown.

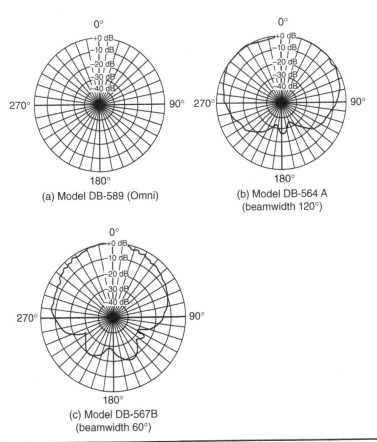

(a) Model DB-589 (Omni)

(b) Model DB-564 A
(beamwidth 120°)

(c) Model DB-567B
(beamwidth 60°)

Figure 3.1.7.2.1.1 Normalized horizontal patterns of antennas.

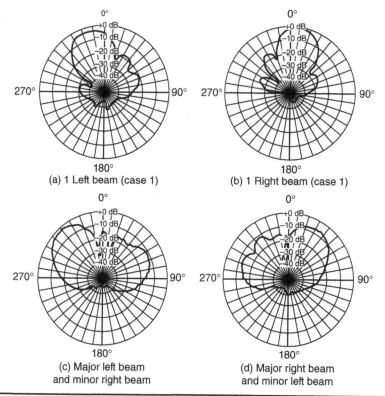

0°
(a) 1 Left beam (case 1)

0°
(b) 1 Right beam (case 1)

0°
(c) Major left beam
and minor right beam

0°
(d) Major right beam
and minor left beam

Figure 3.1.7.2.1.2   Normalized smart antenna horizontal pattern.

3.1.7.2.2 Simulation Result   A parameter named "C/I satisfaction percentage" is used to present our simulation result. The C/I satisfaction percentage is defined as the percentage of area with a satisfactory C/I value in a given cell. In an analog system, a satisfactory C/I value is greater than or equal to 18 dB. The coverage area is obtained by measuring the received carrier power C. The interference I is obtained by taking the signals from the neighboring sites. Therefore, even the smart antenna covers a larger area but receives a higher interference. As a result, the C/I satisfaction percentage may not be high. Figure 3.1.7.2.2.1 shows that the C/I satisfaction percentage improves as the path loss exponent increases. This is because using a low path loss exponent for the received signal at mobile with a certain C/I can cover only a small area near the base station. Also, the C/I satisfaction percentage improves as the number of sector increases due to the additional gain from directional antennas. In Fig. 3.1.7.2.2.2(a), the worst case in a smart antenna system[21] is shown in the figure (a). The worst case in Lee's microcell system[18] is shown in Fig. 3.1.7.2.2.2(b).

The simulation plots the results of smart antenna cell, Lee's microcell, and the other three systems. From the simulation, if we do not specify the environmental areas (free space, open, suburban, urban, and metropolitan), the smart antenna system has the best C/I performance, followed by six-sector cell, Lee's microcell, three-sector cell, and omni-cell, which has the lowest C/I performance. The advantage of using smart antenna

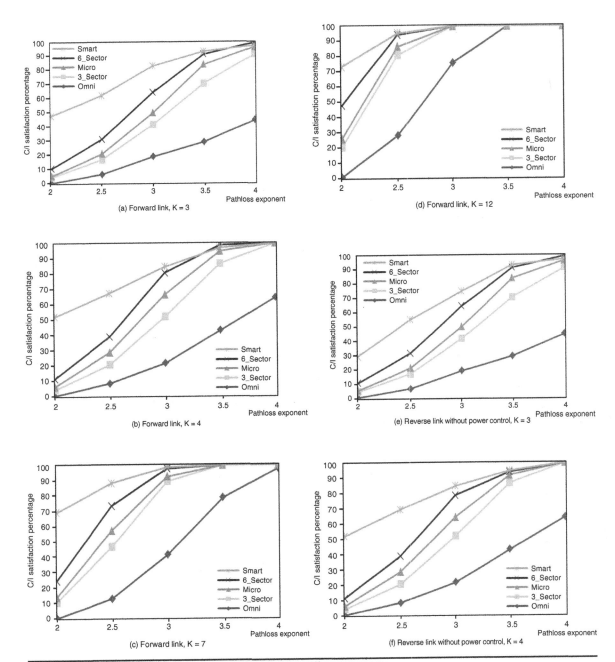

FIGURE **3.1.7.2.2.1** *C/I* performance comparison among different systems.

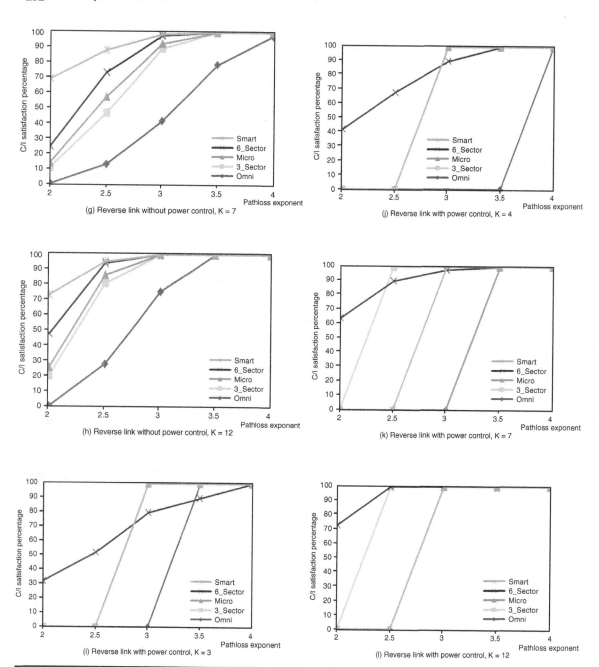

FIGURE 3.1.7.2.2.1   *(Continued)*

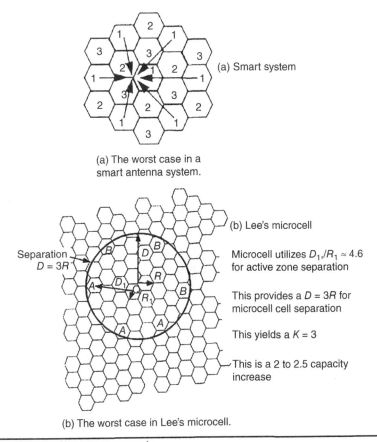

(a) Smart system

(a) The worst case in a
smart antenna system.

(b) Lee's microcell

Separation
D = 3R

Microcell utilizes $D_1/R_1 \approx 4.6$
for active zone separation

This provides a $D = 3R$ for
microcell cell separation

This yields a $K = 3$

This is a 2 to 2.5 capacity
increase

(b) The worst case in Lee's microcell.

FIGURE 3.1.7.2.2.2   Worst case scenario.

system over other systems is the most significant improvement of C/I coverage at a low path loss ($\gamma = 2$). It is in a free space area. Because of the high gain provided by the smart antenna, at a high path loss ($\gamma = 4$), the six-sector cell out performs the smart antenna system. Since the smart antenna system has very good C/I performance with a low path loss exponent, the cellular operator can have both the good C/I performance and the higher trunking efficiency in the system in those areas with $\gamma = 2$.

From the experiment, we used the measured data collected in Los Angeles. The drive routes with longitude and latitude are shown in Fig. 3.1.7.2.2.3. Through the signal propagation analysis using Lee's model, we found that the path loss exponent in Los Angeles area is very close to 4, as shown in Fig. 3.1.7.2.2.4. Figure 3.1.7.2.2.1 shows from the plots for forward and reverse links, with the pass lost exponent $\gamma = 4$, at a 95 percent C/I satisfaction, the frequency reuse factors are 7, 4, 3, 3, and 3 for the omni-cell, three-sector cell, Lee's microcell, six-sector cell, and smart antenna systems, respectively. Given a frequency spectrum bandwidth of 12.5 MHz, their capacities are listed in Table 3.1.7.2.2.1.

We can see from Table 3.1.7.2.2.1 that the smart antenna system and Lee's microcell have the highest capacity. Table 3.1.7.2.2.1 gives a statistical comparison only between Lee's microcell and the smart system. Now let us compare these two in a worst case.

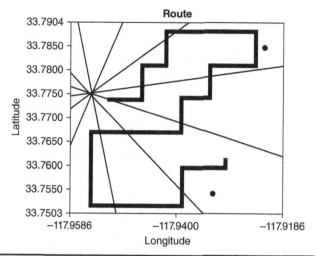

FIGURE 3.1.7.2.2.3   Drive route in Los Angeles.

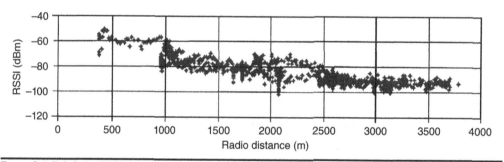

FIGURE 3.1.7.2.2.4   RSSI versus radial distance.

Parameter	Omni	3-Sec	Lee's	6-Sec	Smart
Bandwidth (MHz)	12.5	12.5	12.5	12.5	12.5
# of channels	417	417	417	417	417
Reuse factor	7	4	3	3	3
Channel/cell	59.57	104.3	139	139	139
Channel/cell/sector	59.57	34.75	—	23.17	—
Erlang/sector	49.2	26.1	—	15.9	—
Erlang/cell	49.2	78.3	126	95.4	126
Erlang/cell/MHz	3.94	6.26	10.08	7.63	10.08

TABLE 3.1.7.2.2.1   Capacity Comparison Between Different Systems at a 2%
Blocking Rate

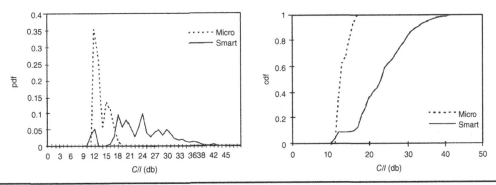

FIGURE 3.1.7.2.2.5   Comparison of C/I distribution at the worst point.

In the worst case, Lee's microcell is about 10 dB better than the smart antenna system. However, due to the randomness of the frequency assignment between different sectors or zones within a cell, the C/I at the worst position is not a constant. The distributions of C/I for the two systems are drawn in Fig. 3.1.7.2.2.5. We can see that the C/I distribution for the smart antenna system is more spread out than Lee's microcell.

For a smart antenna, its worst case is that all beams of the same frequency are pointing to one vertex of the center cell shown in the figure. It is not difficult to see that it is the same performance as in the omni-cell. And, at the vertex of the center cell,

$$\left(\frac{C}{I}\right)_{smart} = \frac{R^{-4}}{\sum_{i=1}^{6} D_i^{-4}} = \frac{R^{-4}}{2R^{-4} + 2\cdot(\sqrt{7}R)^{-4} + 2\cdot(\sqrt{13}R)^{-4} + (4R)^{-4}} = 9.24 \text{ dB} \qquad (3.1.7.2.2.1)$$

Similarly, for Lee's microcell, the worst position is in zone Q of the center of the cell. We have

$$\left(\frac{C}{I}\right)_{Lee's\ microcell} = \frac{(R_1)^{-4}}{3(4.6R_1)^{-4} + 3(5.5R_1)^{-4}} \approx 100 \text{ or } 20 \text{ dB}\quad \text{(worst case)} \qquad (3.1.7.2.2.2)$$

3.1.7.2.3 Conclusion   A simulator was built to handle the analysis of C/I. The C/I satisfaction percentages are plotted and compared among the omni-cell, three-sector cell, Lee's microcell, six-sector cell, and smart antenna cell. We find that the smart antenna cell has the best C/I performance at the low path loss exponent. The six-sector cell outperforms the smart antenna cell when the path loss exponent is high. In an urban area, with a given path loss exponent $\gamma = 4$ and a given C/I satisfaction percentage of 95 percent, the capacities among different systems are also compared. We find that the smart antenna cell and Lee's microcell have the best spectrum efficiency. Besides, Lee's microcell performs better than the smart antenna system in the worst case by about 10 dB.

3.1.8 Prediction Data Files
Depending on the various parameters applied to the Lee macrocell model, the signal strength is predicted for each point on a radial from the base station to the mobile. The calculation of the signal strength at each point on each radial on a terrain map uses the digitized terrain elevation data, land-use data, and antenna data stored in the file. The path loss values of the propagation prediction are stored in a data file.

3.1.8.1 Coverage Files[5]

A file stores the predicted path loss values along each radial for a given set of elevation, terrain, and model parameters. Whenever possible, files are shared across sectors and tiers at a single geolocation to save disk space and file calculation time. When antenna heights or model parameters differ between co-located sectors/tiers, the additional final files are created.

A coverage file has two components:

1. A header containing various identification fields, including the format version, date and time stamp, prediction area extents, parameter values used, and the number of points along each radial.

2. Path loss values stored in n arrays of p values, where n is the number of radials used in the calculation and p is the number of elements along each of the radials. Note that $p =$ (radial distance/radial distance unit) $+ 1$.

The time required for calculating a coverage file as well as the file size are determined by three calculation parameters as follows:

1. The radial distance (from 1 to 100 miles), which can be specified by the individual user on a sector-by-sector basis

2. The number of radials (either 360, 720, 1440, or 2880), which can be specified by the individual user on a sector-by-sector basis

3. The distance between points along the radial (radial distance unit)

3.1.8.2 RSSI Grid Files

An RSSI (Received Signal Strength Indicator) file stores the received signal strength value (in dBm) for each bin in the predicted area. An RSSI file is the result of the conversion of a radial-based coverage file into a grid-based block of values for a given bin size. Several options and formats of the stored final file values can be used to provide an aligned sample space for the different tiers involved.

Note that the distance between arc seconds differs in the latitudinal and longitudinal directions, depending on the latitudinal position on the earth. Thus, the bin sizes given in the table are approximate, and the bins are rectangular, not square.

Storage space requirements for grid files increase four times when bin size is reduced in half. To help conserve disk space, choose the largest bin size suitable for the intended analysis. If desired, you may create plots using a size of multiple bins.

Grid files are generally created only when needed to support data visualization or other bin-based analyses. The grid file is based on the bin size and does the interpretation to attain better granularity. For example, there can be four grids in one of the bins, and the data for each grid are interpreted based on the data in each bin. There are many different ways of doing the interpretation.

Arc sec	24	12	6	3	1.5	0.75	0.375	0.1875
Meters	720	360	180	90	45	22.5	11.25	5.625
Feet	2400	1200	600	300	150	75.0	37.5	18.75

TABLE **3.1.8.2.1**   Sample of Supported Bin Sizes

3.2   Fine-Tuning the Lee Model[22]

Both terrain contour and human-made structures strongly affect the received mobile radio signal strength. But it is difficult to separate the effects of the natural environment and human-made structures on the received data for two reasons. First, in a natural environment, the ground is never flat; second, human-made structures in each area are different. Finding the propagation characteristics, such as the intercept signal level at a given range (1 km or 1 mile away from the base station) and the slope of path loss along the radio path, is a challenging task. This section introduces a method that can separate the effects of human-made structures from those of the terrain contour with a high degree of confidence. The propagation characteristics differ from different areas, and the proper ones are used as valuable inputs to the propagation prediction software program. The theoretical shadow loss and the effective antenna height gain are also compared with the large amounts of measured data when a mobile is either blocked or not blocked by terrain from the base station. Finally, a means of fine-tuning the propagation model due to the terrain effect by feeding back the measured data is discussed.

The increasing demand for cellular services offers a great challenge in designing cellular systems. To be able to design a good cellular system and fine-tune intricate parameters demands a good design tool. In the meantime, an accurate propagation prediction tool is a key to successfully designing a cellular system, especially in the early stages of cellular and PCS cell design. The alternative to the prediction tools is to drive every route in the interested areas before deciding on the locations of base stations, and this is both costly and tedious. A more practical and cost-effective solution would be to collect a sufficient amount of measurement data ahead of the system deployment and use the statistics gained from these values as an input to fit the prediction tool. It is necessary to collect the measured data so as to enhance the predicted outputs and at the same time reveal the characteristics of the region under consideration.

Since every area has its own unique building structures, terrain configuration, and morphology, it is very difficult to separate the effects caused by the natural obstructions of the area from the human-made obstructions of the measured radio signal data.[5,13] In general, the parameters can be classified in two categories:

Impact of Human-Made Structures.   The effect of human-made constructions can be translated to a path loss curve of slope and a 1-mile (or 1-km) intercept value, which become the input parameters for the prediction model. Thus, every region will have its own characteristic slope and intercept values due to the different human-made constructions.[2]

Impact of the Natural Terrain and Variation.   Variations in terrain, valleys, mountains, and so on. result in strong or weak signal reception, depending on the effective antenna height gain or loss in the nonobstructive case and on the diffraction loss in the obstructive case due to terrain contours.[1]

The methodology of separating the effects of human-made structures from those of terrain contours using the pre-collected measurement data is discussed here. Then two deduced parameters are obtained as an input in the prediction tool. These two deduced parameters are the correct slope and the 1-mile intercept value for the area of interest.

In the real environment, the ground level is never flat, the structures above ground level are of different types, and the accuracy of the collected measurement data is never 100 percent accurate. We cannot expect the measured data to be totally reliable, but the accuracy of the terrain database does play a major role in the prediction results.

We will first explain the method of how to separate the factors of human-made structures from the factors of natural terrain contours. Then examples will be presented regarding how to apply this method to real system applications. Finally, suggestions for future enhancements will be discussed.

3.2.1 The Terrain Normalization Method

Before discussing normalization of the terrain, it is necessary to know exactly how the terrain contour affects the prediction. The path loss curve constitutes a 1-mile intercept P_0 and the slope of the path loss γ, as shown in Eq. (3.2.1.1). The method used to find P_0 and γ for each geographical area is described below.

From Eq. (3.2.1.1), it is shown how the gains and losses vary with terrain along the mobile direction. The purpose at this stage is to factor out the terrain variation and to show the effects of human-made structures on the prediction. To achieve this, we need to either nullify the terrain effect or do terrain normalization.

Slope γ and intercept P_0 are determined due to different human-made environments. Gains or losses calculated based on the antenna height gain use terrain contour. As mentioned in the previous section, the measured data are affected by both human-made structures and terrain contours. If we can separate the terrain variation effects on the received signal from the human-made effects, then we can obtain the right set of values for the slope and the 1-mile intercept for the human-made effects.

Now let us explain what exactly the terrain normalization does on the measured data points. Every measured data point is considered individually, and the following steps are carried out:

Step 1: Collect drive test data

Step 2: Screen the valid measured data

Step 3: Run terrain normalization on the measured data

Step 4: Calculate the slope and intercept based on data from step 3

In order to obtain meaningful statistics, a large amount of data needs to be collected. The measured data need to be carefully screened because bad data can lead to a wrong conclusion. The data collected in tunnels, elevated highways, or below the noise floor need to be discarded.

The terrain normalization method in step 3 is discussed in more detail here. For every measured point, the following steps need to be performed.

First, extract a terrain radial from a cell site from the terrain base to match the measured point. Then calculate the prediction along the radial. Depending on whether or the path is being clear or shadowed, calculate both the gain or the loss at that point and the adjustment factor:

$$P_r = P_0 - \gamma \log\left(\frac{r}{r_o}\right) + G_{effh} - L(v) + \alpha$$
$$\underset{\text{Gain}}{\phantom{G_{effh}}} \quad \underset{\text{Loss}}{} \quad \underset{\text{Adjustment}}{}$$

$$(3.2.1.1)$$

By nullifying the effects of gain or loss and adjusting the measured data M from a different setting, the fully normalized points from the measured data are then obtained:

$$M_{\text{norm}} = M - G_{effh} + L(v) - \alpha \qquad (3.2.1.2)$$

where M_{norm} is the normalized signal from the measurement.

Normalization is done only on the valid measured points. After all measured points have been normalized based on terrain, use the mean square best-fit algorithm to calculate the slope and intercept for the cell site.

After inserting the proper path loss slope and the 1-mile intercept in Eq. (3.2.1.1), the signal strength predictions can be found by adjusting G_{effh} or $L(v)$ from a terrain contour map.

3.2.2 Measurement Data Characteristics

The measured data from a single cell were collected in the Italy Ivrea area. There are a total of 10 drive routes. This cell site had an elevation of 246 m above sea level with a 50-m transmitter height and an ERP of 45.3 W.

The measured data were collected by driving away and toward the cell site. Figure 3.2.2.1 shows the actual physical locations of these 10 routes, which demonstrated most of the characteristics of the measured data at this cell site.

Because the morphology might be drastically different in making the same setup of cell sites, it is important to be able to apply different slopes and intercepts for different areas covered by the same setup of cell sites. This situation can easily be handled and solved by applying this treatment.

When a cell site is located in a hilly area, there are some known terrain problems. Also, morphology data, such as tunnels, elevated highways and bridge information, are not always available. Usually, the measured data were collected without screening these restrictions. Although there are questions regarding how accurate the measured data are, the amount of sampled data is believed to be large enough to be statistically

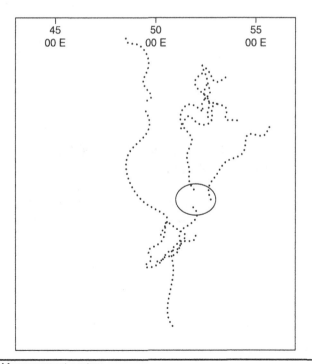

FIGURE 3.2.2.1 Measurement routes.

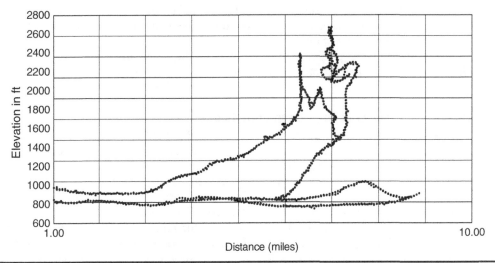

FIGURE 3.2.2.2   Terrain elevation (ft) of measurement data versus distance on different routes.

meaningful. The goal is to make the best use of the drive test data by applying the feedback to enhance the propagation model. Deriving a set of path loss slopes and 1-mile intercepts, and then applying these parameters back to the propagation model for an accurate prediction in a certain area is the objective of this section. Figure 3.2.2.2 presents all measured points and their corresponding elevations along the radial line from the cell site. It shows that the measured data were collected in an area where the elevation of the terrain is generally going up. Without applying the terrain normalization process, the slope and intercept values are misleading and cannot be applied to an accurate prediction of the signal coverage.

After applying the terrain normalization process, a more reasonable slope and 1-mile intercept were achieved and should be applicable for those areas similar to the area where the drive test was performed.

3.2.3   Comparison of Measured and Predicted Curve for the Nonobstructive Case

Considering the nonobstructive situation under real terrain conditions shown in Fig. 3.2.3.1 and selecting a case of sloping up, the terrain contour does affect the signal received by the mobile. There is an effective antenna gain because of the position of the mobile on an elevated slope. The effective antenna gain G_{effh} is given in Eq. (3.1.2.3.1).

All the field points that are collected over a nonobstructive path from the cell site can be identified from the drive test data based on the terrain data provided in Fig. 3.2.3.2. The effective antenna height gain at each data point along the nonobstructive path is then calculated and plotted in Fig. 3.2.3.3, which also shows the graph of signal strength versus radial distance for these points.

If a set of morphology data involving buildings is available, the same procedure can be applied to handle the effect on both blocking and nonblocking by buildings. Again, this procedure can be useful only if the terrain database in the field and the measurement data from the test equipment are accurate.

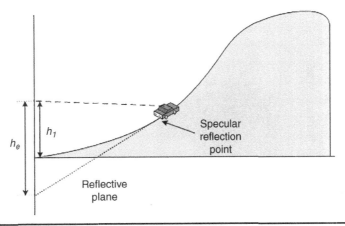

Figure **3.2.3.1**   LOS.

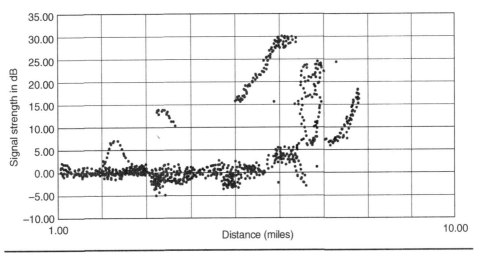

Figure **3.2.3.2**   LOS versus radial distance.

Figure 3.2.3.3 compares the theoretical effective antenna gain with the effective antenna gain derived from the drive test data after normalization. We can see that the actual effective antenna gain can go higher than the theoretical value. The Lee model can be easily fine-tuned with the measurement data and enable those data to be flexible and accurate.

Figure 3.2.3.4 shows another situation when the terrain blocks the mobile. We will go through the same exercise for the terrain blocking situation in Sec. 3.2.4

3.2.4   Comparison of Measured and Predicted Curves for the Obstructive Paths

In the previous section, we covered the effects of gains and losses due to terrain variation where the mobile is in the nonobstructive path from the base station. Besides, the varying terrain contour along the path of the mobile traveling and the scattering loss

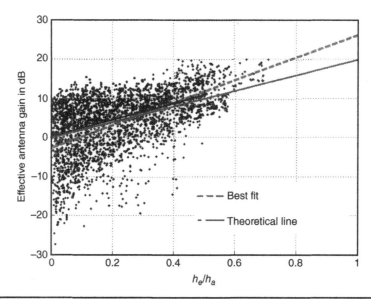

FIGURE 3.2.3.3   Effective antenna gain between best-fit and theoretical prediction.

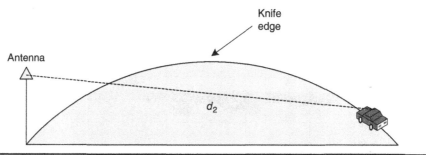

FIGURE 3.2.3.4   Mobile blocked by terrain.

due to the buildings add one more factor that contributes significantly to the gains or losses of the received signal. Now let us consider a more complex case of a mobile under hidden conditions, or a mobile blocked by the terrain contour of a hill. Due to a hill, there is a loss in the signal strength because of diffractions caused by the obstruction. The diffraction loss can be calculated from the knife-edge diffraction loss curve shown in Fig. 1.9.2.2.1.2. The diffraction factor, v, is also defined in Fig. 1.9.2.2.1.2. In this section, where we have to adjust the loss in the drive test data due to a shadowing situation, the new normalized field data then contains only the human-made effect and can be used to find the slope and the 1-mile intercept. Figure 3.2.4.1 shows the signal strengths of all spots, some of which are blocked from the cell site and some not. The theoretical diffraction loss at each spot under the shadowing situation is calculated. Figure 3.2.4.2 presents three sets of data: the theoretical shadowing loss curve; the measured data, which are in the shadowing situation; and the new best-fit shadowing loss curve. The measurement data were plotted on the parameter v scale. Each data point

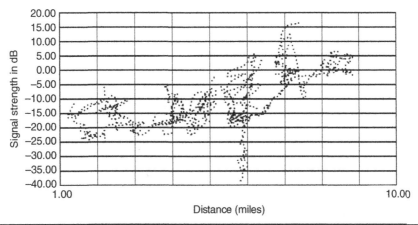

Figure 3.2.4.1   Diffraction loss versus radial distance.

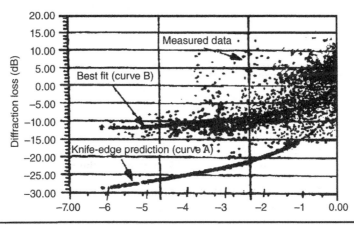

Figure 3.2.4.2   Diffraction loss comparison between best-fit and knife-edge prediction.

was first calculated its v according to the propagation condition. Also, each data point was subtracted from its path loss before plotting on the v scale in Fig. 3.2.4.2. The measured data points always have less diffraction loss than the theoretical diffraction loss, as shown in the figure. A best-fit shadowing loss curve is shown in the figure.

For making a correction to predict the shadow loss more accurately, we have introduced three methods, as shown below.

3.2.4.1   A Method of Using Max G_{effh}

Using Max G_{effh} to correct the prediction loss in the shadow condition has been described in Eq. 3.1.2.4.2.

3.2.4.2   An Empirical Method

As we can see from the measured data points, the best-fit shadowing loss curve would be used for adjusting the parameter v because the theoretical knife-edge diffraction loss prediction curve is not fitted to the measured data, as shown in Fig. 3.2.4.2. For making

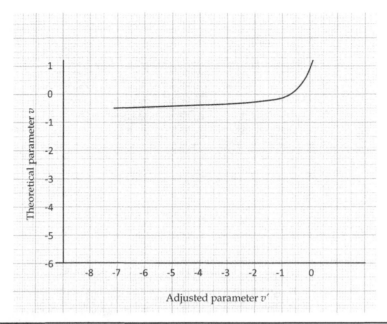

FIGURE **3.2.4.3** The adjusted v' versus the theoretical v.

an adjusting v' for predicting the diffraction loss from the real measured data, we have proposed an adjusted curve, shown in Fig. 3.2.4.3. The adjusted v' is taken by following the best-fit curve and would be more likely to match the real measured data points.

3.2.4.3 A Method for the High-Knoll Condition

The basic concept of dealing the high-hill case is that at the top of a high knoll, the signal is very strong. The signal, beyond the knoll, will be weaken due to the shadow. Then the signal strength, not the average path loss signal, will be subtracted from the knife-edge diffraction loss. Therefore, we implement the correction of knife-edge diffraction loss for the following two cases: one for a single knoll and one for double knolls.

3.2.4.3.1 Single-Knoll Case (see Fig. 3.2.4.4)

1. In a Negative-Slope Condition (Fig. 3.2.4.4(a))

The receiving power at a distance r is expressed as

$$P_{r\text{(area-to-area)}} = P_{r_0} - \gamma \log\left(\frac{r}{r_0}\right) - A_f + \alpha \qquad (3.1.2.1)$$

where r_0 is either 1 mile or 1 km, and the path loss slope γ will be found from Fig. 3.1.2.1.2, depending on whether it is an open or a suburban area.

The power at distance P_{r_1} can be obtained from Sec. 3.1.2.1:

$$P_{r_1} = P_{r\text{(area-to-area)}} \quad \text{at } r_1 + G_{effh} \qquad (3.2.4.3.1)$$

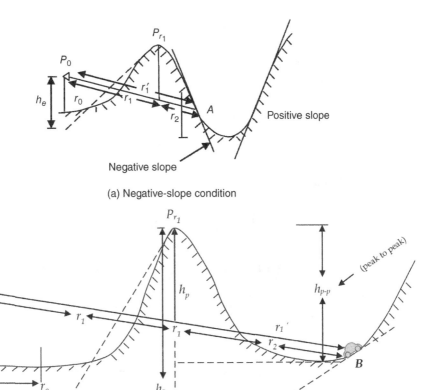

(a) Negative-slope condition

(b) Positive-slope condition

FIGURE 3.2.4.4 Single-knoll coordinate.

where G_{effh} is shown in Eq. (3.1.2.3.1). Then the received power $P_{r_1'}$ at A where $r_1' = r_1 + r_2$, becomes

$$P_{r_1'} = P_{r_1} + 10 \log \left(\frac{r_1'}{r_1} \right)^{-\gamma} + L \text{ (knife-edge diffraction loss at } A) \qquad (3.2.4.3.2)$$

2. In a Positive-Slope Condition (Fig. 3.2.4.4[b])

The power P_{r_1} at distance r_1 is the same as Eq. (3.2.4.3.1). At a positive-slope condition, there is a gain that needs to be calculated. First is to measure from the peak of the knoll to the valley of the knoll, h_{p-p}, and then to measure the height from the peak of the knoll to the crossing of the positive slope at the distance r_1, h_{r_1}. The gain G_{effB} is

$$G_{effB} \text{ at point } B = 20 \log \left(\frac{h_{r1}}{h_{p-p}} \right) \tag{3.2.4.3.3}$$

There are two cases ($r_1' = r_1 + r_2$):

Case 1: If $10 \log \left(\dfrac{r_1'}{r_1} \right)^{-\gamma} + G_{effB} < 10 \log \left(\dfrac{r_1'}{r_1} \right)^{-2}$

$$P_{r_1'} = P_{r_1} + 10 \log \left(\frac{r_1'}{r_1} \right)^{-\gamma} + G_{effB} + L \text{ (knife-edge diffraction loss)} \tag{3.2.4.3.4}$$

Case 2: If $10 \log \left(\dfrac{r_1'}{r_1} \right)^{-\gamma} + G_{effB} \geq 10 \log \left(\dfrac{r_1'}{r_1} \right)^{-2}$

$$P_{r_1'} = P_{r_1} + 10 \log \left(\frac{r_1'}{r_1} \right)^{-\gamma} + L \text{ (knife-edge diffraction loss)} \tag{3.2.4.3.5}$$

3.2.4.3.2   Double-Knoll Case (see Fig. 3.2.4.5)

1. In a Negative-Slope Condition (Fig. 3.2.4.5[a])

The power P_{r_1} at distance r_1 is the same as Eq. (3.2.4.3.1). The path loss from the first peak to the second peak of the knoll would be assumed the free space loss. Then the received power at point A, where $r_1'' = r_1 + r_2 + r_3 = r_1' + r_3$, is

$$P_{r_1''} = p_{r_1} + 10 \log \left(\frac{r_1'}{r_1} \right)^{-2} + 10 \log \left(\frac{r_1''}{r_1'} \right)^{-\gamma} + L \text{ (double knife-edge diffraction loss)} \tag{3.2.4.3.6}$$

2. In a Positive-Slope Condition (see Fig. 3.2.4.5[b])

We are following the same derivation as shown in the single-knoll case. In this double-knoll case, there is a gain G'_{effB} which can expressed as

$$G'_{effB} = 20 \log \left(\frac{h'_{r_1}}{h_{(p-p)_2}} \right) \tag{3.2.4.3.7}$$

where h_{r_1} and $h_{(p-p)_2}$ have been shown in Fig. 3.2.4.5(b).

There are two cases:

Case 1: If $10 \log \left(\dfrac{r_1''}{r_1'} \right)^{-\gamma} + G_{effB} < 10 \log \left(\dfrac{r_1''}{r_1'} \right)^{-2}$

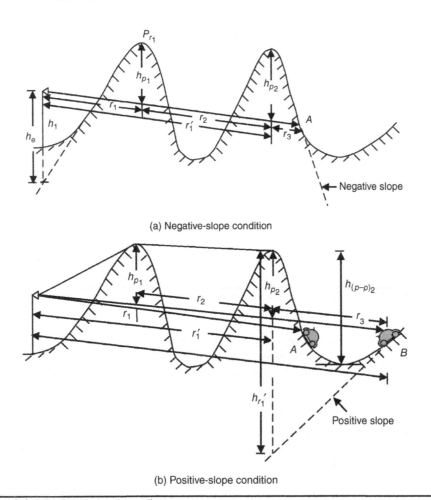

(a) Negative-slope condition

(b) Positive-slope condition

Figure 3.2.4.5 Double-knoll coordinate.

$$P_{r_1''} = P_{r_1} + 10 \log\left(\frac{r_1'}{r_1}\right)^{-2} + 10 \log\left(\frac{r_3}{r_1'}\right)^{-\gamma} + G_{effB} + L \text{ (double-knife-edge diffraction loss)}$$

$$(3.2.4.3.8)$$

Case 2: If $10 \log\left(\dfrac{r_3}{r_1'}\right)^{-\gamma} + G_{effB} \geq 10 \log\left(\dfrac{r_3}{r_1'}\right)^{-2}$

$$P_{r_1''} = P_{r_1} + 10 \log\left(\frac{r_1'}{r_1}\right)^{-2} + 10 \log\left(\frac{r_3}{r_1'}\right)^{-\gamma} + L \text{ (double-knife-edge diffraction loss)} \quad (3.2.4.3.9)$$

where $r_1'' = r_1' + r_3$

3.2.4.3.3 Measured Data versus Predicted Values   Figure 3.2.4.6 shows the measured data taken in San Diego along 25th Street while the site was located at 25th Street and F Street. The transmit power was 1 W, the antenna gain was 6 dB, the antenna height was 20 ft, and the carrier frequency was 850 MHz. The mobile antenna height was 6 ft

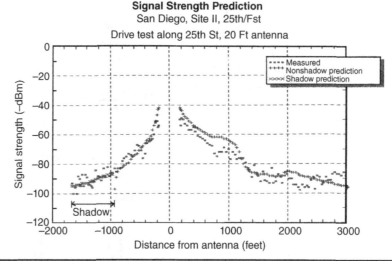

FIGURE 3.2.4.6   Measured data versus prediction values in San Diego.

with a 3-dB-gain dipole. The drive test was started from the north of 25th Street, and passed the site and continued to the south of the street. In this test run, the test range was short and close to the antenna site; therefore, we had to use the microcell prediction model shown in Sec. 4.2.1.2. In the nonshadow area, the prediction values followed were the LOS path loss plus the building block loss. Also, the effective antenna gain was applied due to the terrain contour changes with respect to the antenna site. The shadow section is also shown in the figure. In the shadow section, the shadow loss prediction from the high-knoll condition was used. Comparing the measured data with the predicted values, we can see a amount of agreement.

3.2.5   Conclusion

It is critical for the prediction tool to be accurate by separating the two parts, natural contours and human-made effects, from the collected measurement data. After the part of natural contours is removed, the collected measured data can be used to calculate the correct slope and the 1-mile intercept of the propagation model. Especially in irregular terrain, the terrain varies drastically within a short distance. Without applying this terrain normalization method, it is difficult to come up with a correct slope and 1-mile intercept for the propagation model. This normalization method also allows users to collect measurement data with more flexibility and less effort. Users need to be aware of this fact while applying this method. Several dominant factors can change the result drastically. For example, if the terrain database is not accurate, the poor data will produce inaccurate results. Filtering out the poor or invalid data is crucial for an accurate prediction. Also, the approach of using measured data to further characterize the human-made environment in specific areas is critical to cellular engineers. For example, the theoretical curve of the knife edge is too pessimistic in estimating the loss due to the knife edge. Therefore, the data that are affected by the nature terrain contour (in either the nonobstructive or the obstructive case) must be normalized out before obtaining the 1-mile intercept level and the path loss slope affected by the human-made environment.

3.3   Enhanced Lee Macrocell Prediction Model

3.3.1   Introduction

This chapter describes the enhancements added to the Lee macrocell prediction model, which has been well recognized by the wireless industry as one of the most accurate propagation prediction models.[23-25] This section discusses innovative approaches dealing with a pile of rough digital samples of terrain data and the enhancements to the Lee model during the validation process.

In general, the Lee model is composed of two parts: the impact of human-made structures and the impact of the natural terrain variation.[1,2,26] Other authors discuss innovative algorithms for calculating effective antenna gain[27,28] and diffraction loss[29,30] as well as for enhancing the Lee model.[23,25,31] This section focuses on the natural (terrain) factor. The new algorithm presented in this section is quite different from the others. It integrates the calculation on both LOS and shadow loss. First, in the LOS scenario, it addresses the issue of the big swing of effective antenna gain due to noncontinuous terrain data. Second, in the obstructive situation, the effective antenna gain is integrated with shadow loss. Both single- and multiple-knife-edge scenarios are discussed. The new algorithm is developed based on the analysis of measured and predicted (calculating the theoretical shadow loss and effective antenna gain) data. The new algorithm involves more calculation but improves the accuracy of the predicted value.

This algorithm was implemented and verified using field terrain and measured signal data from a variety of environments in different countries, including Italy, the United States, Spain, Japan, South Korea, Taiwan, and Romania.

3.3.2   The Algorithm

As we have mentioned, the new algorithm covers two areas. First, it addresses the big swing of effective antenna gain due to noncontinuous terrain data. There are several ways to handle this, such as averaging the terrain data along the radial path around the reflection point, averaging the effective height gains from all potential reflection points, and so on. Based on data from the large volume of measured data, the biggest gain must compare with the value from free space loss as a cap value.

Second, effective antenna gains need to be integrated with knife-edge diffraction loss. This section proposes an integrated solution that effectively combines both the effective gain and the shadow loss for each knife-edge situation. Section 3.3.2.1 discuss single knife-edge integration, and Sec. 3.3.3 addresses the integration of multiple knife edges.

3.3.2.1   Effective Antenna Gain Algorithm

In order to work on the original effective antenna height gain (in the LOS case) calculation,[1,2] we have to find the reflection point on the ground first. If the ground is not flat, there are two reflection points between the base station and the mobile receiver. Connecting the image of the transmit antenna height at the base station below the ground to the actual receiving antenna at the mobile, the intercept point on the ground is the first reflection point. Connecting the actual transmit antenna height of the base station to the image of the mobile receiving antenna below the ground, the intercept point on the ground is the second reflection point. Between the two points, the one that is close to the mobile receiver is chosen as the effective reflection point to be used. Take a tangential line on the curved terrain at the reflection point and extend the line to intercept at the transmitting antenna mast. The height from the intercept point on the transmitting antenna mast to the antenna top is the effective antenna height.

The calculation of the effective antenna gain is restrained by the noncontinuous terrain data. Every grid along the terrain radial might result in a different effective antenna gain due to this kind terrain data. Sometimes the effective antenna gain becomes huge regardless of the elevation of the mobile and the position of the reflection point from the base station transmitter.

The new algorithm introduced earlier calculates the effective antenna gain based on all the reflection points and selects the one with the biggest gain.

3.3.2.2   Integrated Solution: Single Knife Edge

This section discusses how the Lee model deals with the natural situation. The shadow loss and effective antenna gain were calculated separately and applied to calculate the received signal strength.

If the mobile is at the LOS condition, the received signal strength is

$$P_r = P_0 - \gamma \log^{\frac{r}{r_0}} + G_{effh} \qquad (3.3.2.2.1)$$

If the mobile is behind a single knife edge, the received signal strength is (see Eq. (3.1.2.5))

$$P_r = P_0 - \gamma \log^{\frac{r}{r_0}} + \text{Max } G_{effh} - L_D \qquad (3.3.2.2.2)$$

where Max G_{effh} is the maximum gain assuming the mobile is on top of a hill.[2] Figure 3.3.2.2.1 shows using the algorithm on identifying the valid knife edges and calculating effective antenna gains.

3.3.2.3   Integrated Solution: Multiple-Knife Edge

In this section, the integration of the diffraction loss with the effective antenna height gain when the mobile is blocked by multiple knife edges is discussed.

Figure 3.3.2.3.1 shows the algorithm on how to identify the valid knife edges and how to integrate these with effective antenna gains.

Under a multiple-knife-edge condition, the major contributions were the following:

1. Path loss

2. Diffraction loss due to all knives (e.g., sum of the losses from all true knives).

3. Effective antenna height gain from the first knife edge and the h_r/h_{pp} gains from the rest of the knife edges.

Figure 3.3.2.3.1 explains how to integrate the effective antenna gain in the knife-edge calculation, especially in the multiple-knife-edge scenarios. h_r/h_{pp} is the effective antenna height gain while the mobile is blocked. The formulas for knife-edge losses were discussed in previous sections.

3.3.3   Measured versus Predicted Data

All the field points are collected and used to calculate the path losses along the radial paths from many cell sites, where the LOS and blocked situations are identified. Those field points are stored in the drive test database.

Figure 3.3.3.1 shows all measured LOS data compared with the theoretical prediction curve, the best-fit curve, and the curve generated from the new algorithm. As shown, the

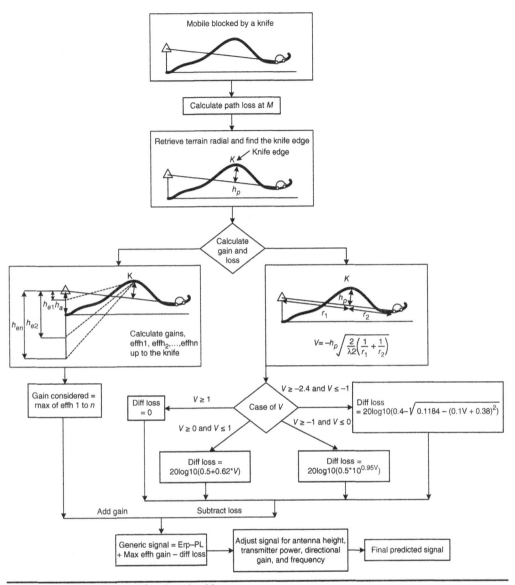

FIGURE 3.3.2.2.1 Single-knife-edge algorithm.

effective antenna gain added into the new algorithm provides a slight and unnoticeable improvement. Therefore, the original prediction method should be used:

$$\text{Signal} = \text{ERP} - PL + \text{Max}_{200\,fi}(effh)\text{gain}_{1\text{knife}} - \text{DiffLoss}_{1\text{knife}}$$

$$+ \sum_{j=2}^{N} \left\langle \text{Max}_{2000\,fi}\left[\frac{\text{hrj}}{\text{hppj}}\right]\text{gain} - \text{DiffLossKnife} \right\rangle \qquad (3.3.3.1)$$

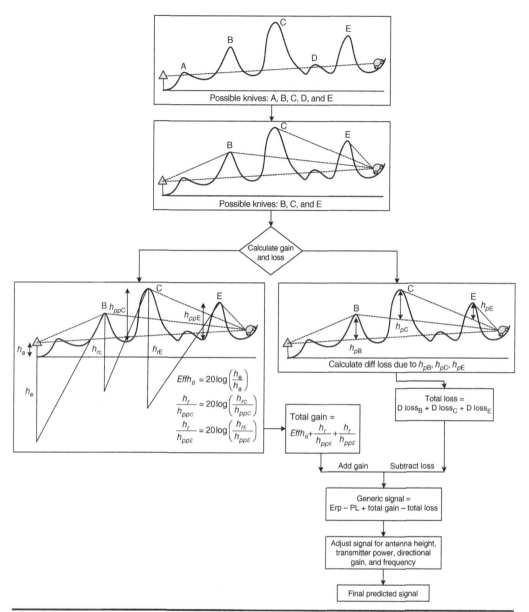

FIGURE 3.3.2.3.1   Multiple-knife-edge algorithm.

Figure 3.3.3.2 shows the original diffraction loss curve, the best-fit curve, and the new algorithm for a new adjusted v' found from the best-fit curve shown in Fig. 3.2.4.3 for the diffraction loss prediction.

To demonstrate the importance of the accurate prediction model, Fig. 3.3.3.3 compares the two coverage plots—one from the predictions using the original diffraction

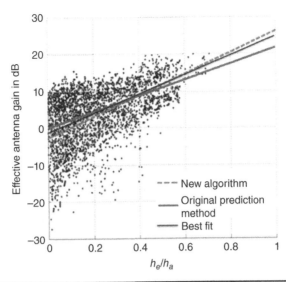

FIGURE **3.3.3.1** Comparison of the predicted curves from the original prediction method and the new algorithm method for the LOS condition. (A color version of this figure is available at www.mhprofessional.com/iwpm.)

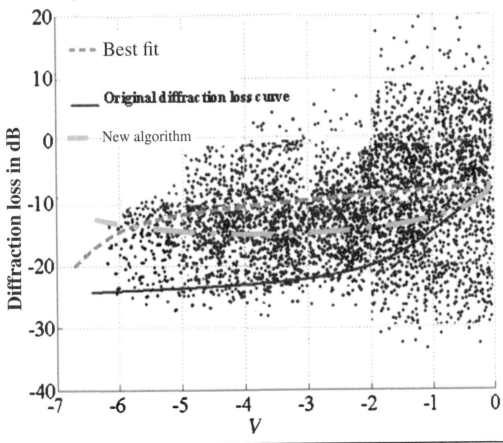

FIGURE **3.3.3.2** Comparison of the predicted curves from the original prediction method and the new algorithm method for the shadow condition.

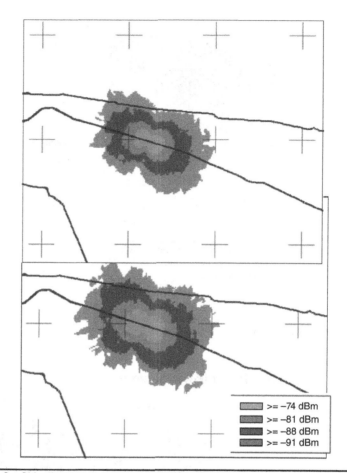

	>= −74 dBm
	>= −81 dBm
	>= −88 dBm
	>= −91 dBm

FIGURE 3.3.3.3   Old versus new coverage plots. (A color version of this figure is available at www.mhprofessional.com/iwpm.)

loss algorithm (top) and another from the predictions using the new diffraction loss algorithm (bottom). The predicted coverage is much larger based on the new algorithm. The original diffraction loss algorithm is more conservative, as we point out in Fig. 3.3.3.2.

3.3.4   Conclusion

It is important to have an accurate propagation model that deals with all the different environments and limitations. This section presents several innovative and possible revolutionary approaches in the Lee model that can better handle issues created by the rough terrain sampling data. These enhancements enable the Lee model to be more flexible and more accurate.

The model works only as long as the input data are accurate. In Chap. 6, we will look into the input data that affect both urban areas and in-building propagation modeling.

3.4 Longley–Rice Model

In the 1960s, Longley and Rice introduced the Longley–Rice model, which is used mainly for predicting the signal strength over a longer distance, says 10 mi or more. The Longley–Rice model is a well-established, general-purpose propagation model. Basically, this model presents methods for calculating the median transmission loss relative to the free space transmission loss based on the characteristics of the propagation path.

The Longley–Rice model is essentially a computer implementation of many techniques described in Ref. 32. This model uses terrain information to compute terrain roughness and radio horizons automatically. The users must supply other environmental variables, such as average climate conditions, soil conductivity, and so on. While these factors can be set to custom values, the program includes default or average values that are applicable in most cases.

In this section, we merely introduce the program with associated formulas of the Longley–Rice model. The reader can search for details from the reference.[32]

3.4.1 Point-to-Point Model Prediction

When a detailed terrain path profile is available, the path-specific parameters can be easily determined, and the prediction is called a point-to-point model prediction. This prediction tool is often applied in computer program and for a range usually over 10 miles. The method is applicable for radio frequencies above 20 MHz. The point-to-point prediction procedure requires detailed terrain profiles. From the profiles, one must determine the distance to the respective radio horizon, the horizon elevation angles, and effective antenna heights. These distances, angles, and heights are then supplied as input to the computer program.

3.4.2 Area Model Prediction

If the terrain path profile is not available in the area of interest, the Longley–Rice method provides techniques to estimate the path-specific parameters, and such a prediction is called an area model prediction. This model predicts a long-term median transmission loss based on random paths made in an area where variations in terrain elevation exist. Estimates of variability are provided as the program output. These estimates vary with respect to location and time.

The ranges of system parameters over which the models are applicable are the following:

1. Transmission frequency (MHz): 20 to 20,000

2. Range (km): 1 to 2000

3. Antenna heights (m): 0.5 to 3000

4. Antenna polarization: vertical or horizontal

Figure 3.4.2.1 gives us a brief sketch of a transhorizontal radio path, and some parameters are listed here:

d_{Lb} and d_{Lm} are horizon distances of the antennas.

θ_{eb} and θ_{em} are horizon elevation angles.

θ_e is the angular distance for a transhorizon path.

Δh is a terrain irregularity parameter.

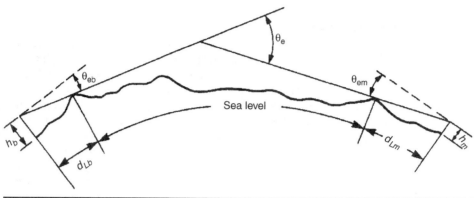

FIGURE 3.4.2.1   Geometry of a transhorizon radio path.

If a terrain data map is available, then the above listed parameters can be determined for any particular path, and the prediction technique operates in a point-to-point mode. If the terrain profile is not available, the program gives techniques for estimating these path-related parameters for use in an area model. The terrain irregularity parameter Δh is a roughness indicator that is related to another parameter $\Delta h(d)$, the interdecile range of heights, estimated at fixed distances along the path. The value of $\Delta h(d)$ increases with path length. For long paths, it reaches an asymptotic value given by

$$\Delta h(d) = \Delta h \left[1 - 0.8 \exp\left(-0.02\, d \right) \right] \tag{3.4.2.1}$$

where   for   water or very smooth plains, $\Delta h = 0\text{–}5$;
       for   plains, $\Delta h \approx 30$;
       for   hills, $\Delta h \approx 80\text{–}150$;
       for   mountains, $\Delta h \approx 150\text{–}300$; and
       for   rugged mountains, $\Delta h \approx 300\text{–}700$.

Longley and Rice define the effective antenna heights with respect to the dominant reflecting plane similar to the effective antenna height defined in Lee model's formula shown in Sec. 3.1.2.3. The smooth earth horizon distance d_L is determined by $d_L = \sqrt{17 h_e}$ in metric units as a function of effective antenna height h_e. The total distance between the antennas and their respective horizons is $d_L = d_{Lb} + d_{Lm}$, where d_{Lb} and d_{Lm} are the smooth-earth horizon distances of the base station antenna and the mobile antenna, respectively.

In the area-to-area model, the statistical estimations of the relevant parameters over irregular terrain are expressed as

$$d_L = d_{Lb} + d_{Lm} \tag{3.4.2.2}$$

$$d_{Lb} = d_{Lsb} \exp\left(-0.07 \sqrt{\frac{\Delta h}{h_e}} \right) \tag{3.4.2.3}$$

$$d_{Lm} = d_{Lsm} \exp\left(-0.07 \sqrt{\frac{\Delta h}{h_e}} \right) \tag{3.4.2.4}$$

where h_e is either h_{eb} or h_{em} as appropriate. h_e should be 5 m or greater. The total distance between the antennas and their respective horizons is $d_L = d_{Lb} + d_{Lm}$. The value of θ_{eb} (radians) can be obtained as

$$\theta_{eb} = \frac{0.0005}{d_{LSb}}\left[1.3\left(\frac{d_{LSb}}{d_{Lb}} - 1\right)\Delta h - 4h_{eb}\right] \tag{3.4.2.5}$$

The sum of the elevation angles θ_e is

$$\theta_e = \theta_{eb} + \theta_{em} \text{ or } -\frac{d_L}{8495} \tag{3.4.2.6}$$

where the larger value is selected for θ_e.

The angular distance for a transhorizon path is

$$\theta = \theta_e + \frac{d_i}{8495} > 0 \tag{3.4.2.7}$$

where d_i is the distance of the transmission path in kilometers.

For computing diffraction loss, the distances d_1 and d_2 to two ideal knife edges over the horizon distances can be expressed as

$$d_1 = \begin{cases} d_{LS} & d_1' \le d_{LS} \\ d_1' & d_1' > d_{LS} \end{cases} \tag{3.4.2.8}$$

$$d_1' = d_L + 0.5\left(\frac{72165000}{f_c}\right)^{\frac{1}{3}} \tag{3.4.2.9}$$

and

$$d_2 = d_1 + \left(\frac{72165000}{f_c}\right)^{\frac{1}{3}} \tag{3.4.2.10}$$

where d_1, d_1', and d_2 are in kilometers.

In the diffraction loss formula, the v parameters appropriate to knife edges at distances d_1 and d_2 are

$$v_{b,i} = 1.2915\theta_{ebi}\left[\frac{f_c d_{Lb}(d_i - d_L)}{d_i - d_{Lm}}\right]^{\frac{1}{2}} \tag{3.4.2.11}$$

$$v_{m,i} = 1.2915\,\theta_{emi}\left[\frac{f_c d_{Lm}(d_i - d_L)}{d_i - d_{Lb}}\right]^{\frac{1}{2}} \tag{3.4.2.12}$$

with $i = 1$ and 2 (two knife edges).

The diffraction losses L_1 and L_2 are then estimated using

$$L_1(\mathrm{dB}) = L(v_{b,1}) + L(v_{m,1}) \qquad (3.4.2.13)$$

$$L_2(\mathrm{dB}) = L(v_{b,2}) + L(v_{m,2}) \qquad (3.4.2.14)$$

The diffraction loss L_D at a mobile located at a distance d from the base station is

$$L_D(\mathrm{dB}) = d \cdot m_d + L_0 \qquad (3.4.2.15)$$

where

$$m_d = \frac{L_1 - L_2}{d_2 - d_1} \qquad (3.4.2.16)$$

and

$$L_0 = A_{fo} + L_2 - d_2\, m_d \qquad (3.4.2.17)$$

where an empirical clutter factor A_{fo} is determined as

$$A_{fo}(\mathrm{dB}) = \min(A'_{fo}, 15)\,\mathrm{dB} \qquad (3.4.2.18)$$

where A'_{fo} is given by

$$A'_{fo}(\mathrm{dB}) = 5\log_{10}\!\left[1 + h_m h_b f_c \sigma(d_{LS}) \times 10^{-5}\right]\ \mathrm{dB} \qquad (3.4.2.19)$$

and $\sigma(d_{LS})$ is given by

$$\sigma(d_{LS}) = 0.78h\,(d)\exp\{-0.5[\Delta h(d)]^{\frac{1}{4}}\}\ \ \mathrm{m} \qquad (3.4.2.20)$$

The Longley–Rice model has used the empirical formulas based on the measured data to predict the path loss. This model has been widely used in US government projects.

3.5   Summary

This chapter focuses on macrocell modeling. Although two models were discussed, the key is to understand those approaches that different models take to solve the common problems. The input from measured data should be carefully examined before use. The characteristics of the mobile environment and the algorithms from the different models that deal with these issues are discussed.

The Lee model has been around for a long time. The earlier version of Lee model has been used by AT&T and named the ADMS planning tool.[33] The later version has been used internally in Pactel and AirTouch markets. Some new algorithms and enhancements for the Lee macrocell prediction model are discussed in this chapter. The results are based on measurement data collected from all over the world to prove that the Lee model is one of the most efficient and accurate macrocell prediction models.

The Carey (Part 22) propagation model[34] started in Chap. 2 is based on the pertinent sections of the U.S. FCC Rules and Regulations. This method is essentially a simplified statistical method of estimating field strength and coverage based only on a station's effective radiated power and height above average terrain. Many details of terrain information have been ignored. Therefore, it can be used only for a generally for designing cellular systems. It cannot be used to design a realistic cellular system in a specified area. The main use of this model is for license applications or other submissions to the FCC that specifically require the use of the methods described in Part 22 of the FCC Rules. The Carey model is also used for other administrative requirements, such as certain frequency coordination procedures.

The Bullington,[35] Okumura,[36] Lee,[1] and Longley–Rice[32] models described both in Chaps. 2 and 3 are more analytical and consider a number of other factors, such as individual obstructions (either terrain or human-made), and terrain roughness. Okumura is often used in urban environments and includes correction factors for various area types, such as urban and suburban. Bullington considers individual obstructions and computes losses, such as for terrain obstructions, and ridges. The Longley–Rice model is a general model that considers radio horizons and various environmental conditions. The Lee model has been enhanced with much drive test data and can deal with more situations with more accuracy and is a better point-to-point prediction model. In the past, the first author and his colleagues had used their hand calculation on the 5×8 mile terrain contour maps from the Defense Map Agency to get the predicted signal strengths along mobile routes. Therefore, the basic Lee model is very simple to use and has developed the ability to integrate measured data back to the model to continue to improve its accuracy. The Lee model also can evolve from macrocell to microcell, covering dense urban and in-buildings with an integrated solution.

To sum up, we can see that all the propagation models were developed based on the requirement of deployed information, the available data, and the technology. When the cellular network first started, the goal was to provide a large coverage, and the focus was only on terrain contour. As capacity demand rises, especially in urban areas, the human-made environment becomes more important. Once wireless services become more popular, the dense urban microcell prediction models take center stage. These models will be introduced in Chap. 4. Many old existing theories that were prohibited before have now reinvented themselves; such as FDTD and ray tracing. Then in-building coverage, such as intrabuilding and interbuilding penetration from outdoor cells, becomes necessary. In-building models will be introduced in Chap. 5. The macrocell models were developed very early, then microcell and in-building (or picocell) models came out later. Table 3.5.1 shows a timetable of the different models.

Although there are many different ways to implement these different models, the basic principles do not change, and the fundamental challenges among them are still the same. We need to collect measured data to fine-tune the models.

Name	Short-Name	Category	Year
Friis' Freespace	friis	Foundational	1946
Young's Model	Young	Measurement	1950
Bullington Model	Bullington	Nomograms	1950
Carey Model	Carey	Terrain	1950
Egli	egli	Basic	1957
Longley Rice	Longley-Rice	Terrain	1960
Okumura	Okumura et al.	Basic	1968
Edwards-Durkin	edwards	Basic/Terrain	1963
Allsebrook-Parsons	allsebrook	Basic/Terrain	1977
Blomquist-Ladell	blomquist	Basic/Terrain	1977
Lee Macrocell	Lee	Terrain	1978
Hata-Okumura	Hata	Basic	1980
Longley-Rice Irregular Terrain Model	Longley-Rice	Terrain	1982[22]
ITM	itm	Basic	1988
Walfisch-Bertoni	bertoni	Basic	1991
Lee Microcell	Lee	Building/Terrain	1991[23]
Flat-Edge	flatedge	Basic	1993
COST-Hata/Cost-231	cost231	Basic	1993
Walfisch-Ikegami	walfisch	Foundational	1994
Lee Picocell	Lee & Lee	Building	1994
Two-Ray (Ground Reflection)	two.ray	Basic	1997
Hata-Davidson	davidson	Basic	1998
Ereeg-Greenstein	ereeg	Supplementary	1999
Directional Gain Reduction Factor (GRF)	grf	Basic	2000
Rural Hata	rural.hata	Terrain	2001
ITU Terrian	itu	Basic	2001
Standford University Interium (SUI)	sui	Basic	2002
Green-Obaidat	green	Basic	2002
ITU-R/CCIR	itur	Basic	2003
ECC-33	ecc33	Supplementary	2006
Riback-Medbo	Riback-Medbo	Basic	2007
ITU-R 452	itur452	Basic	2008
IMT-2000	imt2000	Supplementary	2009
deSouza	deSouza	Basic	2010
Effective Directivity Antenna Model (EDAM)	Edam	Basic	2010

TABLE 3.5.1 Timetable of Different Models

3.5.1 Ways of Implementation of Models

If we look at the models described in Chaps. 2 and 3 from the ways they are implemented, we can categorize them into four different approaches:

Empirically Based—Young, Okamura, and Hata use the measurement-based graphical method that predicts the received field strength as a function of various terrain features and system parameters. Young's measurement data were taken at the microwave frequency range from 150 to 3700 MHz on the East Coast of the United States.[37] Okamura's measurement data[36] were taken around 150 to 800 MHz in Japan.

Terrain Data—An accurate terrain database for calculating the transmission path loss over irregular terrain based on the statistical properties of the terrain profile is needed. The detailed point-to-point version rather than the area version is studied in this chapter because it provides better prediction accuracy. A portion of the Longley–Rice procedure is used in conjunction with International Radio Consultative Committee (CCIR) propagation curves based on a method published by Durkin in 1977.[38] Durkin's computer program generates field strength contours around a base station by computing path loss estimates along radial lines. Two other prediction techniques, one from the Joint Radio Committee (JRC) of the Nationalized Power Industries of the United Kingdom and the other from Palmer of the Canadian Communications Research Centre (CRC)[39] are also operating on computer programs using terrain database. The Lee model is a point-to-point model and has used mixed theoretical derivations and statistical approaches in predicting the signal strengths on U.S. Defense Map Agency (DMA) topographic data ($1° \times 1°$ tape) with a satisfactory prediction as compared with the measured data.[22]

Plane Earth—Egli was the first to publish a technique incorporating the plane earth model, estimating received power by treating the earth as a plane surface and adding a terrain clutter factor to account for surface obstructions. Egli[40] estimates the median path loss in decibels by summing the plane earth reference loss and the median terrain factor. Later, Murphy[41] presented a statistical method for predicting transmission loss over the irregular terrain using empirical data from the plains and mountains of Colorado. A third method was developed by Allsebrook and Parsons in the United Kingdom.[42] They designed a prediction technique specifically for urban environments using data in three British cities. The Allsebrook–Parsons model was introduced in Chap. 2.

Monograms—Bullington provides a short manual procedure for making estimates of the average received power using graphs and monograms. His study also contains a brief summary of vehicular radio propagation.

3.5.2 Features Among Models

Different models have different features. A summary of model features is shown in Table 3.5.2. There are three major categories: input parameters, propagation factors, and output treatment parameters. Under each category, is a list of items. We take each model and check whether covers the items. N means not treated, L means limited treated, and E means extend treated. The simpler models that can be found in the table are Carey[34] and CCIR 567[43] introduced in Chap. 2. Depending on the purpose of using a particular model, some parameters may not be important. The Lee model has covered most features and has tried to make a more accurate prediction.

Model	Input parameters									Propagation factors							Output treatment parameters			
	Antenna height				Terrain data	Building data	Foliage data	Hill shape	Distance	Free space	Diffraction by smooth earth	Reflection from earth surface	Reflection from hills	Diffraction by hills	Atmospheric refractivity	Building penetration	Loss deviation	Location variability	Time fading	Median transmission loss
	Above average terrain	Above street level	Effective height	Mobile (1–3 m)																
Bullington	N	E	L	E	L	N	N	L	E	E	E	L	N	E	N	N	N	L	L	E
Egli	E	N	L	E	N	N	N	N	E	N	N	L	N	N	N	N	N	E	N	E
Carey	E	N	N	N	N	N	N	N	E	N	N	N	N	N	N	N	N	L	L	E
Longley–Rice pt.-to-pt.	N	E	E	E	E	N	N	L	E	E	E	E	N	L	E	N	N	E	E	E
Longley–Rice Area	N	E	L	E	N	N	N	N	E	E	E	E	N	L	E	N	E	N	E	E
Okumura	N	E	E	E	N	L	N	N	E	E	E	E	N	E	N	N	L	E	N	E
Tirem	E	E	E	E	E	N	N	L	E	E	E	E	N	E	E	E	L	E	E	E
Lee	E	E	E	E	E	E	E	E	E	E	E	N	E	E	E	E	N	E	E	E
CCIR Report 567	E	L	N	N	N	E	N	N	E	E	E	N	N	E	N	N	N	E	E	E
Bertoni and Walfisch	N	E	L	E	N	E	N	N	E	E	N	N	N	E	N	N	N	N	N	E

Rating Scale: N = not treated;   L = Limited treatment;   E = extensive treatment

TABLE 3.5.2 Summary of Model Features

References

1. Lee, W. C. Y. *Wireless and Cellular Telecommunication*. 3rd ed. New York: McGraw-Hill, 2006, Chap. 8.
2. Lee, W. C. Y. *Mobile Communications Design Fundamentals*. 2nd ed. New York: John Wiley & Sons, 1993. Chap. 2.
3. Lee, W. C. Y. *Mobile Communications Engineering*. 2nd ed. New York: McGraw-Hill, 1998. Chaps. 3 and 4.
4. "Lee's Model, Appendix VI, Propagation Special Issue." *IEEE Transactions on Vehicular Technology* 37, no. 1 (February 1988): 68–70.
5. *Phoenix Handbook*, "A Propagation Signal Strength Prediction Tool Based on Lee Model." Internal Publication, Pactel Corp. (became AirTouch after 1996), 1991–2000.
6. Lee, W. C. Y. *Mobile Communications Engineering*, Ibid. p.142.
7. Lee, W. C. Y. *Wireless & Cellular Telecommunication*, Ibid., p. 382.
8. Lee, W. C. Y. "Estimate of Local Average Power of a Mobile Radio Signal," *IEEE Transactions on Vehicular Technology* 34 (February 1985): 22–27.
9. Lee, W. C. Y. *Wireless & Cellular Telecommunication*, Ibid. p. 382.
10. U.S. Geological Survey 7.5 minute maps and terrain elevation data tapes (a quarter-million scale; 250,000:1 map tapes).
11. Defense Map Agency, 3-second arc (roughly 200 to 300 ft depending on the geographic locations) tape and a planar 0.01-in tape (an elevation at intervals of 0.01 inch, i.e., 208 ft).
12. Vogler, Lewis E. "An Attenuation Function for Multiple Knife-Edge Diffraction." *Radio Science* 17, no. 6 (November–December 1982): 1541–46.
13. Bullington, P. "Radio Propagation at Frequencies above 30 Megacycles." *Proceedings of the IRE* 35, no. 10 (October 1947): 1122–36.
14. Lee, D. J. Y. "4D Morphology Model with Feedback Loop." IEEE Vehicular Technology Conference, 56th Proceedings. VTC 2002-Fall 2002, pp. 839–43.
15. Lee, W. C. Y. *Wireless and Cellular Telecommunication*, Ibid. Sec. 8.3.2.
16. Ibid, Sec. 9.7.3
17. Lee, W. C. Y. "Cellular Mobile Radiotelephone System Using Tilted Antenna Radiation Patterns." U.S. Patent 4,249,181, February 3, 1981.
18. Wang, J. S. W., and Wang, I. "Effects of Soft Handoff, Frequency Reuse and Non-ideal Antenna Sectorization on CDMA System Capacity." Proceeding of the 43rd IEEE VTC, Secaucus, NJ, May 1993, pp. 850–54.
19. The six patents of Lee's microcell system, listed in reference 27 of Chap. 1.
20. Lee, W. C. Y. "A Computer Simulation Model for the Evaluation of Mobile Radio Systems in the Military Tactical Environment." *IEEE Transactions Vehicular Technology* 32, no. 2 (May 1983): 403–9.
21. Lee, D. J. Y., and Xu, Ce. " The Capacity and Trunking Efficiency of Smart antenna System." Proceedings of the Vehicular Technology Conference, Phoenix, May 1997, pp. 612–16.
22. Lee, W. C. Y., and Lee, D. J. Y. "Fine Tune Lee Model." Personal, Indoor and Mobile Radio Communications, 11th IEEE International Symposium, Vol. 1, PIMRC, London, September 2000, pp. 406–10.
23. de la Vega, David, Lopez, Susanna, Gil, Unai, Matias, Jose M., Guerra, David, Angueira, Pablo, and Ordiales, Juan L. "Evaluation of the Lee Method for the

Analysis of Long-Term and Short-Term Variations in the Digital Broadcasting Services in the MW Band." 2008 IEEE International Symposium on Broadband Multimedia Systems and Broadcasting, March 31 2008–April 2, 2008, pp. 1–8.

24. Evans, Greg, Joslin, Bob, Vinson, Lin, and Foose, Bill. "The Optimization and Application of the W. C. Y. Lee Propagation Model in the 1900 MHz Frequency Band." *IEEE Transactions on Vehicular Technology*, March 1997, pp. 87–91.

25. Kostanic, Ivica, Rudic, Nikola, and Austin, Mark. "Measurement Sampling Criteria for Optimization of the Lee's Macroscopic Propagation Model." IEEE 48th Vehicular Technology Conference, May 18–21, 1998, Vol. 1, pp. 620–24.

26. Lee, W. C. Y., and Lee, D. J. Y. "The Propagation Characteristics in a Cell Coverage Area." IEEE 47th Vehicular Technology Conference, Vol. 3, Phoenix, May 1997, pp. 2238–42.

27. Ostlin, E., Suzuki, H., and Zepernick, H. J. "Evaluation of a New Effective Antenna Height Definition in ITU-R Recommendation P.1546-1." Proceedings of Asia-Pacific Conference on Communications, October 2005, pp. 128–32.

28. Green, D. B., and Obaidat, M. S. "An Accurate Line of Sight Propagation Performance Model for Ad-Hoc 802.11 Wireless LAN (WLAN) Devices." IEEE International Conference on Communications, Vol. 5, April 28–May 2, 2002, pp. 3424–28.

29. Zhao, Xiongwen, and Vainikainen, Pertti. "Multipath Propagation Study Combining Terrain Diffraction and Reflection." *IEEE Transactions on Antennas and Propagation* 49, no. 8 (August 2001): 1204–9.

30. Giovaneli, C. L. "An Analysis of Simplified Solutions for Multiple Knife-Edge Diffraction." *IEEE Antennas and Propagation* 32, no. 3 (January 2003): 297–301.

31. Lee, W. C. Y., and Lee, D. Y. J. "Enhanced Lee Model from Rough Terrain Sampling Data Aspect." IEEE VTC, Fall 2010, Anchorage, Canada, pp. 1–5.

32. Longley, A. G., and Rice, P. L. "Prediction of Tropospheric Radio Transmission over Irregular Terrain: A Computer Method." ESSA Technical Report, ERL 79- ITS 67, 1968.

33. Lee, W. C. Y. *Lee's Essentials of Wireless Communication.* New York: McGraw-Hill, 2001, Chap. 1.

34. Carey, R. "Technical Factors Affecting the Assignment of Frequencies in the Domestic Public Land Mobile Radio Service." Washington, DC: Federal Communications Commission. Report R-6406, 1964.

35. Bullington, Kenneth. "Radio Propagation for Vehicular Communications." *IEEE Transactions on Vehicular Technology* 26, no. 4 (November 1977): 295–308.

36. Okumura, Y., Ohmori, E., Kawano, T., and Fukuda, K. "Field Strength and Its Variability in the VHF and UHF Land Mobile Radio Service." *Rev. Elec. Comm. Lab.* 16, no. (9/10), (1968): 825–73.

37. Young, W. R. "Comparison of Mobile Radio Transmission at 150, 450, 900 and 3700 MC." *Bell System Technical Journal* 31 (1952): 1068–85.

38. Durkin, J. "Computer Prediction of Service Areas for VHF and UHF Land Mobile Radio Services." *Transactions on Vehicular Technology* 26, no. 4 (November 1977): 323–27.

39. Palmer, F. H. "The CRC VHF/UHF Propagation Prediction Program: Description and Comparison with Field Measurements." *NATO-AGARD Conference Proceedings* 238, (1978): 49/1–49/15.

40. Egli, J. J. "Radio Propagation above 40MC over Irregular Terrain." *Proceedings of the IRE* (1957): 1383–91.

41. Murphy J. P. "Statistical Propagation Model for Irregular Terrain Paths Between Transportable and Mobile Antenna." NATO-AGARD Conf. Proc. No. 70, 1970, pp. 49/1–49/20.

42. Allsebrook, K., and Parsons, J. D. "Mobile Radio Propagation in British Cities at Frequencies in the VHF and UHF Bands." *IEEE Transactions on Vehicular Technology* 26, no. 4 (1977): 95–102.

43. CCIR. "Methods and Statistics for Estimating Field Strength Values in the Land Mobile Services Using the Frequency Range 30MHz to 1GHz." CCIR XV Plenary Assembly, Geneva, Report 567, Vol. 5, 1983.

CHAPTER 4

Microcell Prediction Models

4.1 Introduction

The use of microcells to increase the capacity of cellular mobile communication, especially in dense urban areas, is an attractive tactic. A number of experimental and theoretical studies have been undertaken regarding propagation in the urban microcellular system. However, predicting the local mean of this microcell requires building layout data in a 2D or 3D format on aerial photographical maps. Usually, the more detailed the data, the higher the cost and the more accurate it is to form the model. Balancing the cost and resolution of building layout data with the accuracy of the model is always a challenging task.

Microcell prediction is a very important tool in RF design in dense urban areas, especially for both cellular and WLAN applications. Generally, a microcell is defined as a cell located in dense urban area, providing less than 1 mile or 1 km of coverage (in radial) on a generally flat terrain with a low transmitted power (less than 1 W ERP) and embedded within a cluster of buildings.

When the size of the cell is small, less than 1 mile or 1 km, the street orientation and individual blocks of buildings have significant effects on signal reception. These varying street orientations and individual blocks of buildings do not make any noticeable difference in reception when the signal is well attenuated at a distance over 1 mile or 1 km based on the measured data. However, when the cells are small, a signal receiving multiple reflections from the individual buildings along the path, before arriving at the mobile unit, is weakened. These buildings directly affect the received mobile signal strength and are considered part of the path loss. The general path loss prediction is an average process. Therefore, we do not need high-resolution 2D and 3D building data, which are expensive and difficult to obtain and to process correctly. Even though detailed building data are available, the associated parameters might make the implementation and accuracy of the model more complicated and less efficient.

There are many microcell prediction models for dense urban-area propagation predictions.[1-9] This chapter starts with the introduction of the Lee microcell model, how it works, and what are its enhancements. Then other microcell prediction models will be presented and discussed. The first few sections focus on the Lee microcell propagation model.[1, 10] The Lee model predicts the local mean from the statistical properties of the average signal strength, or long-term fading, based on physical parameters, such as antenna height, frequency, and building thickness. The model starts with 2D building data, and has been verified against many dense urban areas around the world with good results. It was enhanced first to deal with the terrain effect in dense urban areas,

187

as many major cities reside on top of varying terrains. For example, in Los Angles, San Francisco, Tokyo, and Seoul, the terrain undulation over a few feet has a great impact on the accuracy of the model. The Lee model was also enhanced to deal with attributes, such as water and foliage. The water in Amsterdam and the trees in Atlanta have a profound impact on the accuracy of the model.

Later, the Lee microcell prediction model will be used to determine the path loss statistical properties of the cell site based on multiple breakpoints. This microcell path loss prediction model predicts the signal strength based on both theory and experiments, that is, multiple-break-point formulas, cell site parameters, and measured data. As the prediction model will be shown later, the results of this model are very promising.

The Lee model provides a baseline for integrating the microcell to the macrocell so that coverage, interference, and handoffs among microcells and macrocells can be properly simulated and integrated. This model has been validated with the measured data from drive tests performed in various countries and cities, with different transmitter heights, unique frequencies, and separate cell site parameters as well as different kinds of mobile environments. The initial findings are that this empirically and theoretically based microcell model performs well in all areas under various conditions. Furthermore, this model can be reinforced based on the collected measured data.

Many other microcell prediction models, including their empirical, theoretical, and physical approaches and site specifics, are discussed as well. Note that the FDTD model and the TLM model for the microcell and the picocell will be discussed in Chaps. 5 and 6.

4.2   The Basic Lee Microcell Prediction Model

4.2.1   Basic Principle and Algorithm

4.2.1.1   Near-In Distance in the Microcell Prediction Model

Near-in distance has been derived from the simple plane earth model, which has been described in Sec. 1.9.1.3. Actually, it is a two-ray model, and the Lee model has been derived from this fundamental model. Because in the microcell region, the distance between the base station and the mobile is much longer than the antenna height of each of them, the incident angle between the reflected wave and the ground is very small. In this model, the signal s received at the mobile as shown in Fig. 4.2.1.1.1:

$$s = \sqrt{P_0} \left(\frac{1}{4\pi d/\lambda} \right) \cdot \exp(j\phi_1) [1 + a_v \exp(-j(\phi_1 - \phi_2))]$$

$$= \sqrt{P_0} \left(\frac{1}{4\pi d/\lambda} \right) \cdot \exp(j\phi_1) [1 + a_v e^{-j\Delta\phi}] \qquad (4.2.1.1.1)$$

The received power P_r at the mobile is $|s|^2$. It was expressed in Eq. (1.9.1.3.1) as

$$P_r = P_0 \left(\frac{1}{4\pi d/\lambda} \right)^2 |1 + a_v e^{-j\Delta\phi}|^2$$

where $a_v = -1$ for small incident angle.

The phase difference between two wave paths is shown in Eq. (1.9.1.3.6) as

$$\Delta\phi = \beta \, \Delta d \approx \frac{2\pi}{\lambda} \frac{2h_1 h_2}{d}$$

and the received signal P_r is expressed as a function of $\Delta\phi$, shown in Eq. (1.9.1.3.2), as

$$P_r = P_0 \frac{2}{(4\pi d/\lambda)^2}(1-\cos\Delta\phi)$$

From the above equation, we can find that when $\Delta\phi = \pi$, the received signal P_r becomes maximum:

$$P_r = P_0 \frac{2}{(4\pi d/\lambda)^2}(1-\cos\Delta\phi) = \text{max} \qquad \text{when } \Delta\phi = \pi \qquad (4.2.1.1.2)$$

Let $\Delta\phi = \pi$ in Eq. (1.9.1.3.6):

$$\Delta\phi = \frac{2\pi}{\lambda}\frac{2h_1 h_2}{d} = \pi \qquad (4.2.1.1.3)$$

From Eq. (4.2.1.1.3), we can find the near-in distance d_f[10]:

$$d_f = \frac{4h_1 h_2}{\lambda} \qquad (4.2.1.1.4)$$

where h_1 = height of base station antenna, h_2 = height of mobile antenna, and λ = wavelength in meters. The near-in distance was calculated and described in Lee's book.[2] The criterion for defining the near-in distance is when the phase difference between the direct wave and the reflected wave is 180°, as shown in Fig. 4.2.1.1.1.

Within the near-in distance, the received signal is still very strong and is not disturbed by the reflected wave. We may consider the signal path loss following the free space path loss in this reagin. The near-in distance is used for the microcell model and can be used for the macrocell model but does not apply for the in-building model. This is because the angle of incidence of the reflected wave in the in-building environment is not small. The derivation of the distance in which the received signal is still strong for the in-building model will be shown in Chap. 5. We call it the *close-in distance* to distinguish it from the near-in distance.

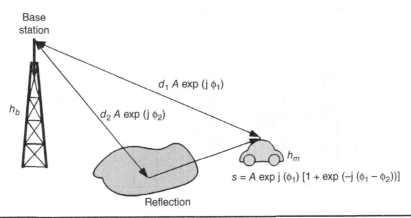

FIGURE 4.2.1.1.1 The criterion of near-in distance.

4.2.1.2 The Basic Model[1]

The microcell model is used for a distance of up to 1 km or 1 mile. When the size of a cell is small, less than 1 km, the street orientation and individual blocks make a difference in signal reception. These buildings that directly affect the received signal strength level are taken into account for the path loss. Although the strong received signal at the mobile unit is coming from the multipath reflected waves outside the buildings, not from the waves penetrating through the buildings, there is a correlation between the attenuation of the signal and the number of building blocks along the radio path. The larger the number of building blocks with the size of the blocks, the higher the attenuation of the signal. The propagation mechanics within the microcell are shown in Fig. 4.2.1.2.1. The microcell prediction formula over a flat terrain is

$$P_r = P_t - L_{LOS}(d_A, h_1) - L_B + G_A + G_a \quad (\text{dBm})$$

$$= P_{LOS} - L_B + G_A + G_a$$

$$= P_{OS} + G_a \tag{4.2.1.2.1}$$

where P_t is the ERP in dBm and $L_{LOS}(d_A, h_1)$ is the line-of-sight (LOS) path loss at distance d_A with the antenna height h_1. The general formula of the theoretical $L_{LOS}(d_A, h_1)$ at a location A, which is a distance d_A away from the base station (shown in Fig. 4.2.1.2.2), can be obtained as

$$L_{LOS} = 20 \log \frac{4\pi d_A}{\lambda} \quad (\text{free space loss}) \quad d_A < d_f$$

$$= 20 \log \frac{4\pi d_f}{\lambda} + \gamma \log \frac{d_A}{d_f} \qquad d_A > d_f \tag{4.2.1.2.2}$$

where d_f is the near-in distance, $d_f = 4h_1 h_2 / \lambda$ derived in Sec. 4.2.1.1, h_1 is the base station antenna height, and h_2 is the mobile antenna height. γ is the path loss slope. Figure 4.2.1.2.2 shows the signal strength $P_{LOS} = P_t - L_{LOS}(d_A, h_1)$ for the LOS conditions as the curve (a) indicated at the carrier frequency of 850 MHz, the transmitting power of ERP = 1 W (20 dBm), the antenna height $h_1 = 20$ ft, and the gain $G_a = 3$ dB. The curve (a) is an empirical curve.

The L_B in Eq. (4.2.1.2.1) is the loss due to the transmitted signal blocked by the buildings measured by the building blockage length B shown in Fig. 4.2.1.2.3.

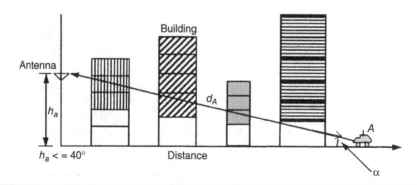

FIGURE 4.2.1.2.1   The propagation mechanics of low antenna heights at the cell site.

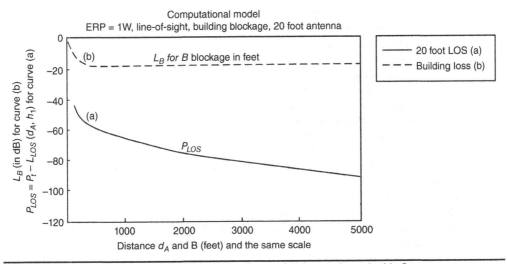

FIGURE 4.2.1.2.2 Empirical curves of L_B and P_{LOS}. Curve of G_A is not shown in this figure.

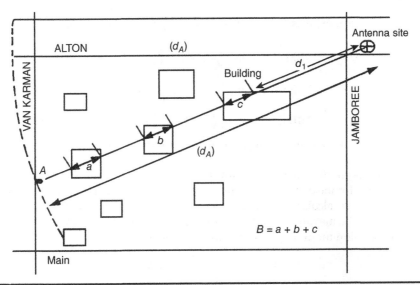

FIGURE 4.2.1.2.3 Building block occupancy between antenna site and location A. Here d_1 is the distance between the base station and the building (Building c).

In Fig. 4.2.1.2.3, the distance between the base station and the first building, d_1 is shown. G_A in Eq. (4.2.1.2.1) is a gain when $d_1 > 200'$ and $G_A = 0$ when $d_1 \leq 200'$.

The total block length is $B = a + b + c$. The value of L_B for a blockage of B can be obtained as follows:

$$L_B(B = a+b+c) = P_{LOS}(d_A, h_a = h_1) - P_{os}(\text{at } A) \qquad (4.2.1.2.3)$$

The empirical curve of L_B is obtained by first calculating the sum of lengths of building blocks between the antenna site and location A, as shown in Fig. 4.2.1.2.3 and plotted on Fig. 4.2.1.2.2 as curve (b), where P_{os} is the measured local mean at point A.

Therefore, we will use an empirical curve P_{os} which can be found by taking P_{LOS} from curve (a) and adding the block loss L_b in curve (b) shown in Fig. 4.2.1.2.2 as our predicted curve to compare with the measured data. Again G_A will be considered when $d_1 > 200'$. When $d_1 \leq 200'$, $G_A = 0$.

The blockage length B shown in Fig. 4.2.1.2.3 is the sum of the widths of all buildings, which are located in line with the transmitted signal. The curve (b) in Fig. 4.2.1.2.2 shows that rapid attenuation occurred while B is less 500 ft. When B is greater 1000 ft, a nearly constant value of 18 dB is observed. The corner turning effects were found during the test. When the mobile turned from one street to another street, the signal strength changed noticeably. Because the blockage length of the buildings varies quite differently during the turning of the mobile from one street to another street, this scenario can be taken care of in the model. The curve b is used for obtaining L_B in Eq. (4.2.1.2.1) in the microcell prediction model.

In the microcell environment, the antenna height h_a at the base station and antenna gain G_a relationship follows a 9-dB/oct rule was observed.[11] Therefore, G_a in Eq. (4.2.1.2.1) is shown as follows:

$$G_a = 10\log\left(\frac{h_a}{h_1}\right)^3 = 30\log\frac{h_a}{h_1} \qquad (4.2.1.2.4)$$

The numerical expression of two curves in Fig. 4.2.1.2.3 can be depicted as follows:

$$
\begin{aligned}
P_{LOS} &= P_t - 77 \text{ dBm} - 21.5\log\frac{d}{100'} + 30\log\frac{h_1}{20'} & 100' \leq d < 200' \\[2mm]
&= P - 83.5 \text{ dBm} - 14\log\frac{d}{100'} + 30\log\frac{h_1}{20'} & 200' \leq d < 1000' \qquad (4.2.1.2.5) \\[2mm]
&= P - 93.3 \text{ dBm} - 36.5\log\frac{d}{100'} + 30\log\frac{h_1}{20'} & 1000' \leq d < 5000' \\[2mm]
&= \text{use the macrocell prediction model for} & 5000' \leq d
\end{aligned}
$$

Readers may notice that the symbol d is for the distance in the microcell model and that the symbol r is for the distance in the macrocell model. This makes it easier to recognize that the prediction calculation is for the microcell or the macrocell model from the equations.

We set the maximum building blockage loss to 20 dB instead of 18 dB in the following expression for giving a conservative estimation:

$$
\begin{aligned}
L_B &= 0 & 1' \leq B \\[2mm]
&= 1 + 0.5\log\frac{B}{10'} & 1' \leq B < 25' \\[2mm]
&= 1.2 + 12.5\log\frac{B}{25'} & 25' \leq B < 600' \qquad (4.2.1.2.6) \\[2mm]
&= 17.95 + 3\log\frac{B}{600'} & 600' \leq B < 3000' \\[2mm]
&= 20 \text{ dB} & 3000 \leq B
\end{aligned}
$$

The predicted signal strength P_{OS} shown in Eq. 4.2.1.2.1 is

$$P_{OS} = P_{LOS} + L_B + G_A = P_{LOS} + L_A \qquad (4.2.1.2.7)$$

The measurements for the case of $d_1 \leq 200'$ were carried out along various routes, as shown in Fig. 4.2.1.2.4. The measurement on the zigzag route (R2) in San Francisco was

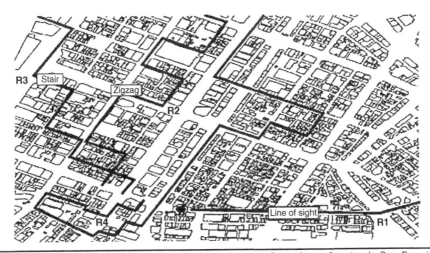

FIGURE 4.2.1.2.4   A sample building layout with various configurations of routes in San Francisco.

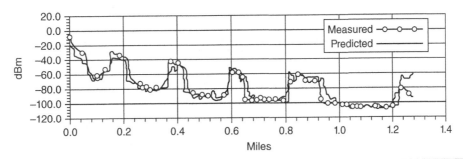

FIGURE 4.2.1.2.5   Measured and predicted versus mobile distance from the zigzag drive—microcell model predictions compared to measurements versus delta.

shown in Fig. 4.2.1.2.5 with a frequency at 1.9 GHz, $h_1 = 13.3$ m, and the power at ERP $= 1$ W. The measured and predicted curves agree fairly well with each other,[12,13] as shown in the figure. In Sec. 4.2.1.3, there are more comparisons between the measured and the predicted on different scenarios of runs. Those data were used to develop Lee microcell system.[14]

4.2.1.3   Handling Different Scenarios

In predicting the signal strengths in the microcell,[1] two different parameters deal with two different scenarios. Along the radio path, one is the range d_1 of the first close building to the antenna site, and the other is the total length B of the sum of the building blockage thickness.

4.2.1.3.1   The Range d_1   Empirical Curve: The range d_1 is the range between the first building and the antenna site along the radio path. From the measured data, we have found that when the range d_1 is small, the new path loss L_A (see Eq. 4.2.1.2.7) due to the buildings will be higher and remain higher after the distance d_A of the mobile unit reaches 1000 ft and beyond. When d_1 is large, the new path loss L_A becomes smaller and remains smaller after d_A reaches 1000 ft and beyond. An empirical formula can be found for the different range d_1 when $B \geq 500$ ft, as shown in Fig. 4.2.1.3.1. In the figure, the new path loss L_A due to the buildings is about 18 dB when d_1 is less than 200 ft. Then L_A will

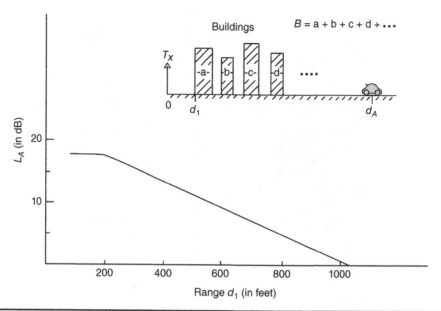

Figure 4.2.1.3.1   An empirical curve for predicting L_A with Range d_1 when $B \geq 500$ ft.

be reduced as the range d_1 increased linearly. The new path loss L_A with different values of B and d_1 can be found in Fig. 4.2.1.3.6.

Measured Data   The measured data were taken at Irvine and San Diego, California, with a frequency of 850 MHz, $h_1 = 20$ ft, ERP = 1 W (30 dBm), and antenna gain at the base $G_a = 3$ dB. Figure 4.2.1.3.2 shows the LOS scenario ($d_1 = > 1000'$) along Main Street in Irvine. The predicted curves were found by taking the P_{LOS} curve of Fig. 4.2.1.2.2 adjusted by the effective antenna height gain G_{effh} due to the ground contour. Figure 4.2.1.3.3 shows the measured data in San Diego. It was at a site of $d_1 \leq 200$ ft and $G_A = 0$. The predicted curve P_{OS} (at A) is drawn as subtracting the building blockage L_B curve from the P_{LOS} curve, shown in Fig. 4.2.1.2.2. The predicted and the measured curves were in agreement. Figure 4.2.1.3.4 shows the mobile unit run on Jamboree Road from north to south, while the measured data were collected starting from the Barranca Parkway intersection (on the left of the figure) and ending at the Michaelson Road intersection (on the right of the figure). The antenna site was located on Alton Avenue. The range d_1 on the north side (left side of the figure) was $d_1 = 500$ ft and on the south side (right side of the figure) was $d_1 \leq 200$ ft. The predicted curves (P_{OS} (at A)) were found from Fig. 4.2.1.3.1 for different ranges of d_1. Figure 4.2.1.3.5 shows the measured data collected on Alton Avenue from the Jamboree Road intersection to Red Hill Avenue. The range is $d_1 = 700$ ft, as shown in the figure. The predicted P_{OS} (at A) is found from Fig. 4.2.1.3.6 for $d_1 = 700$ ft. The predicted curve and the measured data agree with each other well.

4.2.1.3.2   The Total Length B—The Sum of Individual Building Blockage

Empirical Curve   The total length B is the sum of individual building blockage along the radio path. In some scenarios, the length B can be less than 500 ft. If this happens, the new path loss L_A will be less than shown in the measured data. Therefore, an empirical curve was generated for a different length B and a different d_1, as shown in Fig. 4.2.1.3.6. The predicted P_{OS} (at A) can be obtained when the values of B and d_1 are known.

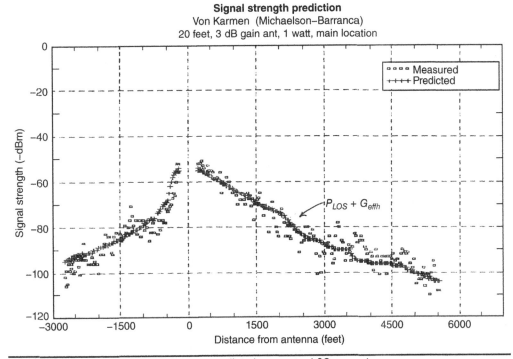

FIGURE 4.2.1.3.2 Measured data versus predicted curve at a LOS scenario.

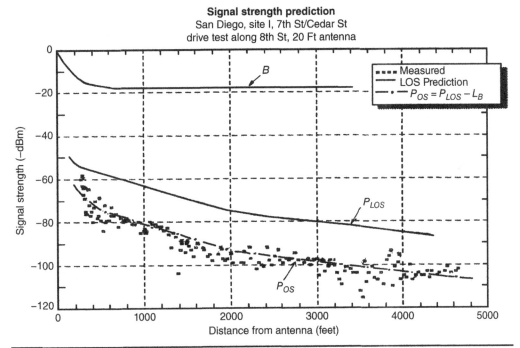

FIGURE 4.2.1.3.3 Measured data versus predicted curve at $d_1 \leq 200$ ft.

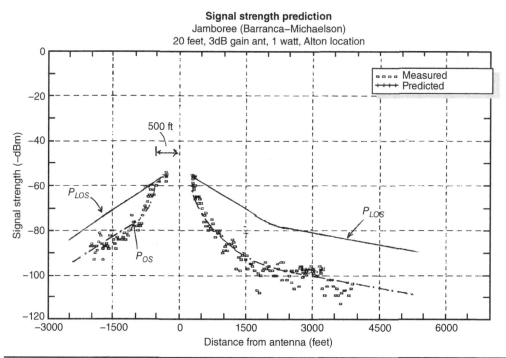

FIGURE 4.2.1.3.4 Measured data versus predicted curve at $d_1 \leq 200$ ft (right-side data) and at $d_1 = 500$ ft (left-side data).

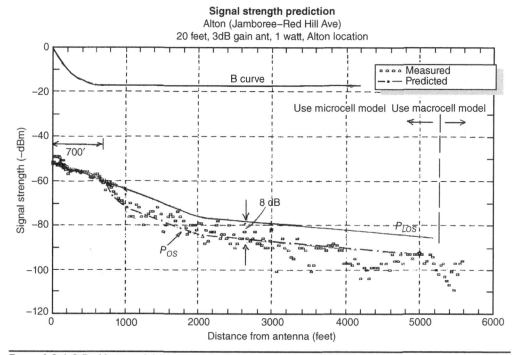

FIGURE 4.2.1.3.5 Measured data versus predicted curve at $d_1 = 700$ ft.

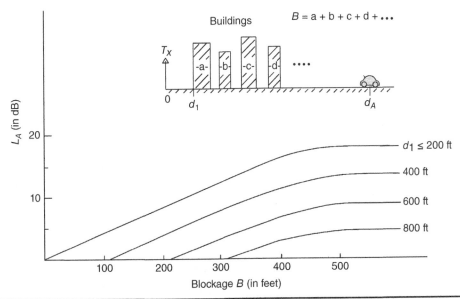

FIGURE 4.2.1.3.6 An empirical curve for predicting L_A with blockage B and d_1.

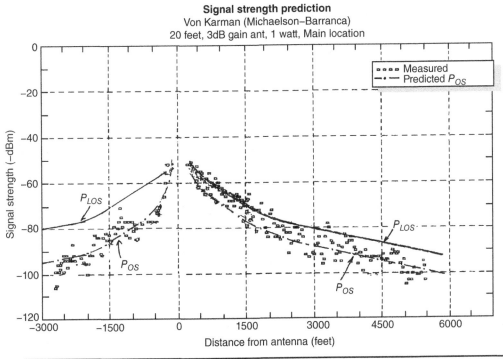

FIGURE 4.2.1.3.7 Measured data versus predicted curve when $B = 200$ ft (right-side data) and $B = 350$ ft (left-side data).

Measured Data   Figure 4.2.1.3.7 shows the data that were collected while the mobile unit ran on Von Karman Avenue from south to north. The antenna site was on Main Street, which was a relatively quiet street. The value of B blockage on the north side was $B = 200$ ft, corresponding to $L_A = 8$ dB. On the south side, $B = 350$ ft, corresponding to $L_A = 15$ dB. The new path losses L_A due to the blockage B are found from Fig. 4.2.1.3.6. The predicted P_{OS} (at A) after adding L_A was obtained from two different blockages of B in two different areas, then plotted in Fig. 4.2.1.3.7 for comparing with the measured data. They fairly agree with each other.

4.2.1.4   A Simplified Algorithm

Since digitalizing building layout maps of a city is a tedious job, we may simplify this by digitizing only the streets, not the buildings, as shown in Fig. 4.2.1.4.1. The methodology is as follows:

1. Digitize the streets (see Fig. 4.2.1.4.1).

2. Identify the street blocks.

3. Calculate the percentage of building occupation within each street block (density).
   $$P_i = \text{density of } i\text{th street block} = \frac{\text{total area occupation by the building blocks}}{\text{area of street block}}.$$
   Indicate P_i for each street block as a ratio shown in the figure.

4. Take the LOS path loss curve $L_{LOS}(d_A, h_1)$ from Fig. 4.2.1.2.2 curve a.

5. At point A, there are three street blocks—a, b, and c (see Fig. 4.2.1.4.1).

6. Calculate the total equivalent block length B_{eq} by considering the density of each street block shown in Fig. 4.2.1.4.1:
   $$B_{eq} = a \cdot p_a + b \cdot p_b + c \cdot p_c$$
   $$= a \cdot (0.347) + b \cdot (0.31) + c \cdot (0.41) \tag{4.2.1.4.1}$$

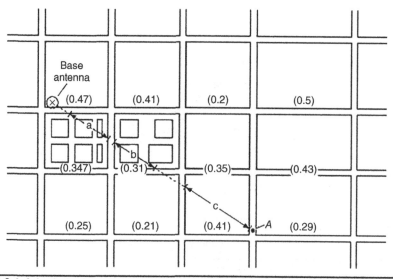

Figure 4.2.1.4.1   A simple method of estimating building blocks.

7. Find L_B from Fig. 4.2.1.2.2 curve b or from Eq. (4.2.1.4.1) for the value of B_{eq}.

8. The predicted signal strength is expressed similarly to Eq. (4.2.1.2.1) as

$$P_r(\text{at } A) = P_t - L_{LOS}(d_A, h_1) - L_{B_{eq}} + G_A + G_a \qquad (4.2.1.4.2)$$

4.2.2   Input Data for Microcell Prediction

4.2.2.1   The Consideration of Input Data

Having reliable input data is always a challenge for the prediction model to work. If the data are too coarse, the prediction cannot be accurate. However, detailed and accurate data are not easy to get and are too expensive. Sometimes, even with data with very high granularity, the prediction model may actually lose some accuracy and efficiency. For the Lee microcell model to work, we need accurate building, terrain, and attribute (e.g., water and foliage) data. As explained earlier, the building block data are the primary input to the Lee microcell model, such as in dense urban areas, due to the fact that buildings impact propagation loss the most. Some flat-city microcell models are available. In most major dense urban cities, the terrain factor cannot be ignored. Even a terrain variation of 3 to 5 m will impact the result quite drastically. This is based on the drive test data from San Francisco, Tokyo, Seoul, and Los Angles, to name just a few. As far as attribute (morphology) data are concerned, water and foliage data are two major attributes that are needed to provide accurate predictions. In most cases, there are not many trees in dense urban areas. However, water can make quite a difference for a city like Amsterdam. Again, this is based on the analysis of measured data.

One issue to be noted here is that sometimes GPS does not work accurately in a high-rise dense urban area. Recorded measured data need to be handled carefully as input data so that these data and measured data can be lined up correctly to ensure the accuracy of the prediction.

The flexibility of the Lee microcell model is that although building data are hard to acquire, the model works well with street map data. It can simply assume that the area bounded by four streets is a building block. This provides an efficient way to assess first-order assessment on the coverage and interference at an urban base station antenna. As shown in Fig. 4.2.2.1.1, the street data were actually used for the verification and benchmark with the Lee microcell model. The street information detailing certain street widths is stored in the database for an area shown in Fig. 4.2.2.1.2. Since the data include street classification, users can easily specify widths for varying kinds of streets (major, minor, alley, and highway). Also, based on street data, a building database can be created.

4.2.2.2   Collecting and Leveraging Measured Data

Deriving the building loss curve and specifying the propagation characteristics of a predicted area from the measured data is crucial. Especially, each city most likely has different building materials, structures, and distributions of buildings.

To collect drive test data, it is more accurate to use ETAK and GPS data together, for GPS might not be accurate, especially in dense urban areas. The ETAK data can be used to correct the GPS error. The drive routes need to be predefined and explored to make sure that they preset the characteristics of the predicted areas. Typically, routes selected for the validation of the model are LOS, zigzag, staircase, and random. A typical drive test route is shown in Fig. 4.2.2.2.1. Since many tall buildings are around, during the

FIGURE **4.2.2.1.1** Building block data.

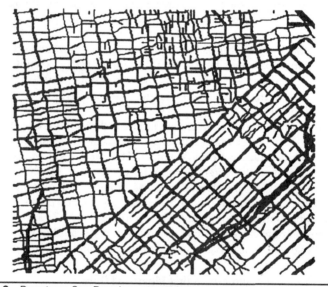

FIGURE **4.2.2.1.2** Downtown San Francisco street data.

measurement, we want to make sure that the antenna pattern at the base station will not get distorted, and the drive test needs to be kept at a constant speed to obtain a local mean later by averaging every 20 to 40 λ length of data along the test run.

The following process is used to derive the building loss curve:

1. Manage the measured database on GPS and ETAK data
2. Filter the unreliable data

FIGURE 4.2.2.2.1   Drive test route and data analysis.

3. Identify LOS data points
   Validate near-in distance
   Validate the slope after break points
   Basically identify the microcell LOS curve
4. Identify NLOS data points
   Calculate building thickness
   Create the building loss curve

Building height is one of the parameters in many urban microcell prediction models. 3D building data are shown in Fig. 4.2.2.2.2. If 3D building data must be used, the data need to be very accurate to ensure the benefit of using these data. It would complicate both the process and the ability to achieve high accuracy. It might be beneficial to validate a specific propagation model, but it comes with more prone to disarray the data

FIGURE 4.2.2.2.2   Downtown San Francisco building data in 3D.

base and yet with limited gain in prediction accuracy, especially when applied to the whole city. However, based on the drive tests conducted in many cities, it appears that building height does not play a critical role in the accuracy of the model, as in most cases antenna height is about the same or just a bit higher than the surrounding buildings to ensure that the coverage is within 1 km. This is another benefit of the Lee microcell model, as it simplifies a 3D problem to a 2D one without sacrificing accuracy.

Figure 4.2.2.2.3 shows the San Francisco building data in 2D. These are the input data for the Lee microcell prediction model. The data can cover the area under test using street map data as the baseline for prediction. As shown in the figure, there is a big open space without any building in the middle. If by using the street data, we assume that these are buildings, the accuracy of the model will suffer. A way to deal with open space is to implement it in the program. Again, the basic requirement for any model to work accurately is to keep the input data reliable to a certain degree.

Figure 4.2.2.2.4 shows the Lee microcell model coverage plot. Again, the same coverage threshold is shown in one color, and the threshold colored red demonstrates that building data are most important input data for the model to function accurately.

An accurate microcell prediction model needs to consider the effects of terrain contour and building factors on urban microcell radio propagation. An inexpensive and accurate way to execute this prediction is to utilize readily accessible street and terrain data as proposed in this section. Various comparisons were made within the

Figure 4.2.2.2.3   Downtown San Francisco building data in 2D. (A color version of this figure is available at www.mhprofessional.com/iwpm.)

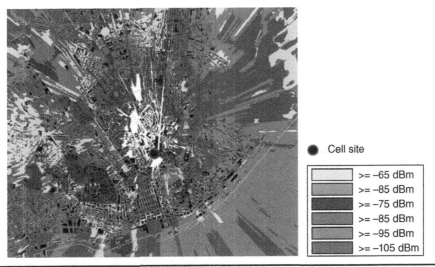

Cell site

	>= −65 dBm
	>= −85 dBm
	>= −75 dBm
	>= −85 dBm
	>= −95 dBm
	>= −105 dBm

Figure 4.2.2.2.4   Downtown San Francisco Lee microcell coverage plot. (A color version of this figure is available at www.mhprofessional.com/iwpm.)

San Francisco Bay Area and discussed in this section. The model calculations for the combination of attenuation from buildings and the gain or loss from terrain (i.e., diffraction loss and effective antenna gains) encountered along the path of the mobile are fairly accurate. The street map data utilized can be employed either as the base for creating the building database or as a method of fine-tuning to improve the accuracy of the microcell model.

Historically, there has been a need to increase the capacity of cellular mobile communication systems in urban areas. The Lee microcell model has been one of the most effective methods in achieving this goal. Most of the models and investigations performed within this realm have concentrated on analyzing propagation under the effect of building blockage in smaller coverage areas, assuming that the terrain is flat.[1-9] It is thought that within smaller regions, the terrain does not vary to a large extent in producing any major effect on propagation. However, the microcell predictions considering only flat terrain become much simpler but not necessarily more accurate.

This section explains how to forecast a signal at different locations by considering building loss, effective antenna gain, and diffraction losses due to terrain. The approach discussed here is to anticipate the signal strength in the microcell environment by considering the following factors: each loss encountered along the radio propagation radial separately, building loss, effective antenna height gain (due to terrain sloping upward), and diffraction loss (due to the shadow condition). Comparison with experimental results originating from the measurements of downtown San Francisco shows that this method provides reasonable prediction outcomes in various terrain fluctuations.

Building data are still expensive to get, but the need for achieving accurate microcell prediction is increasing. Creating building data by using street map data is a cheap yet effective method. Therefore, street data can be used as input for microcell prediction. This approach is inexpensive and accurate due to the proliferation of buildings located in dense urban areas. Users also have the option of specifying height and width of the selected buildings so that the resulting 3D database can be created at nominal cost.

FIGURE 4.2.2.2.5   Sample terrain data in the Los Angeles area.

This street data can be further utilized not only as a base for creating a building database but also as a method of fine-tuning, which is necessary to improve the accuracy of the microcell model. The ETAK data can also be employed as a base to build building data with low granularity. Building data with fine granularity can then be structured by combining ETAK data and street map data.

Figure 4.2.2.2.5 displays a sample of variation in terrain in the Los Angeles region. These data are readily accessible via the Internet. As shown in Fig. 4.2.2.2.6, there is a significant amount of terrain undulation within the region, and the terrain effect cannot be ignored in many major urban cities.

Figure 4.2.2.2.6 shows the sample of the water attribute file in the Sacramento area. The water attribute is highlighted in blue. When the propagation prediction runs through this area, certain correction factors need to be applied to fine-tune the Lee model for accuracy. The water enhancement formula is discussed in Sec. 3.1.6 and also in a later section of this chapter, and Lee has developed a generic approach to deal with this attribute parameter. It was named the Lee morphology model, and it provides a generic way to apply correction factors to the model.

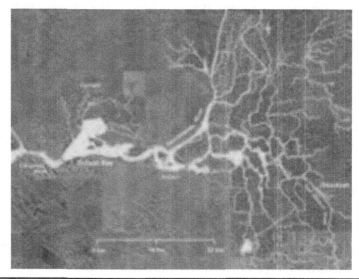

FIGURE 4.2.2.2.6   Sample water data in the Sacramento area. (A color version of this figure is available at www.mhprofessional.com/iwpm.)

By integrating building, terrain, and attribute data, such as water, the Lee microcell prediction model can work with these inputs and come up with a much more adaptable and accurate propagation model.

These data are readily accessible through the Internet. However, to integrate these data into the model, calculation is not a trivial task.

Cities in different countries most likely will have different building loss curves, as they are built with different materials. The Lee model also developed an algorithm to derive the building loss curve L_b. Measurement integration is one of the key features of the Lee model to ensure the model's accuracy and adaptability. All impacted parameters need to go through the measurement integration process so that data are normalized in each category. Then the model can be applied with the new derived parameters to secure better accuracy.

Let us give more details on how the model works. The effect of buildings on propagation will be addressed first, followed by an analysis of the effect of terrain on microcell prediction.

4.2.3 The Effect of Buildings on Microcell Prediction

A received signal can be in LOS either without any building blockage or with significant obstruction. These two factors are discussed in more detail in Lee.[1] The flat-earth assumption is utilized in this section as well.

Let us begin by considering a basic microcell scenario in which there is no building obstruction and the existing terrain loss occurring on the radio path is L_{LOS}.

4.2.3.1 The Basic LOS Situation

Considering a very basic case of a microcell where there are no buildings obstructing the LOS and assuming a flat-terrain scenario to start with, the only loss that occurs on the radio path is L_{LOS}. Figure 4.2.3.1.1 provides a map of San Francisco for searching the LOS situation.

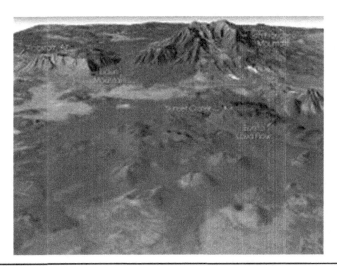

FIGURE 4.2.3.1.1   San Francisco terrain.

The received signal strength under flat terrain is as follows:

$$P_r = P_t + G_t - L_{LOS} + G_r \tag{4.2.3.1.1}$$

where P_r = received signal strength, P_t = transmitted power, G_t = transmitter antenna gain, L_{LOS} = loss under LOS, and G_r = receiving antenna gain. L_{LOS} is computed as follows: First, calculate the near-in distance (see Sec. 4.2.1.1) from the following equation:

$$d_f = \frac{4h_1 h_2}{\lambda} \tag{4.2.3.1.2}$$

where d_f = near-in distance (Eq. [4.2.1.1.4]), h_1 = transmitter height, h_2 = receiver height, and λ = Wavelength. If the mobile's position at distance d is within the near-in area, $d \le d_f$,

$$L_{LOS} = 20 \log \frac{4\pi d}{\lambda} \tag{4.2.3.1.3}$$

for $d > d_f$

$$L_{LOS} = 20 \log \frac{4\pi d_f}{\lambda} - \gamma \log \frac{d}{d_f} \tag{4.2.3.1.4}$$

where γ = slope and d = distance between transmitter and receiver.

Considering Figure 4.2.3.1.2 under true terrain conditions, taking a case of terrain sloping up the terrain contour does have an effect on the signal received by the mobile. There is an effective antenna gain that is due to the position of the mobile on an elevated slope. The signal received under this condition would be

$$P_r = P_t + G_t - L_{LOS} + G_r + G_{ef} \tag{4.2.3.1.5}$$

The effective antenna gain G_{effh} is given as

$$G_{eff} = 20 \log \frac{h_e}{h_a} \tag{4.2.3.1.6}$$

where at the base station antenna h_e = effective height, and h_a = actual height. Equation (4.2.3.1.5) is the same as used in Lee macrocell prediction model (see Sec. 3.1.2.4).

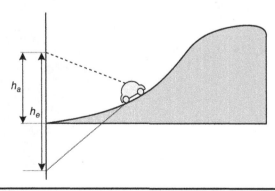

Figure 4.2.3.1.2   LOS.

4.2.3.2 The Non-LOS Situation

Now focusing on the losses due to building loss, we need to eliminate the terrain effects. To do this, we need to have a building layout of the area under consideration. Figure 4.2.2.1.1 shows a sample building's layout, which is in the standard AutoCAD DXF file format. The prediction tool is developed for the analysis of the model and takes the vector-based building data as the input. In reality, the buildings are not on a flat surface, but so that we can accurately find the effect of the building losses on the received signal, we consider the surface to be flat. The model currently is based on 2D building data. There is no consideration of the height of the building; rather, the thickness of the building is the main phenomenon for predicting the losses.

The detailed explanation of the loss due to building blocks has been discussed.[1, 6] The blocking is caused by the thickness of buildings B_a, B_b, and B_c and their respective radial cutting thickness being a, b, and c as shown in Sec. 4.2.1.2. The blockage L_B can be expressed as

$$L_{B(a+b+c)} = P_{LOS} - P_b \qquad (4.2.3.2.1)$$

where P_{LOS} = LOS signal and P_{bld} = the signal received after passing the building blocks (the measured local means).

4.2.4 The Terrain Effect

4.2.4.1 Nonshadow Region

Now we are considering the realistic LOS case with the elevated terrain, where the radio path is blocked by the buildings, as shown in Fig. 4.2.4.1.1.

The mobile is blocked by buildings 1 and 2, but the terrain contour contributes an antenna effective height gain. Both factors affect the signal received at the mobile receiver.

The received signal due to both the LOS gain and the losses from the building thickness blockage is expressed as

$$P_r = P_t + G_t + G_a - L_{LOS} + G_r + G_{effh} - L_B \qquad (4.2.4.1.1)$$

where P_r = received signal strength, P_t = transmitted power, G_t = transmitter antenna gain, G_a = antenna height gain, L_{LOS} = loss under LOS, G_r = receiving antenna gain, G_{effh} = effective antenna height gain, and L_B = building block transmission loss.

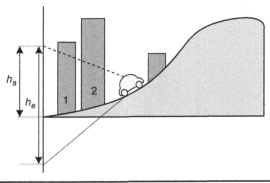

FIGURE 4.2.4.1.1   LOS terrain and building blockage.

4.2.4.2 Shadow Region

We have already considered the signal received over the elevated terrain along the LOS where there are effective antenna height gains or losses due to terrain variation. Now a more complex case of having a mobile under a hidden condition or the mobile being blocked by a knife-edge case is considered in this section.

The basic algorithm of detecting hidden conditions, where the direct path from the mobile is obstructed by a hill, has been described in Sec. 3.3.2. The mobile signal has occurred in an obstructed condition by terrain, as shown in Fig. 4.2.4.2.1.

The received signal P_r takes a generalized form as the typical output in the prediction analysis tool:

$$P_r = P_t + G_t + G_a - L_{LOS} + G_r + G_{effh} - L_B - L_D \qquad (4.2.4.2.1)$$

where effective antenna gains G_{effh} can be calculated based on the terrain contour (see Sec. 3.1.2.3). Loss L_B is due to building blocks (see Sec. 4.2.1.2), and diffraction loss L_D due to terrain (see Sec. 3.1.2.4).

Now we have to demonstrate how to use Eq. (4.2.4.2.1) to deal with the real measurement conducted in various cities, as shown in Fig. 4.2.4.2.2. The current situation is such that the mobile is far away from the base station transmitter in an elevated terrain situation and 10 buildings are blocking the radio path propagation.

As shown in Fig. 4.2.4.2.3, considering only the loss due to building blockage on a flat terrain, the sum of the 10 buildings' thickness is computed, and the building blockage loss is L_B. It is worth mentioning at this stage that the height of the building will not make a significant difference because the incident angle θ of the direct wave is very small.

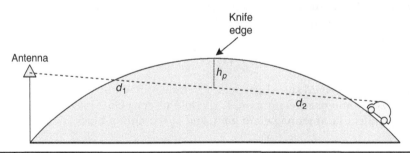

Figure 4.2.4.2.1   Mobile blocked by terrain.

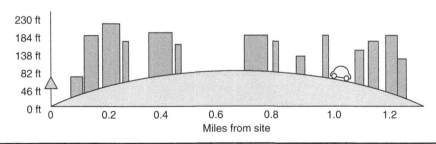

Figure 4.2.4.2.2   The true transmitter and mobile situation.

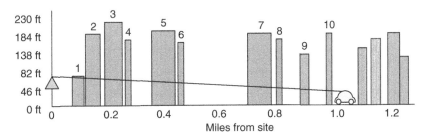

FIGURE 4.2.4.2.3 Building blockage loss.

Building height might be an important issue if the antennas are placed on the top of the building, but this issue is not discussed in this section.

If L_{LOS} is the LOS loss of the direct wave reaching the mobile due to the transmitter of power P_{erp} and only the building blockage is considered, then the received signal at the mobile would be

$$P_r = P_{erp} - L_{LOS(at\ Mobile)} - L_B \qquad (4.2.4.2.2)$$

Introducing the elevated terrain contour along the radial path from the transmitter to the mobile and considering only the terrain effect, it is found in Fig. 4.2.4.2.4 that the mobile is blocked. The knife-edge height h_p is calculated, and the corresponding diffraction loss L_v due to the knife edge h_p can be obtained from Fig. 3.1.2.3.2.

Figure 4.2.4.2.5 is a caption of the tool that provides the gains and losses along the radial due to the terrain contour and the effective antenna height gains/losses. Those gains/losses will be added to the received signal P_r shown in Eq. 4.2.4.2.3.

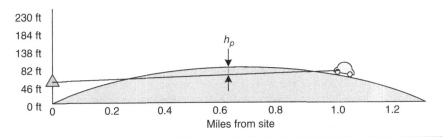

FIGURE 4.2.4.2.4 Diffraction loss.

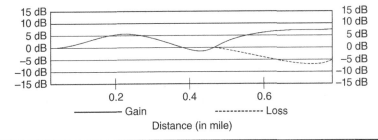

FIGURE 4.2.4.2.5 Gain and loss curves.

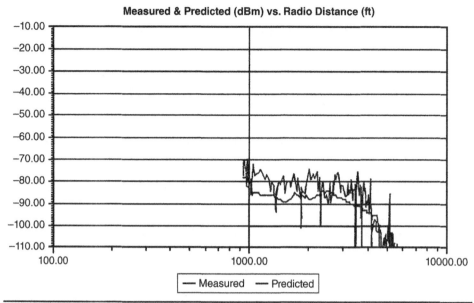

Figure 4.2.4.2.6 Predicted versus measured data without integrating with terrain (downtown San Francisco).

The actual received signal at the mobile is

$$P_r = P_{erp} - L_{LOS} - L_D + G_{effh} - L_B \qquad (4.2.4.2.3)$$

If the prediction model is not integrated with G_{effh} due to the terrain contour, we observe a big difference in the predicted versus the measured data. We have explained the reason for this using Eq. (4.2.4.2.3) in Sec. 3.1.2.4. Figure 4.2.4.2.6 shows measured versus predicted data without considering the terrain elevation. Figure 4.2.4.2.7 shows measured versus predicted data in which both terrain and building characters are integrated into calculation. We can see that if we consider only the building blockage, accuracy suffers drastically. Once we put in the elevated terrain calculation, the predicted and the measured data follow each other nicely. This happens in most big cities, and we do notice that even when terrain elevation varies slightly, the measured data vary greatly. The reason for this is discussed in Chap. 3.

4.2.5 Prediction Model with Four Situations

A model constituting a microcell prediction under a real terrain situation and the effect of elevated terrain contour on urban microcell radio propagation has been introduced in Sec. 4.2.4. Various comparisons were done to show that the model is appropriate under the LOS condition, the obstruction due to terrain contours (i.e., under shadow conditions), and the situation of building blockage. A wide-scale analysis of the San Francisco, Tokyo, Amsterdam, and Rome terrain proved that the model calculation

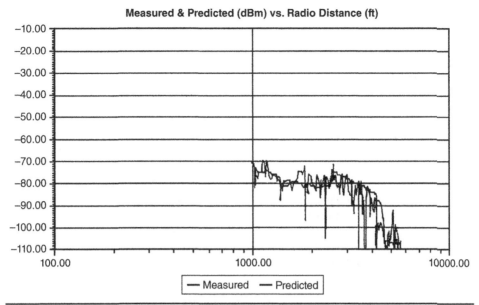

Figure 4.2.4.2.7 Predicted versus measured data integrating with terrain.

of considering diffraction loss and effective antenna gains encountered over the radio path of the mobile is fairly accurate.

There has always been a need to increase the capacity of cellular mobile communication systems in urban areas, and using the microcell[10, 15] has been one of the positive solutions to achieve this. Most of the models have concentrated only on analyzing the propagation under the LOS condition and reflections due to buildings. The terrain is assumed flat because the microcell has a far smaller coverage area.[16] It has always been assumed that within a small region, the terrain elevation cannot vary to a large extent and cannot produce a major effect on propagation. Of course, considering a flat terrain makes the predictions much simpler but not necessarily accurate and correct. This section provides more evidence that the fluctuations of terrain contour play a major role. How to predict the signal strength at a desired location considering building loss, LOS gain, and diffraction losses due to terrain will be discussed.

The approach is to predict the signal strength on urban streets by considering building loss, effective antenna height gain due to terrain elevation, and diffraction loss due to the shadow condition separately along the radio propagation path. A near-in distance criterion under the LOS condition is also considered for predicting signal strength. Comparison of experimental results from measurements of downtown San Francisco shows that this method gives reasonable prediction results for various terrain elevations. Figures 4.2.2.2.2 and 4.2.2.2.3 show the San Francisco building data in 2D and 3D. Figure 4.2.3.1.1 shows the terrain in 3D covering the terrain variations within 2 miles of the cell site. The benefit of the Lee model is that both the terrain and the building data are considered in the prediction.

The following four cases result from the combination of terrain factors and building data:

	Terrain	Building
Case 1	Nonblocking	Nonblocking
Case 2	Blocked (shadow case)	Nonblocking
Case 3	Nonblocking	Blocked
Case 4	Blocked (shadow case)	Blocked

Cases 1 and 2 are handled the same way as in the macrocell prediction model. Case 3 is discussed next.

Let's assume a scenario of Case 3 in which the LOS signal is affected by the terrain elevation but not in the shadow case and also that the buildings block the radio path. In Fig. 4.2.4.1.1, the radio path is blocked by buildings 1 and 2 and experiences an effective antenna height gain in the received signal instigated by the surrounding terrain. The received signal at the mobile, under the Case 3 condition, can be expressed as

$$P_r = P_t + G_t - L_{LOS} + G_r + G_{effh} - L_B \tag{4.2.5.1}$$

where P_r = received signal strength, P_t = transmitted power, G_t = transmitter antenna gain, L_{LOS} = loss under LOS, G_r = receiving antenna gain, G_{effh} = effective antenna height gain, and L_B = building block transmission loss.

For case 4, the current scenario dictates the received mobile signal, which can be separated from the effect by the shadow of varying terrain contour and the building blockage. First considering only the loss due to building blockage, the sum of the building thickness along the radio path is computed, and the building blockage loss is L_B. To simplify the calculation of L_B, we can find L_{eq} from Eq. (4.2.1.4.1).

Also, L_{LOS} is the path loss of the direct wave reaching the mobile from the transmission of power P_{erp}. Then the received signal at the mobile would be

$$P_r = P_{erp} - L_{LOS} - L_B \tag{4.2.5.2}$$

The next step is to introduce the terrain contour fluctuations along the radial path from the transmitter to the mobile and to consider only the terrain effect. The knife-edge height h_p with the corresponding diffraction loss L_v due to h_p has been described in Sec. 3.1.2.4.

4.2.6   Characteristics of the Measured Data

San Francisco, as an example, exhibits a variety of building layouts and a fluctuating terrain contour. Figure 4.2.6.1 displays the building layout of downtown San Francisco. A selected portion of San Francisco is also shown in Fig. 4.2.6.2 for the sake of demonstrating the significant impact of signal enhancement from the terrain elevation. There are 200 to 300 ft of undulation of terrain contour.

The comparison of results, such as measured versus predicted, on the various routes leads to a difference between the two sets of data around 6 to 10 dB, as shown in the next section.

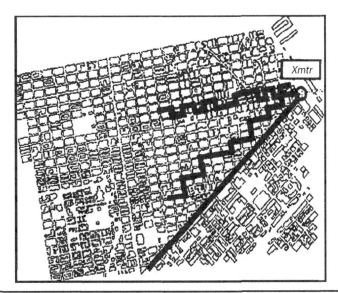

FIGURE 4.2.6.1 San Francisco measurement route.

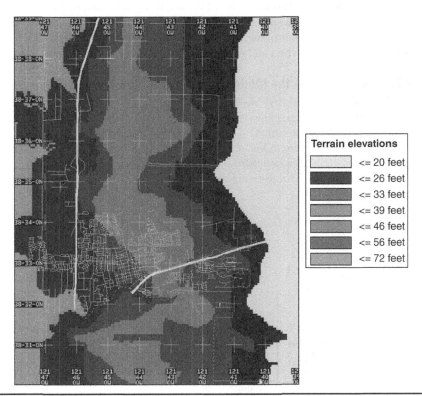

FIGURE 4.2.6.2 Downtown San Francisco terrain. (A color version of this figure is available at www.mhprofessional.com/iwpm.)

The route types were as follows:

1. Along a straight line and maintaining LOS
2. Zigzag kind of route
3. Route following a staircase pattern
4. Random routes

Figure 4.2.6.2 displays the variation of terrain contour in the San Francisco region. There is a lot of undulation of the terrain, and this is within a small region of 2 miles.

Figure 4.2.6.3 shows the map when building and terrain data are combined. The terrain elevation plays a crucial role in ensuring the accuracy of the prediction in the dense urban area. The drive test in Tokyo shows that with a difference, of only a few feet, the propagation prediction can have a standard deviation 6 to 8 dB from the measurement data, as shown in following section.

Figure 4.2.6.3 displays the variation of terrain contour as seen between the transmitter and the receiver. The received signal is affected from terrain elevation as well as buildings blockage. It is critical to cover these two parameters in this scenario. As discussed later, other parameters (water, foliage, and so on) should also be handled by the model. The Lee model covers these parameters, which are easily integrated into the model.

All the terrain profiles are drawn for the city of San Francisco from the same transmitting site but in different radial angles, as shown Fig. 4.2.6.4. Based on the different terrain profiles, the Lee model provides the predicted results at each mobile location along the radio path. The validation of the model is introduced in the next section.

4.2.7   Validation of the Model: Measured versus Predicted

The Lee model is validated based on the measurements collected in the real elevated terrain. Different configurations of runs taken in San Francisco are shown in Fig. 4.2.1.2.4. Comparisons of the measured data with the predicted results along various routes in San Francisco are shown in Fig. 4.2.7.1 for the zigzag-run case (the same figure has been shown in Fig. 4.2.1.2.5 when introducing the basic principle), in Fig. 4.2.7.2 for the LOS-run case, in Fig. 4.2.7.3 for the stair-run case, in Fig. 4.2.7.4 for the random-run case 1, and

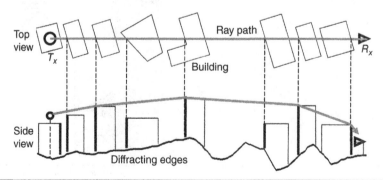

FIGURE 4.2.6.3   Top and side view of combined terrain and building data.

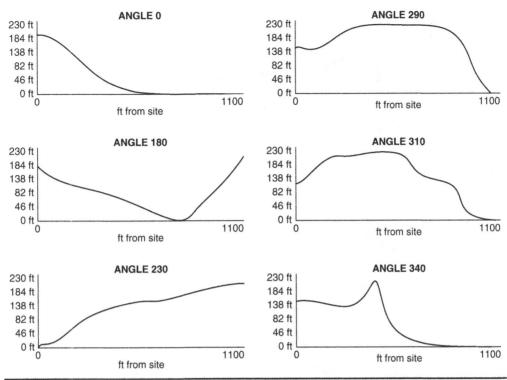

Figure 4.2.6.4 Terrain radial profile in downtown San Francisco.

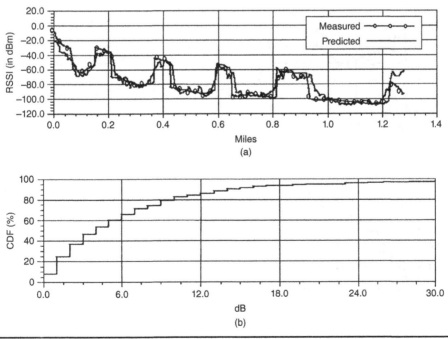

Figure 4.2.7.1 Zigzag-run case. (a) RSSI of measured versus predicted. (b) CDF of deviation between measured and predicted.

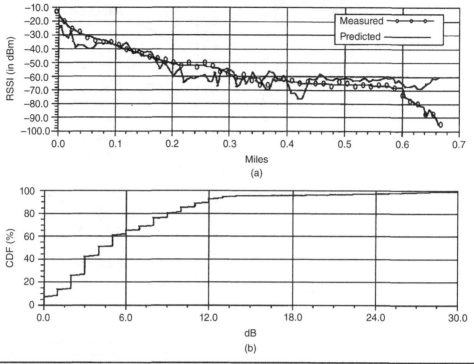

FIGURE 4.2.7.2 LOS-run case. (a) RSSI of measured versus predicted. (b) CDF of deviation between measured and predicted.

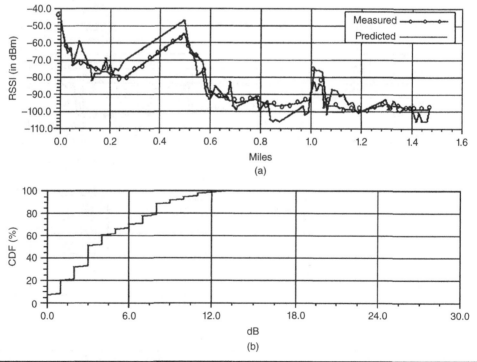

FIGURE 4.2.7.3 Stair-run case. (a) RSSI of measured versus predicted. (b) CDF of deviation between measured and predicted.

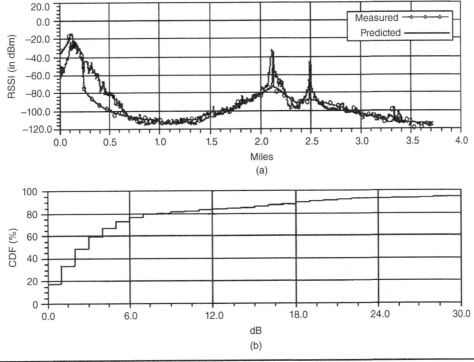

FIGURE 4.2.7.4 Random-run case 1. (a) RSSI of measured versus predicted. (b) CDF of deviation between measured and predicted.

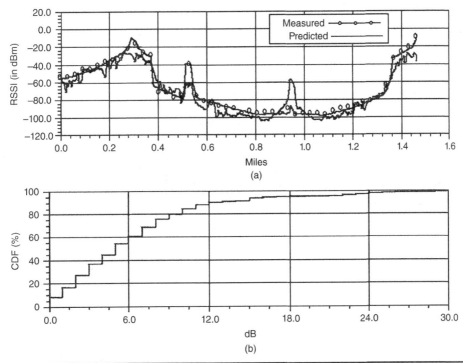

FIGURE 4.2.7.5 Random-run case 2. (a) RSSI of measured versus predicted. (b) CDF of deviation between measured and predicted.

in Fig. 4.2.7.5 for the random-run case 2. Each figure shows the received signal strength indicator (RSSI) and the cumulative distribution function (CDF) on a particular run. A large scale (in miles) is used to plot RSSI in each figure. We find that a deviation between the measured and predicted at 60% data is about 6 dB in all five figures. The following four figures showing the RSSI received on a particular street in San Francisco are plotted on a small scale (in feet). Figure 4.2.7.6 is the LOS run on California Street at 909 MHz, and Fig. 4.2.7.7 is the zigzag run on Drum-Sacramento Street at 1937 MHz. Fig. 4.2.7.8 shows the same run as Fig. 4.2.7.7 but at 909 MHz. Figure 4.2.7.9 is the stair run on Front Street at 844 MHz. The CDF of deviations at 60 percent of data in the last four figures are 6, 7, 6.5, and 5 dB, respectively. This shows that the prediction model is a fairly accurate tool to use.

For a city like San Francisco, if terrain is not a part of the parameters to be considered, the results will be misleading and thus derive a wrong conclusion. Just like if the water is not considered a parameter in Amsterdam, the results will be twisted as well. The attributes are very important in leading to a correct conclusion and developing models.

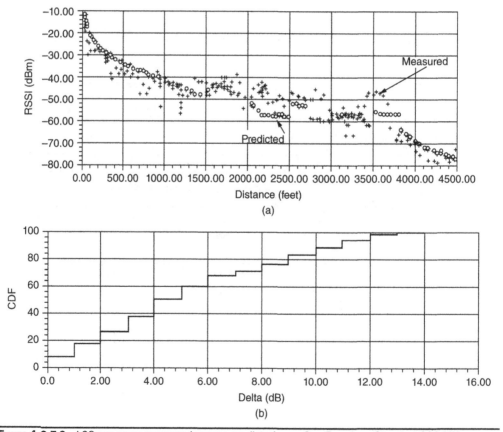

FIGURE 4.2.7.6 LOS-run case: measured versus predicted on a San Francisco Street at 909 MHz. (a) RSSI of measured versus predicted. (b) CDF of Deviation between measured and predicted.

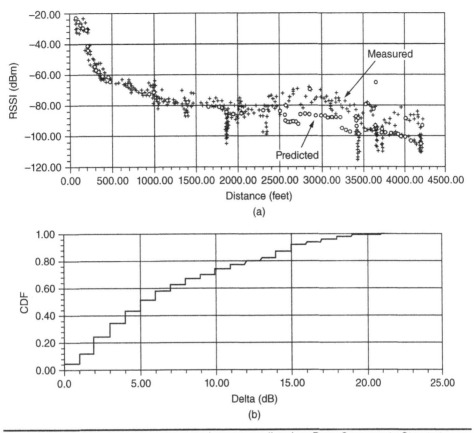

Figure 4.2.7.7 Zigzag-run case: measured versus predicted on Drum-Sacramento Street at 1937 MHz. (a) RSSI of measured versus predicted. (b) CDF of the deviation between measured and predicted.

4.2.8 Integrating Other Attributes into the Model

The Lee microcell model has successfully integrated building and terrain attributes to provide an accurate propagation model. There are many other important attributes in the dense urban area, such as, water, trees, parks, and open areas shown in Fig. 4.2.8.1. The model needs to be able to integrate capabilities to handle all these different attributes. The Lee macrocell model has developed the ability to integrate these attributes, such as water, and other morphologies into the model.

The grid along the radio radial in the map can be divided into a smaller unit, a "bin." Each bin has its own unique attributes associated with it. During the calculation of the prediction, the associated "attributes" in each bin can be considered and the adjusted to the final prediction value. This capability enables the Lee microcell model to handle many different environments as well as many unique challenges in the complex urban area.

The attribute file can be overlaid on the terrain and building information to enable the model to be more flexible in handling different characteristics of the environment.

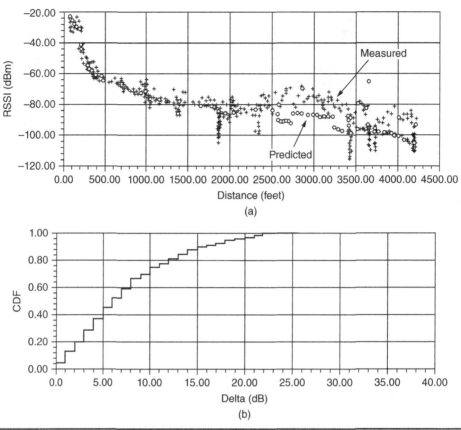

FIGURE 4.2.7.8   Zigzag-run case: measured versus predicted on Drum-Sacramento Street at 909 MHz. (a) RSSI of measured versus predicted. (b) CDF of the deviation between measured and predicted).

The Lee microcell model does not leverage the building height data because usually the transmitter antenna height is below the surrounding building; therefore, the building height is, on average, much higher. In reality, the building distribution is very nonuniform, and the street spacing and layout are unique in every city. All these factors will impact the accuracy of the propagation prediction. For example, if a few unique buildings in a certain area are much higher than of her buildings in the coverage area, we will need to categorize the area with unique building IDs so that the model has the flexibility to add some extra parameters in predicting that area. If the street is very narrow or has multiple paths through which the signal can be received from the same cell site, the model can leverage these "data" or this "knowledge" to further fine-tune the formula.

Figure 4.2.8.2 shows the 3D view from the cell site. If the cell site is much higher than the surrounding buildings, the propagation path will be much different than if the cell site is about the same height of the surrounding buildings. Normally, a microcell base station antenna is mounted at about the average height with the surrounding buildings. However, with complex building deployment becoming more popular, the coverage from inside the buildings needs to go to the in-building (picocell) models, and

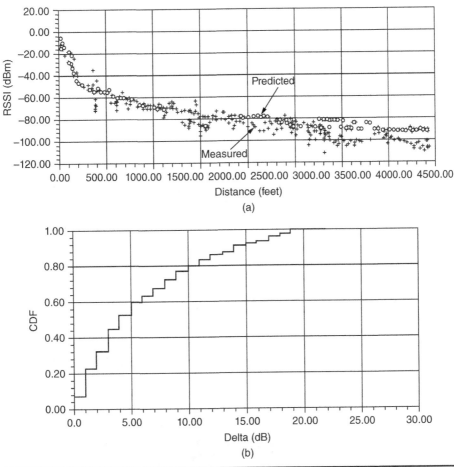

FIGURE 4.2.7.9 Stair-run case: measured versus predicted on Front Street at 844 MHz. (a) RSSI of measured versus predicted. (b) CDF of the deviation between measured and predicted.

these picocell sites can be much lower than the surrounding buildings. This will be discussed in more detail in the next chapter, as one must consider some in-building situations more carefully so that coverage, cost, and capacity can be balanced.

Figure 4.2.8.3 shows the building height clusters in different colors. Depending on the distribution and height of these buildings, the propagation characteristics over the radio path vary. The propagation model needs to have the ability to integrate these characteristics in a proper and effective manner. Based on the work from the Lee macrocell model with the automorphology factors described in Sec. 3.3, these factors and functions can be easily added to the Lee Microcell Model.

From a measurement integration perspective, some cities have both old and modern buildings coexisting in the same vicinity. Applying one empirical curve for the building loss might not provide an accurate result. The measurement data need to be integrated into the attribute file to make the prediction more accurate. The model can thus derive the building loss curve based on the building material. The model can add

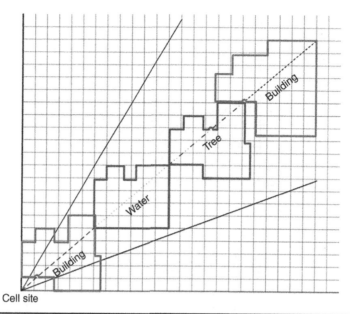

Cell site

FIGURE 4.2.8.1 Attribute files.

FIGURE 4.2.8.2 3D view from the cell site.

extra loss/gain at certain "attributed" areas. This will definitely improve the accuracy of prediction. Furthermore, with accurate signal coverage, interference can be further controlled.

Figure 4.2.8.4 shows an addition to the generic approach by applying attributes that enable the model to be more accurate. Along this radial path, the model needs to be able to handle many attributes. The radio wave travels over and/or through the buildings, the treetops, and the water, and in an open area encountering the terrain effect. The generic path loss can be used as the baseline, but other propagation factors can be easily added into the propagation prediction with the help of the attribute files.

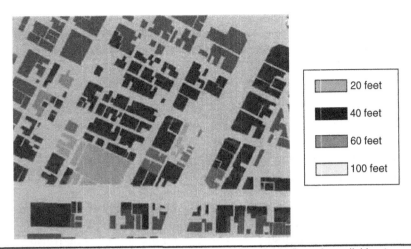

FIGURE 4.2.8.3   Building height clustering. (A color version of this figure is available at www.mhprofessional.com/iwpm.)

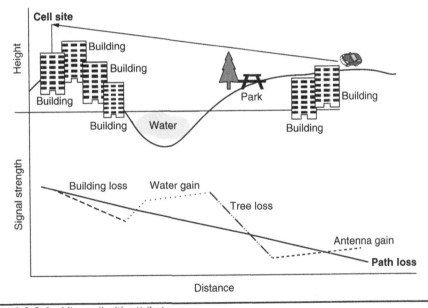

FIGURE 4.2.8.4   Microcell with attributes.

4.3   Integration of the Microcell Prediction Model and the Macrocell Prediction Model

Lee has developed the macro, micro, and pico propagation models. The key advantage of the Lee models is that they can be integrated and work seamlessly together so that one integrated set of three models can cover all radio planning needs, from pico- (in-building) to micro- to macrocells. Many field data have been collected and used to verify the models as well as to understand the key issues that are to be addressed in this section. From the measured data, we clearly understand the impact of the nature and thus enhance the models so that the models interact among themselves and smoothly transition from one to another.

In this section, we will focus on the transition of models from microcell to macrocell.[17,18] The microcell prediction can work closely with the concept of using multiple radio radials. We have to ensure that the characteristics of microcell propagation are simulated through the multiple radials in the database of the macrocell model.

4.3.1   The Algorithms for Integrating the Two Models

The Lee model for microcell prediction has been enhanced to support this multiple breakpoint macrocell propagation prediction model, as mentioned in Chap. 3. The 1-mile intercept is used as the final breakpoint of this multiple-breakpoint microcell propagation model. Note that the symbol for distance is the Lee macrocell model uses r, and the Lee microcell model uses d. It is for making a distinguishability between the two groups of formulas.

A. The Lee Macrocell formula used for distances greater than 1 mile is

$$P_r = P_0 - \gamma \log\left(\frac{r}{r_0}\right) + G_{effh} + L_D + \alpha \quad \text{for } r > r_0 \qquad \text{in dBm} \qquad (4.3.1.1a)$$

When the signal is obtained over a flat ground with a standard setting ($\alpha = 0$), then

$$P_r = P_0 - \gamma \log\left(\frac{r}{r_0}\right) \qquad\qquad \text{for } r > r_0 \quad \text{in dBm} \qquad (4.3.1.1b)$$

where P_0 = a power at r_0 in dBm;
  r = distance from the transmitter in meters;
  r_0 = distance measured in feet or meters (the default distance is 1 mile or 1.6 km);
  γ = macrocell slope in dB/dec;
  G_{effh} = effective height gain = 20 log (h_e/h_a) in dB, when in shadow $G_{effh} = 0$;
  h_e = effective antennas height in meters[1];
  h_1 = height of base station antenna in meters;
  L_D = shadow loss in dB if blocked (when in LOS, $L_D = 0$; and
  α = standard adjustments in dB for those parameters having different values from the standard set.

B. The received signal for the area within a radius of the near-in distance d_f is expressed as

$$P_r = P_{st} - 20 \log\left(\frac{4\pi d}{\lambda}\right) + L_D + G_{effh} + \alpha \quad \text{in dBm} \qquad \text{for } d \le d_f \text{ in dB} \qquad (4.3.1.2a)$$

C. The radio signal received at the near-in distance is

$$P_{d_f} = P_{st} - 20 \log\left(\frac{4\pi d_f}{\lambda}\right) + L_D + G_{effh} + \alpha \qquad \text{in dBm} \qquad (4.3.1.2b)$$

where P_{st} = standard radiated power in dBm;
 G_{effh} = effective antenna height gain in dB, G_{effh} = 0 for $d = d_f$;
 L_D = diffraction loss in dB;
 d = distance from transmitter to mobile in meters;
 λ = wavelength in meters; and
 α = standard adjustments in dB.

D. The received signal in the range d_f to d_1, assuming that no measured data exist between d_f and d_1, is

$$P_r = P_{d_f} - \left(\frac{P_{d_f} - P_{d_1}}{d_1 - d_f}\right)(d - d_f) \quad \text{in watts} \quad \text{for } d_f < d_1 \text{ and } d_f \leq d \leq d_1 \quad (4.3.1.3)$$

where P_{df} is obtained from (4.3.1.2b) at $d = d_f$ (convert to watts),
 P_{d_1} = the signal power received at d_1 (convert to watts),
$\left(\dfrac{P_{d_f} - P_{d_1}}{d_1 - d_f}\right)$ = the path loss slope in linear scale,

 d_1 = distance less than r_0, d_1 can be adjusted based on the measured data, and
 d_f = Fresnel distance in meters.

As shown in Fig. 4.3.1.1, this microcell model is composed of three breakpoints. The near-in distance shown in Eq. (4.3.1.2b) induces the first breakpoint. The second breakpoint is determined by Eq. (4.3.1.3), and the last breakpoint is the 1-mile intercept in Eq. (4.3.1.1). These equations can be used if no measurement data are available. One of the special characteristics of this model is that the distance $d_1(d_f \leq d_1 \leq r_0)$ can be fine-tuned based on the measured data. When the measured data are available, the breakpoints are calculated and adjusted based on the measured data on a per radial basis.

Also, when the measured data are available in a region, the region around the transmitter will be narrowed to a specific number of radial zones. The measurement data will be assigned to zones and will be based on finding the best line fitting slope to the measurement data to achieve each zone's own path loss curve.

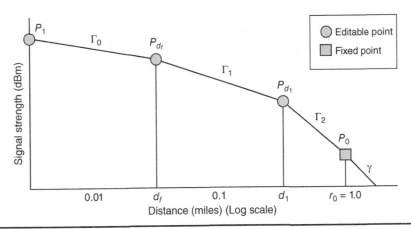

FIGURE 4.3.1.1 A Multiple-break point model.

4.3.2   Treatment of Measured Data

4.3.2.1   Finding the Propagation Path Loss Slopes

When the measurement data exist in the range from d_f to d_1, a best-fit slope Γ_1 is obtained, including the measured power P_{df} at the near-in distance d_f to a distance d_f. Similarly, if the measured data exist in the range from d_1 to r_0 or 1-mile, a best-fit line slope Γ_2 is obtained.

A. Area after the first breakpoint

If the measured data exist in the range d_f to d_1 the received signal P_r can be predicted as

$$P_r = P_{df} - \Gamma_1 \log \frac{d}{d_f} \tag{4.3.2.1.1}$$

where P_{df} = the power at the intercept of the path loss slope Γ_0 (= free space path loss slope) and at the near-in distance d_f in dBm (Eq. [4.3.1.3]) and
Γ_1 = the best-fit slope from the measurement data in dB/dec.

B. Area after the second breakpoint

If the measured data exist in the range from d_1 to r_0 (or 1 mile), then the received signal P_r can be predicted as

$$P_r = P_{d_f} - \Gamma_1 \log \frac{d_1}{d_f} - \Gamma_2 \log \frac{d}{d_1} + G_{effh}$$

$$= P_{d_1} - \Gamma_2 \log \frac{d}{d_1} + G_{effh} \tag{4.3.2.1.2}$$

where Γ_2 = the best-fit slope in the range from d_1 to r_0, where $d_1 < r_0$ (r_0 is usually equivalent to 1 mile);
G_{effh} = effective height gain; and
d_f = near-in distance in meters.

The lines Γ_2 of Eq. (4.3.2.1.2) and γ of Fig. 4.3.1.1a intersect at a point within a range from d_1 to r_0 (at 1 mile). The received signal at d_1 is P_{d_1}. Hence, the received signal strength in the range from d_1 to 1 mile is obtained.

C. Finding the intersection point of two slopes, Γ_1 and γ, at distance d_1 by letting $\Gamma_1 = \gamma$ in Fig. 4.3.1.1, then the distance d_1 can be found as shown below.

We can eliminate the path loss slope Γ_2 if we can find the intersection point of the slope Γ_1 and the slope γ at the distance d_1. The intersection point at the distance d_1 in the range $d_f < d_1 < r_0$ can be obtained by solving Eqs. (4.3.2.1.1) and (4.3.1.1a). We get the distance d_1 at which the intersection point of slopes Γ_1 and γ occurred:

$$\log d_1 = \frac{1}{\Gamma_1 - \gamma} [P_{df} - P_0 + \Gamma_1 \log d_f - \gamma \log r_0] \tag{4.3.2.1.3}$$

When the slopes Γ_1 and γ are known, the distance d_1 can be determined as follows:

$$d_1 = 10 \wedge \left[\left(\frac{1}{\Gamma_1 - \gamma} \right) \{ \Gamma_1 \log d_f - \gamma \log r_0 + P_{d_f} - P_0 \} \right] \tag{4.3.2.1.4}$$

4.3.2.2 Finding the Path Loss Slopes from the Measured Data

The data are stored in corresponding radial zones. The slopes and intercepts are calculated for each zone and each range, that is 0 to d_f, d_f to d_1, and d_1 to 1 mile. The area around the transmitter is typically divided into radial zones, and the zones are typically uniformly distributed. The Fresnel distance is then calculated for the central radial line of each zone.

First, if no measured data exist, then as soon as the predicted slopes γ of Eq. (4.3.1.1) and Γ_1 of Eq. (4.3.2.1.1) are determined, the distance d_1 can be calculated from Eq. (4.3.2.1.4). If measurement data do exist, the data are then distributed into corresponding zones. The data in each zone are then further separated and stored in three ranges: 0 to d_f, d_f to d_1, and d_1 to 1 mile. We may call each range the sub-zone.

We may either subtract diffraction loss or add effective antenna height gain in order to normalize the measured data in each subzone for a flat terrain. When the distance d between the base station and measured point is less than the near-in distance, only diffraction loss is subtracted. When the distance d is greater than the near-in distance, the diffraction loss is subtracted and effective height gain added. Diffraction losses and effective antenna height gain can be calculated at each measurement point. The prediction will start after the normalization of the measurement data.

The line fitting is taken on the normalized measured data in the 0 to d_f range, and we obtain slope Γ_0 and the power intercept P_{df} at the near-in distance d_f. Similarly, the slope Γ_1 can be found from the line fitting on the normalized measured data in the range from d_f to d_1. The slope Γ_2 can be found from the line fitting on the measured data from d_1 to 1 mile (r_0), as shown in Fig. 4.3.1.1 mostly time Γ_2 can be assumed by γ. Since it is possible that the lines Γ_0 and Γ_1 do not intersect at P_{df} at the distance d_f, we may need to modify the distance d_f to be the distance where lines intersect. Similarly, Γ_1 and γ generate a power intercept at d_1, at the second breakpoint between d_f and 1 mile.

The characteristics shown in Fig. 4.3.2.2.1 are the near-in distances after being adjusted, which are different in different radials. The coverage area of the adjusted near-in distance will not be a circle within 1 mile.

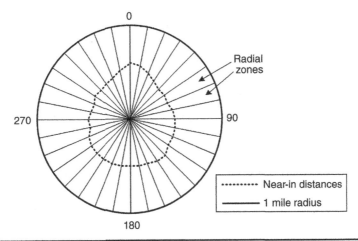

FIGURE 4.3.2.2.1   Radial zones for a specific cell.

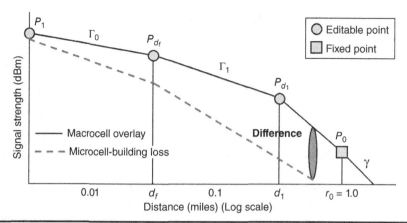

FIGURE 4.3.2.2.2   Delta between macrocells and microcells in an urban area.

For example, as shown in Fig. 4.3.2.2.2, there is a difference in prediction between the microcell and the macrocell coverage. Based on the Lee microcell model applied to the small size of cells, we can ensure that microcell coverage is more accurately defined in small cells. Based on the predicted microcell coverage, a better solution is to engineer the coverage, mitigate the interference, and indicate the handoffs. The integration of microcells and macrocells has served two goals. One is to ensure that without building data, the Lee macrocell model can provide enough flexibility to give accurate predictions for dense urban areas. The other goal is to ensure that the data can be integrated between the microcell and macrocell models so that radio planning can be done much more effectively.

Using the microcell prediction model to predict a small cell when a low-height base station antenna is served, the 1-mile signal strength contour around 360° is shown in Fig. 4.3.2.2.3. In general, the 1-mile signal strength contour coming from the macrocell

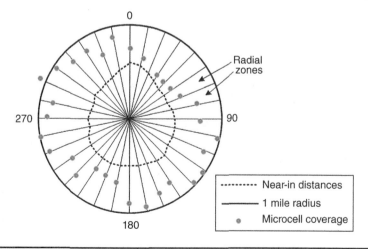

FIGURE 4.3.2.2.3   Macrocell versus microcell radial zones for a specific cell.

prediction model is more or less circular. The difference in using either of the models can make a big difference in dense urban areas when a low-height base station antenna is served in a small cell.

4.3.2.3 Characteristics of Measured Data

Measurement data were collected from various countries/cities and served for selected applications of various frequencies, various transmitting antenna heights, and different cell site parameters in a mobile communication environment. We must also consider certain buildings that have unique shapes, and individual heights and widths, that were built with various materials and different thickness.

The following configurations of drive routes were used to collect various measured data as shown in Fig. 4.3.2.3.1:

1. Measurement under LOS (R_1)

2. Zigzag kind of route (R_2)

3. Staircase kind of route (R_3)

4. Random routes (R_4)

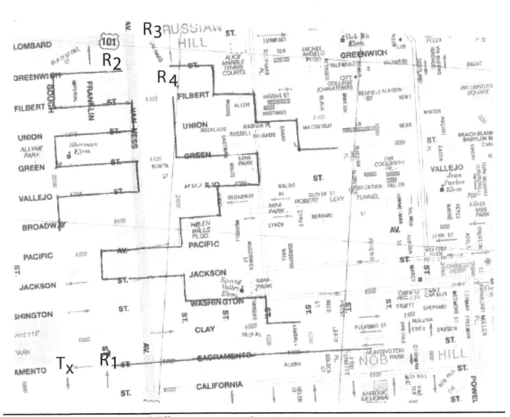

FIGURE 4.3.2.3.1 A sample of different routes on the street map.

The position of a mobile as well as a cell site needs to be extremely accurate. Using only GPS is not recommended in positioning urban microcell measurement data collection. Utilizing both ETAK and GPS is more accurate in collecting measured data. In cities where no ETAK data are available, road markers are used to keep track of the positions between the cell site and mobiles. Comparisons with collected measurement data demonstrate that the model does perform exceptionally well in varying mobile environments with different cell site parameters.

4.3.3   Validation of the Model: Measured versus Predicted

The integrated microcell and macrocell models are validated in both 900- and 1800-MHz frequency bands based on the collected measurement data. Some comparisons with collected data in different countries/cities are shown here for demonstration purposes. Figures 4.3.3.1 to 4.3.3.4 show the RSSI signals along the different radial distances and CDF in percentage of the deviation between measured and predicted. From the four figures (Figs. 4.3.3.1 to 4.3.3.4), their deviations in dB at 60% of data (CDF)

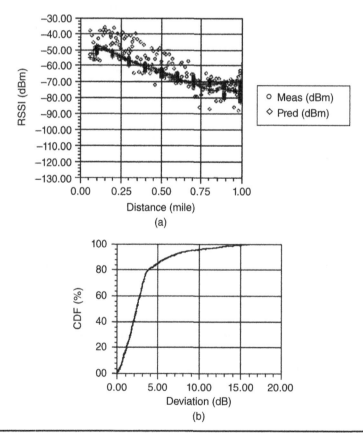

Figure 4.3.3.1   Route A_0044_01. (a) Measured and predicted versus distance. (b) CDF of difference between measured and predicted.

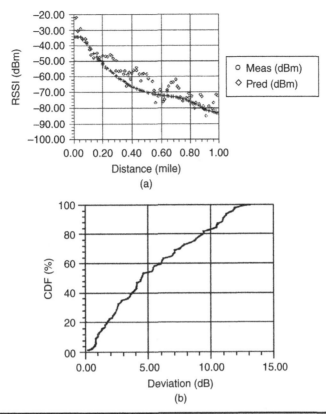

FIGURE 4.3.3.2   Route s_1900_02. (a) Measured and predicted versus distance. (b) CDF of difference between measured and predicted.

are 3, 6, 3.5, and 4.5, respectively. Thus, the predicted values follow the measured signal data fairly accurately, and the standard deviation is approximately 3 to 6 dB on average.

This prediction in the four figures utilizes a statistical approach and a multiple-breakpoint model to calculate the path loss. The method is simple and easy to implement and does not require building data. Validation was done in several different countries/cities, and the prediction results are very encouraging. One area that can still be enhanced is the specification of the average building height for each radial so that the multiple-breakpoint model can more accurately predict the near-in distance.

It is imperative that the two prediction models, macrocell and microcell, can work seamlessly to ensure a smooth transition of coverage. Understanding the characteristics of micro- and macrocells as well as the ability to model them provides a huge advantage to network deployment and engineering. This integrated micro- and macrocell model demonstrates another unique advantage of the Lee models. The integrated model has been used in many different cities, countries, and continents for the success of planning network deployment.

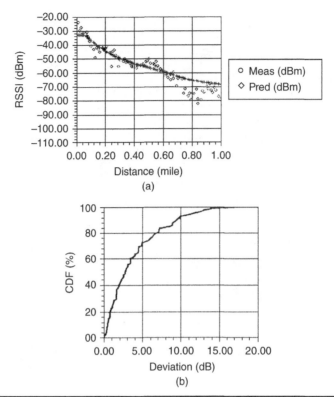

FIGURE 4.3.3.3   Route s_1900_02. (a) Measured and predicted versus distance. (b) CDF of difference between measured and predicted.

4.4   Tuning the Model for a Particular Area

4.4.1   Before Tuning the Lee Microcell Model

The purpose of feeding the field-measured data back to tune the model improves the accuracy of the model.

4.4.1.1   The Multiple-Breakpoint Model

Within Lee's microcell model, the prediction of path loss uses the multiple-breakpoint microcell prediction model with the terrain and building database. It is composed of three breakpoints, P_{df}, P_{d_1}, and P_0, as shown in Fig. 4.3.1.1.

The areas within the first and second breakpoints as well as the intersection point d_1 can be obtained from the measured data and are described in Secs. 4.3.1 and 4.3.2.1. The calculation of near-in distance is shown in Sec. 4.2.1.1.

4.4.1.2   The Algorithms of Integrating the Two Models before Tuning

The algorithms of integrating the two models are described in Sec. 4.3.1. These algorithms are applied without using the measured data. The enhancement is introduced by

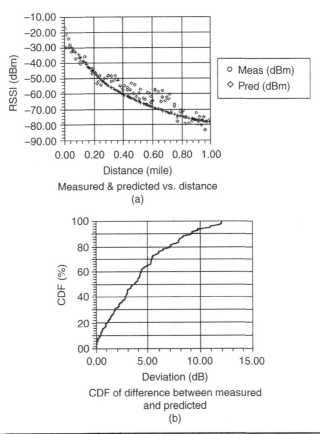

Figure 4.3.3.4   Route E_1937_02. (a) Measured and predicted versus distance. (b) CDF of difference between measured and predicted.

taking the measurement data into the integration model and finding the two breakpoints on the adjusted path loss curve.

4.4.2   The Tuning Algorithm of the Lee Model

4.4.2.1   Tuning the Lee Microcell Model in the First Two Ranges

Next, we will discuss the algorithm based on the optimization[9] to fine-tune the Lee microcell model. The optimization minimizes the difference between the measurement and the prediction by applying the joint local optimization (JLO) procedure to fine-tune the model.

4.4.2.1.1   Linear Optimization   For applying the Lee microcell model within 1 mile, there are two different ranges in the three breakpoints from the equations stated previously:

$$P_r = P_{df} - \Gamma_1 \log \frac{d}{d_f} \qquad (4.4.2.1.1)$$

and

$$P_r = P_{df} - \Gamma_1 \log \frac{d}{d_f} - \Gamma_2 \log \frac{d}{d_1 + d_f} + G_{effh} \qquad (4.4.2.1.2)$$

Let's denote R_m as the measured result from the mobile for each local point and ζ as the difference between our predicted result and the measured result at each local point. Then we have

$$\zeta = |R_m - P_r| \qquad (4.4.2.1.1)$$

4.4.2.1.2 Matrix Optimization In order to match JLO procedure, let us make up our matrices y, ϑ, and A. Note that there is no physical meaning to these matrices. With these newly formed matrices, we can apply the JLO procedure. In matrix y, the first matrix contains only the constant values, such as measured data and distance; in matrix ϑ, the second one contains only the parameters, which can be fine-tuned and have direct effects to the performance of our model; in matrix A, the third one contains only the coefficients related to the parameters, which also are constants. Then we have the following:

1. For the case of the first range

   Three matrixes are

   $$y = [R_{m1} - P_{df} - L, R_{m2} - P_{df} - L, \ldots, R_{mN} - P_{df} - L]^T$$
   $$\vartheta = [\Gamma_1]$$
   $$A = \left[\log \frac{r_1}{d_f}, \log \frac{r_2}{d_f}, \ldots, \log \frac{r_N}{d_f} \right] \qquad (4.4.2.1.2)$$

2. For the case of the second range

   Three matrixes are

   $$y = [R_{m1} - P_{df} - L - G_{effh}, R_{m2} - P_{df} - L - G_{effh}, \ldots, R_{mN} - P_{df} - L - G_{effh}]^T$$
   $$\vartheta = [\Gamma_2]$$
   $$A = \left[\log \frac{r_1}{d_f}, \log \frac{r_2}{d_f}, \ldots, \log \frac{r_N}{d_f} \right] \qquad (4.4.2.1.3)$$

Then we can rewrite Eq. 4.4.2.1.1 into a matrix form as follows:

$$\zeta = y - A\vartheta \qquad (4.4.2.1.4)$$

By now, all we are trying to do is find an optimized solution for ϑ to minimize ζ in Eq. (4.4.2.1.4).

4.4.2.2 Tuning the Lee Microcell Model in the Last Range

The optimization of the predicted results in the first two ranges is described in Sec. 4.4.2.1. In the last range, the third range in the non–near-in zone scenario, the effects of building obstructions are modeled with a correction term L_B (see Eq. (4.2.1.2.3)).

The value of this correction is the loss proportional to the length of the radio path obstructed by buildings. To clarify the loss due to buildings, a scenario considers the radio wave propagating from the transmitter to the receiver, as shown in Fig. 4.4.2.2.1.

The total normalized block length x of the obstructed radio path in Fig. 4.4.2.2.1 is given as

$$x = \frac{(d_A + d_B)}{d_{Bref}} \qquad (4.4.2.2.1)$$

where d_{Bref} is a suitably chosen reference distance. From the study reported in Kostanic et al.,[9] $d_{Bref} = 10$ ft, which is an adjustment factor derived from the measured data. Substituting the normalized length into Eq. (4.4.2.3.1), the building loss factor is calculated as

$$L_B = f(x) \qquad (4.4.2.2.2)$$

where $f(x)$ is a monotonically increasing function of x and can be found in the following equation. Within the original description of Lee microcell model,[1] the format of $f(x)$ is shown in Eq. (4.2.1.1.3). For the implementation of JLO suggested in this section, the form of the function is provided as

$$f(x) = a + bx + c\,\log(x) \qquad (4.4.2.2.3)$$

The coefficients a, b, and c are determined empirically through the analysis of measured data. The three matrixes then can be shown as

$$y = [y_{m1} - P_t - G_r - G_t,\ y_{m2} - P_t - G_r - G_t,\ ...,\ y_{mN} - P_t - G_r - G_t]^T$$
$$\vartheta = [L_b, r]$$
$$A = \begin{bmatrix} 1 & \log\dfrac{d_1}{d_f} \\ \vdots & \vdots \\ 1 & \log\dfrac{d_N}{d_f} \end{bmatrix} \qquad (4.4.2.2.4)$$

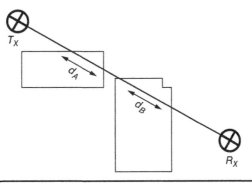

Figure 4.4.2.2.1   Illustration of the building blocks for finding the correction factor.

For all the matrixes for each case or scenario we list above, we can simply follow the JLO procedure to fine-tune the model. In this approach, the values of the model parameters are obtained through minimization of the cost function given by

$$J(\zeta) = \frac{1}{2}\zeta^T \zeta \qquad (4.4.2.2.5)$$

where

$$\xi = (A^T A)^{-1} A^T y \qquad (4.4.2.2.6)$$

The JLO method solves for all parameters of the model simultaneously and generally provides the best fit to the measured data. However, although numerically optimal, the values for the parameters may not be physically intuitive and may vary substantially from cell to cell.

4.4.3   Verification of the Lee Model

The work of fine-tuning the Lee model was published by Kostanic et al.[9] as described in Sec. 4.4.2 and the result is discussed here. Kostanic used a different approach to integrate the measured data with the Lee microcell model. The Lee microcell model and tuning techniques have been implemented by the WIZARD RF propagation software tool of Agilent technologies.[19]

Applying the Lee microcell model and the JFO tuning method, the range of standard deviations was shown to be between 5.3 and 6.6 dB. These low values from the measurement errors rank better than the typical standard deviations observed in dense urban environments, which are often as high as 8 to 10 dB.

Another illustration of the accuracy of the model is presented in Figs. 4.4.3.1 and 4.4.3.2. The predictions are calculated using only initial default microcell propagation parameters and no building losses. These predictions are used for comparison with the predictions from the tuned model. Figure 4.4.3.1 shows a scatter plot of predicted versus measured points for the default (no building losses added) and the tuned models. The ideal zero-mean, no-scatter line is shown as reference. The default model underpredicts by 18 dB when compared with the measured data. As shown in Fig. 4.2.1.2.2, the building loss L_B remains constant as 18 dB after a building block exceeds 250 ft. The basic Lee microcell prediction model in Sec. 4.2 describes the calculation of prediction in microcell environments by adding the building block loss to the default LOS loss. Therefore, by adding 18 dB to the prediction of the default model if the building block data are not available, the results are the same as the prediction from the tuned model, as shown in Fig. 4.4.3.1. The tuned model introduced by the JLO tuning technique also shows the tight fit to the reference line. Both predictions are fairly accurate, but the basic Lee microcell model is much simpler to use than the tuned Lee model by using JLO procedures. Although the prediction from the default model plus the loss due to the building block curve is just as accurate as from the tuned model, it may not have as sound a calculation as the tuned model does. However, the tuned model proves that the default model plus the building block curve, which is the basic Lee microcell model, is the right approach to making the prediction.

A similar comparison can be observed from the histogram of measurement errors shown in Fig. 4.4.3.2. Histograms for both the default model and the tuned Lee microcell model are shown. The difference of 20 dB between two prediction results is the loss due to building blocks, which the default model does not include. Removal of the bias

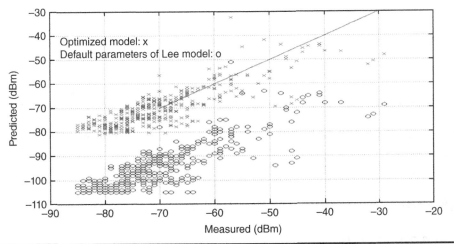

FIGURE 4.4.3.1 Scatter plot of optimized and default model predictions versus measured data.

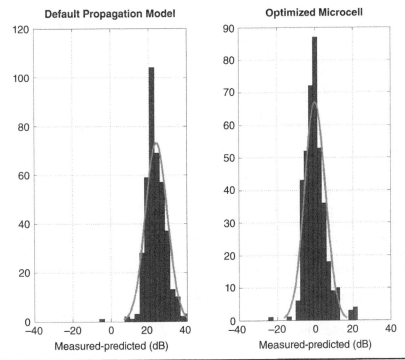

FIGURE 4.4.3.2 Histograms of measurement errors for default and optimized microcell models.

and tighter grouping around zero are shown for the two optimized models. The histo-gram shapes of two predictions are the same, but the difference of about 20 dB is the loss due to the building blocks, as mentioned before. When using the Lee microcell prediction model instead of the Lee default model to compare with the tuned Lee model, the results shown in the two figures will be almost identical.

The performance demonstrated in Figs. 4.4.3.1 and 4.4.3.2 clearly demonstrates the importance of measurement integration and the Lee microcell model as a suitable can-didate for propagation prediction in dense urban environments. When the building data are available, the building block loss L_B will be obtained from the curve shown in Fig. 4.2.1.1.2. When the building data are not available, simply set the block loss $L_B \approx 20$ dB in the Lee microcell model to get a fairly accurate prediction.

4.5   Other Microcell Prediction Models

4.5.1   Introduction

In this section, several approaches of modeling propagation in a microcell environment are introduced. Both theoretical and empirical methods have been used, and ray-tracing techniques have also been investigated. The multipath effects in the prediction results are very important in urban areas, depending on the relative height of the base station antenna and the surrounding buildings.

Generally, all existing microcellular models are valid only in flat urban areas, and little attention has been given to the influence of terrain variations; the effects of vegeta-tion have also been largely ignored. Both of these aspects are important and need to be incorporated specifically into the ray-tracing models.

For the practical application of microcell propagation models, an important trade-off is between the accuracy of the prediction and the computational speed with which the prediction can be obtained. Microcells often have to be deployed in the field very quickly, with little engineering effort. The guiding procedures and rapid statistical planning tools are very important. A very high resolution topographic database is required.

These microcell models are used and run at the start of a system deployment, and then used to create a unique set of predictions and recommendations for deployment. The real-time processes are operating within the base station, with assistance from the mobiles, which can optimize and are used by the system to assess the likely system parameters, such as transmit powers, antenna patterns, and channel assignments, on an ongoing basis. This section describes models other than Lee model that can also yield reasonable prediction accuracy when their parameters are tuned against measurements.

4.5.2   Empirical (Path Loss) Models

Normally, microcell models are based on statistical techniques. The Lee model is one kind of statistical model.

Empirical models are statistical models with lower computational overhead but less accuracy in small urban cells and subjective clutter classification. The parameters of the model are the properties of the buildings along the radio path between the transmit-ter and the receiver. These models are not as accurate for smaller cells under 1 km. However, these models should not require any calibration and are more suited for the urban macro- and microcells.

4.5.2.1 Dual-Slope Model

In a dual-slope model, two separate path loss exponents are used to characterize the propagation, together with a breakpoint distance of a few hundred meters, at which propagation changes from one regime to the other. In this case, the path loss is modeled as

$$\frac{1}{L} = \begin{cases} \dfrac{k}{r^{n_1}} & \text{for } r \leq r_b \\[2ex] \dfrac{k}{\left(\dfrac{r}{r_b}\right)^{n_2} r_b^{n_1}} & \text{for } r > r_b \end{cases} \tag{4.5.2.1.1}$$

or the path loss L in decibels

$$L = \begin{cases} 10n_1 \log\left(\dfrac{r}{r_b}\right) + L_\gamma & \text{for } r \leq r_b \\[2ex] 10n_2 \log\left(\dfrac{r}{r_b}\right) + L_\gamma & \text{for } r > r_b \end{cases} \tag{4.5.2.1.2}$$

where L_γ is the reference path loss at $r = r_b$, r_b is the breakpoint distance between 100 and 500 m, n_1 is the path loss exponent for $r \leq r_b$, and n_2 is the path loss exponent for $r > r_b$.

Typical values for the path loss exponents are found by measurement to be around $n_1 = 2$ and $n_2 = 4$, with breakpoint distances of 200 to 500 m. However, these values vary between different measurements, as discussed in.[20-22]

This model is very similar to the Lee microcell-to-macrocell integration model. The Lee model has more flexibility with multiple breakpoints as well as different radial zones so that the granularity of data points can be further fine-tuned.

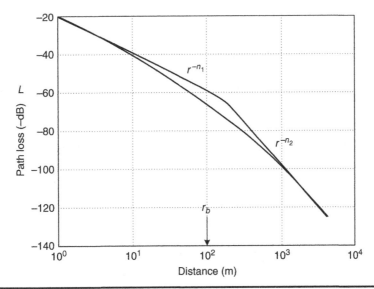

FIGURE 4.5.2.1.1 Dual-slope empirical loss models. $n_1 = 2$, $n_2 = 4$, $r_b = 100$ m, and $L_\gamma = 20$ dB.

4.5.2.2   Kaji and Akeyama Microcellular Model

In this model, two dominant waves are considered: the waves that propagate over the tops of buildings and the waves that propagate along road paths. They are called row waves: "building diffracted waves" (BD waves) and the along-road path (RG).

BD waves propagate in all directions, while RG waves propagate in a linear fashion like plant roots. The total received power is the summation of received power from both BD waves and RG waves. The building distribution in test area is shown in Fig. 4.5.2.2.1. Figure 4.5.2.2.2 shows the following excess path losses for BD waves and RG waves:

$*Ld_1$ = diffraction loss caused by average height buildings at transmitting point,

$*Ld_2$ = diffraction loss caused by average height buildings at receiving point,

$*Ld_3$ = shielding loss caused by buildings of higher than average height,

$*Lg_1$ = shielding loss caused by buildings facing road, and

$*Lg_2$ = diffraction loss at corners of roads.

The strength of waves depends on factors such as base station antenna height, frequency, propagation distance, and building distribution. For BD waves, diffraction losses are large, and the path loss increases noticeably even at small distances. For RG waves, path loss is very small when the road is straight but increases abruptly at the corners of roads. Therefore, RG waves are stronger than BD waves in areas near the transmitting antenna, and BD waves are stronger than RG waves in areas far from the transmitting antenna.

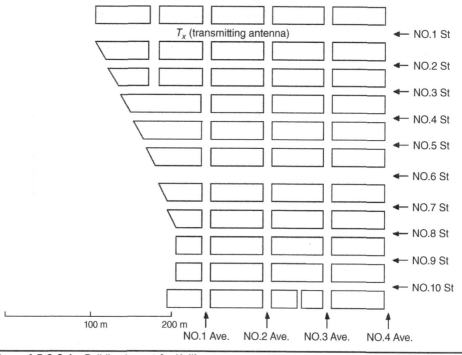

FIGURE 4.5.2.2.1   Building lay out for Kaji's measurement.

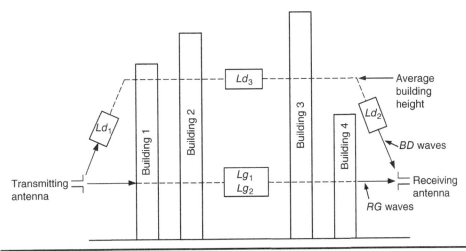

FIGURE 4.5.2.2.2   Multipath propagation model in a high-density urban area.

In the microcell environment, a street-guided wave exists, at least in the immediate vicinity of the transmitter. From a result of their measurements, Kaji and Akeyama[23] suggested an empirical model in the form:

$$S = -20 \log_{10} \left\{ d^a \left(1 + \frac{d}{g} \right)^b \right\} + c \qquad (4.5.2.2.1)$$

for base antenna heights in the range of 5 to 20 m and for distances between 200 m and 1 km. In this equation,

S = the signal level (dB_μV),
d = the distance from the transmitter,
a = the basic attenuation rate for short distances (approximately 1),
b = the additional attenuation rate for distances beyond the turning point,
g = the distance corresponding to the turning point, and
c = the offset factor.

The model has two limiting cases. In case 1, for distances significantly less than g, the attenuation is such that

$$S = -20 \log d^a + c \qquad (4.5.2.2.2)$$

In case 2, for distances greater than g, the attenuation falloff rate tends to be such that

$$S = -20 \log d^{(a+b)} + c + \text{constant} \qquad (4.5.2.2.3)$$

This model was fitted from their measured results at 900 MHz; the values for the various coefficients are given in Table 4.5.2.1.

Antenna Height (m)	a	b	G	c	Sum of Errors Squared
5	1.15	−0.14	148.6	94.5	309
9	0.74	0.27	151.8	79.8	246
15	0.20	1.05	143.9	55.5	577
19	−0.48	2.36	158.3	37.3	296

TABLE 4.5.2.2.1   Results of Fitting the Proposed Propagation Model to Experimental Results at 900 MHz Using a Least Squares Regression Procedure

4.5.3 Physical Models

The empirical models in the previous section can provide reasonable results. However, it is desirable to understand the physical mechanisms underlying them to gain insight and to create physical models with potentially greater accuracy earlier. Depending on the building distributions and antenna location and height, measurement data vary drastically. This is one of the characteristics of microcell propagation.

4.5.3.1 Street Canyon Model[24,25]

This is also known as the urban canyon model. This street canyon model can be also called four-ray model: direct ray, ground reflection, wall 1, and wall 2 reflection. It can be called the six-ray model if double building wall reflections are also taken into account. All the buildings around the mobile can reflect the transmitted signal. A display utilizing the model is shown in Fig. 4.5.3.1.1. It assumes that both the mobile and the base stations are located on a long, straight street and that the buildings on both sides are acting like plane walls.

The characteristics of a four-ray model consider the direct path and all three singly reflected paths from the walls and the ground. The structure of the model follows the two-ray model described in Sec. 1.9.1.3, but the reflections from the vertical building walls involve the Fresnel reflection coefficients, not the reflection coefficient from the ground. The four-ray paths are depicted in Fig. 4.5.3.1.1, and a typical result is shown in Fig. 4.5.3.1.2, in which the multipath fading is more rapid. The average path loss exponent is close to 2 over the whole range up to 10^5 m.

Six possible ray paths are illustrated in Fig. 4.5.3.1.1. They tend to include reflections from more than two surfaces of the walls. These wall reflections are typically weakening the signal.

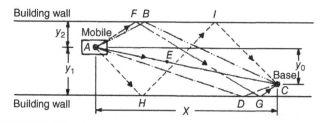

FIGURE 4.5.3.1.1   Street canyon model of LOS microcellular propagation.

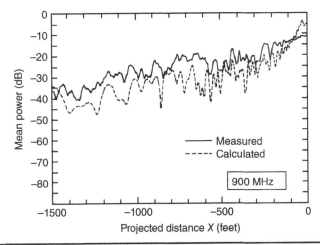

FIGURE 4.5.3.1.2 Comparison of measured power at 900 MHz on Lexington Avenue and calculated mean power based on the six-ray model.

In today's cellular systems, the signal amplitude (especially in urban environments) varies randomly with distance due to multiple scatterings between the mobile and the base. The signal envelope usually follows a Rayleigh distribution,[25] implying many scatterings of comparable strength. Whenever the subscriber (mobile or portable user) and base station are within LOS, the direct path will tend to dominate in strength, and the envelope distribution will be more typically Rician.[26,27] With the increased proximity of the mobile to the base station in LOS microcells, we expect the scatter components of the signal to be weak relative to the direct ray.

Four additional specular wall-reflected rays adequately represent propagation in urban areas. Consequently, the mean power falloff at 900 MHz decays faster than $1/r^2$ in rural environments for distances exceeding 1000 ft, and the asymptotic $1/r^2$ dependence is essentially reached beyond 1500 ft. In urban and semiurban environments, we observe that a $1/r^2$ dependence out to distances exceeding 2000 ft at this frequency. At 11 GHz, we observe a $1/r^2$ dependence of the measured mean power and calculated power in both rural and urban environments out to distances well over 5000 ft.

4.5.3.2 Random Waveguide Model

A random waveguide model is applied to large metropolitan areas that have tall buildings, and where both the transmitting and the receiving antennas are located below the rooftops. The city streets act as a type of wave-guiding structure for the propagating signal.[28,29] Theoretical analysis of propagation in a city street modeled as a 3D multislit waveguide has been proposed.

A new multislit waveguide model with randomly distributed slits, such as gaps between buildings, and screens, such as walls for low antenna heights, in comparison with surrounding building heights along the street level, is presented here.

At long distances, the street canyon LOS models suggest a path loss exponent of 4. This model implicitly considers the multiple reflections on building walls, multiple diffraction on their edges, and reflections from the road surface.

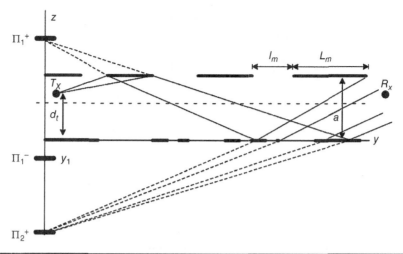

Figure 4.5.3.2.1 A 2D diagram of the waveguide in the *zy* plan. The coordinates of source are $y = y_1$ and $z = d$, and "*a*" is the street width.[28]

Figure 4.5.3.2.1 shows the street in plain view for this model. The distances between the buildings (called slits for the waveguide model) are defined as l_m, where m = 1, 2, 3, ... The buildings on the street are assumed randomly distributed nontransparent screens with lengths L_m with the wave impedance *ZEM* given by

$$Z_{EM} = \frac{1}{\sqrt{\varepsilon}}, \quad \varepsilon = \varepsilon_r - j\frac{4\pi\sigma}{\omega} \tag{4.5.3.2.1}$$

where ε_r is the relative permittivity of the walls and σ is their conductivity in Sm^{-1}.

The model uses a geometrical theory of diffraction calculation to apply on the rays reflected from the walls and diffracted from the building edges. In this approximation, the resulting field can be considered as a sum of the fields arriving at the mobile at a height h_m from the virtual image sources Π_1^+, Π_1^-, and Π_2^+, as shown in Fig. 4.5.3.2.1.

The full field inside the street waveguide can be presented as a sum of the direct field from the source and rays reflected diffracted from the building walls and corners. In order to calculate the full field from the source, we substitute for each reflection from the walls an image source Π_n^+ (for the first reflection from the left-hand walls of the street waveguide) and Π_n^- (for the first reflection from the right-hand walls), where n is the number of the reflections.

The path loss is then approximately[28]

$$L = 32.1 - 20 \log |R_n| - 20 \log \left[\frac{1 - (M |R_n|)^2}{1 + (M |R_n|)^2} \right]$$

$$+ 17.8 \log r + 8.6 \left(\|\ln M|R_n\| \frac{[(\pi n - \varphi_n / a)]r}{\rho_n^{(0)}} \right) \tag{4.5.3.2.2}$$

where it is assumed that $L_m \gg \lambda$ and $l_m \gg \lambda$.

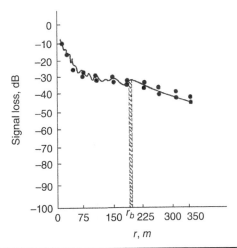

FIGURE 4.5.3.2.2   Field intensity versus distance *r* from transmitter. Solid curve is the calculated data and solid circles are the experimental data.

Here, M is the parameter of brokenness, defined as

$$M = \frac{L_m}{L_m + l} \tag{4.5.3.2.3}$$

so that $M = 1$ for an unbroken waveguide without slits or separations between buildings p is defined as

$$\rho_n^{(0)} = \sqrt{k^2 + (\pi n / a)^2} \tag{4.5.3.2.4}$$

where k is the wave number and n_r represents the number of reflections and $n_r = 0, 1, 2, 3, \ldots$ R_{nr} is the reflection coefficient of each reflecting wall given by

$$R_{nr} = \frac{(\pi n + j \, | \ln M | / a - k Z_{EM})}{(\pi n + j \, | \ln M | / a + k Z_{EM})} = |R_{nr}| e^{j\varphi_n} \tag{4.5.3.2.5}$$

When the street width is larger than the average building heights and both antenna heights, beyond the breakpoint the field intensity attenuates exponentially. This law of attenuation is close to that obtained experimentally in most measurements, where the attenuation mode of field intensity beyond the breakpoint was r^{-q}. The breakpoint happens around 100 to 500 m as shown in Fig. 4.5.3.2.2.

The waveguide multislit model is in agreement with urban canyon model, and predicts the propagation characteristics in LOS conditions along straight streets.

This model is one kind of *physical-statistical* model. It combines physical propagation mechanisms with a statistical analysis of the environment.

4.5.4   Non-LOS Model

The non-LOS (NLOS) empirical model is also known as the diamond-shaped model because the shape of the signal coverage coming out from this model is like a diamond. The NLOS empirical model is derived from a series of microcellular urban radio propagation measurements carried out in the metropolitan areas of the United States,[30-37] where the street orientation is very much a grid-like structure.

4.5.4.1 Propagation Mechanisms in NLOS

In LOS, when the LOS path in a microcell is blocked, signal energy can propagate from the base to the mobile via mainly three mechanisms:

- Diffraction over building rooftops
- Diffraction around building edges
- Reflection and scattering around walls and the ground

Figure 4.5.4.1.1 shows the plan view of buildings arranged in a regular grid structure.[38] Path A shows a building edge created by the motion of a mobile across the shadow boundary. Because this building is isolated, diffraction in the shadow region takes place, and the signal strength drops very rapidly with increasing distance. The other path to reach the mobile from the base station is through path B. The diffraction path propagates through the building roof, no reflected signal exists, and the rooftop-diffracted signal begins to dominate due to the large number of diffractions and causes interference between co-channel microcells. Path C describes the reflection path of the radio wave. In path C, the building is now surrounded by other buildings, which act as reflecting surfaces. In this case, the reflected ray is much stronger than the diffracted ray, so the signal strength remains strong over much longer distances.

In Fig. 4.5.4.1.1, there are many different paths. The short paths A and B are single reflections or diffractions and are main sources of signal strength. The long path C after four sequential reflections becomes very weak at the spot, and the rooftop-diffracted path D then dominates. This variation in propagation mechanism with distance is an additional parameter in the two path loss slopes in the empirical models.

The resultant coverage area in NLOS microcells generally is broadly diamond shaped, as shown in Fig. 4.5.4.1.2. From the measurement, the curved boundaries of signal strength forming the diamonds indicate that the dominant mechanism of propagation into side streets is diffraction rather than reflection.[39]

In more realistic environments, where buildings are not regular in size, advanced planning techniques must be applied, particularly when frequencies are shared between microcell and macrocell layers.[40-42]

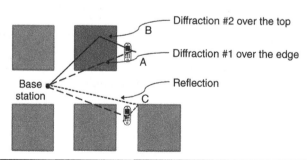

FIGURE 4.5.4.1.1   Street geometry where diffraction dominates.

4.5.4.2 Recursive Model[43]

This is a mathematical model for calculating path losses along streets surrounded by buildings, that are much taller than the height of the antennas. The method is recursive and suited for ray- or path-tracing techniques. The procedure works in cases of perpendicular street crossings, any arbitrary angles in street crossings, and bent streets with linear segments. The method is reciprocal, computer efficient, and simple to use. Before using the procedure, one needs some basic information regarding street orientation and transmitter location. Then the model can handle these cases by choosing appropriate parameter values.

This model is an intermediate model between an empirical model and a physical model. It uses the concepts of GTD/UTD for NLOS propagation at the street intersections where diffraction and reflection points exist. The model breaks down the radio path between the base station and the mobile into a number of segments interconnected by nodes.

An *illusory* distance for each ray path is calculated according to the recursive expressions as

$$k_j = k_{j-1} + d_{j-1} \times q(\theta_{j-1})$$
$$d_j = k_j \times s_{j-1} + d_{j-1}$$

(4.5.4.2.1)

where the initial values of k_0 and d_0 are

$$k_0 = 1 \qquad d_0 = 0$$

(4.5.4.2.2)

In Fig. 4.5.4.2.1, solid lines indicate the path of the propagating wave between the transmitter T_x and the receiver R_x. The path change directions at the nodal point, $j = 1$ and $j = 2$, with the angles θ_1 and θ_2. The illusory distance d_n is calculated for each straight line. In Fig. 4.5.4.2.1, the number n of straight lines along the ray path is 3. The physical distance r_j is the line segment in meters following the jth node.

The path loss is determined by the parameter q_j. The angle at which the path turns at node j is θ_j (degrees). When angle θ_j increases, the illusory distance d_n is increased according to the following equation:

$$q_j(\theta_j) = \left(\frac{\theta_j q_{90}}{90}\right)^\nu$$

(4.5.4.2.3)

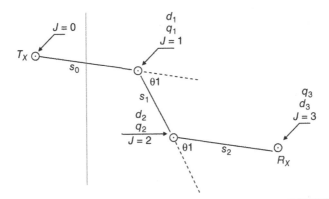

FIGURE **4.5.4.2.1** Recursive model—example of street orientation.

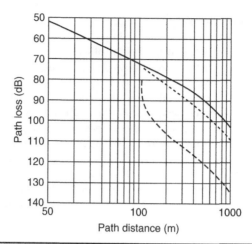

FIGURE 4.5.4.2.2   Loss as a function of the distance from the T_x, $\theta_j = 0$, solid line; $\theta_j = 10$, dotted line; $\theta_j = 90$ dashed line.

The calculated path loss for a configuration according to Fig. 4.5.4.2.1, with $s_0 = 100$ m, $s_1 = 200$ m, $0_1 = 90$ and $0_2 = 0,10$ and 90 degrees is displayed in Fig. 4.5.4.2.2. The parameter values $q_{90} = 0.5$ and $v = 1.5$ is chosen from Eq. (4.5.4.2.3.1). As mentioned earlier, the distance dependency of the path loss has two-slope behavior. The first slope is usually about 2. After a certain breakpoint distance, x_{brk},

$$D(x) = \begin{cases} \dfrac{x}{x_{brk}}, & \text{for } x > x_{brk} \\ 1, & \text{for } x \leq x_{brk} \end{cases} \qquad (4.5.4.2.3.4)$$

Then Eq. (4.5.4.2.3.4) is extended according to

$$L_{dB}^{(n)} = 20 \cdot \log\left[\frac{4 \cdot \pi \cdot d_n}{\lambda} D\left(\sum_{j=1}^{n} s_{j-1}\right)\right] \qquad (4.5.4.2.3.5)$$

where s_j is the physical distance along the path

Equation (4.5.4.2.4) yields a dual-slope path loss curve with a path loss exponent of 2 for distances less than the breakpoint r_b and 4 for longer distances.

The recursive model is simple and applicable to predicting the microcell signal coverage. The major microcell propagation effects, such as dual-slope path loss exponents and street-corner attenuation, are covered in the model.

4.5.5   ITU-R P.1411 Model[44]

This model is a popular model issued by the International Telecommunication Union (ITU). It consists two parts. One deals with LOS paths and the other with NLOS paths. The empirical formulas of this model are explained in this section.

4.5.5.1   Definition of Propagation Situations

In ITU-R P1411, base station BS_1 is mounted above rooftop level in a small macro-cell environment. Base station BS_2 is mounted below rooftop level in a micro- or

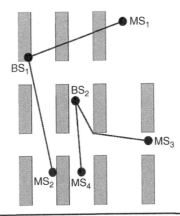

Figure 4.5.5.1.1   Typical propagation situations in urban areas (top view).

picocell environment. In these environments, the signal propagation is occurs mainly within street canyons, as shown in Fig. 4.5.5.1.1. Four situations are considered for the base-to-mobile links:

1. BS_1 propagates over the rooftop to MS_1.
2. BS_1 propagates passing rows of buildings to MS_2.
3. BS_2 propagates passing only one row of buildings to MS_4.
4. BS_2 propagates to MS_3. from diffracted waves.

For mobile-to-mobile links, both ends of the link are below rooftop level, and the models relating to BS_2 may be used.

4.5.5.2 LOS Condition
Under the LOS condition, the model can be applied to two types of LOS paths: the paths BS_1 to MS_2 and BS_2 to MS_4, shown in Fig. 4.5.5.1.1.

The basic transmission loss from the signal propagation in the UHF range can be characterized by two slopes and a single breakpoint.[45] An approximate lower bound is given by

$$L_{LOS,l} = L_{bp} + \begin{cases} 20\log_{10}\left[\dfrac{d}{R_{bp}}\right] & \text{for } d \le R_{bp} \\[2ex] 40\log_{10}\left[\dfrac{d}{R_{bp}}\right] & \text{for } d > R_{bp} \end{cases} \qquad (4.5.5.2.1)$$

where R_{bp} is the breakpoint distance and is given by

$$R_{bp} \approx \frac{4h_b h_m}{\lambda} \qquad (4.5.5.2.2)$$

where λ is the wavelength (m). Equation (4.5.5.2.2) is adapted from the near-in distance d_f shown in Eq. (4.2.1.1.4). An approximate upper bound is given by

$$L_{LOS,u} = L_{bp} + 20 + \begin{cases} 25 \log_{10}\left[\dfrac{d}{R_{bp}}\right] & \text{for } d \le R_{bp} \\ \\ 40 \log_{10}\left[\dfrac{d}{R_{bp}}\right] & \text{for } d > R_{bp} \end{cases} \qquad (4.5.5.2.3)$$

L_{bp} is a value for the basic transmission loss at the breakpoint, defined as:

$$L_{bp} = \left| 20 \log_{10}\left[\dfrac{\lambda^2}{8\pi h_b h_m}\right] \right| \qquad (4.5.5.2.4)$$

4.5.5.3 NLOS Models

Usually under the NLOS condition, NLOS signals can arrive at the BS or MS by diffraction mechanisms or by multipath, which may be a combination of diffraction and reflection mechanisms. There are two cases to be treated in this model: the base station antenna above the rooftop and the base station antenna under the rooftop.

The models are valid under the following conditions:

h_b = 4 to 50 m
h_m = 1 to 3 m
f = 800 to 2 000 MHz
d = 20 to 5 000 m

4.5.5.3.1 Propagation over Rooftops—NLOS[44]   The typical NLOS case (link BS_1 to MS_1, shown in Fig. 4.5.5.3.1.1(a) is a view in a vertical plane) is described by Fig. 4.5.5.3.1.1(b) in a top view. This case is called NLOS1.

The relevant parameters for Fig. 4.5.5.3.1.1(a) are the following:

h_r = average height of buildings (m),
w = street width (m),
b = average building separation (m),
φ = street orientation with respect to the direct path (degrees),
h_b = BS antenna height (m),
h_m = MS antenna height (m),
l = length of the path covered by buildings (m), and
d = distance from BS to MS.

In the NLOS1 case (see Fig. 4.5.5.3.1.1) for rooftops of similar height, the loss between two isotropic antennas is expressed as the sum of (1) freespace loss, L_{bf}; (2) the diffraction loss from rooftop to street, L_{rts}; and (3) the reduction due to multiple screen diffraction past rows of buildings, L_{msd}.

In this model, L_{bf} and L_{rts} are independent of the BS antenna height, while L_{msd} is dependent on whether the base station antenna is at, below, or above the building height.

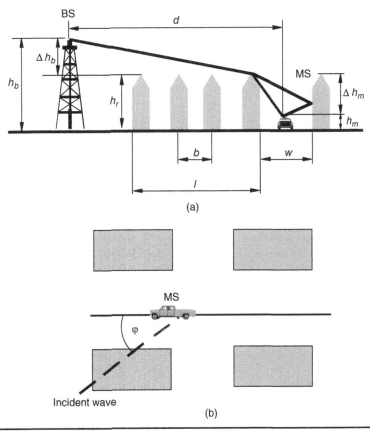

FIGURE 4.5.5.3.1.1 Definition of parameters for the NLOS1 case.

The NLOS path loss L_{NLOS1} can be expressed as

$$L_{NLOS1} = \begin{cases} L_{bf} + L_{rts} + L_{msd} & \text{for } L_{rst} + L_{msd} > 0 \\ L_{bf} & \text{for } L_{rts} + L_{msd} \leq 0 \end{cases} \qquad (4.5.5.3.1.1)$$

The free space loss is given by

$$L_{bf} = 32.4 + 20 \log_{10}\left(\frac{d}{1000}\right) + 20 \log_{10}(f) \qquad (4.5.5.3.1.2)$$

where d = path length (m) and f = frequency (MHz).

The diffraction loss L_{rts} describes the loss after the wave propagating along the multiple-screen path and diffracted into the street where the mobile station is located. It takes into account the width of the street and its orientation:

$$L_{rst} = -8.2 - 10 \log_{10}(w) + 10 \log_{10}(f) + 20 \log_{10}(\Delta h_m) + L_{ori} \qquad (4.5.5.3.1.3)$$

$$L_{ori} = \begin{cases} -10 + 0.354\varphi & \text{for } 0° \leq \varphi < 35° \\ 2.5 + 0.075(\varphi - 35) & \text{for } 35° \leq \varphi < 55° \\ 4.0 - 0.114(\varphi - 35) & \text{for } 55° \leq \varphi < 90° \end{cases} \qquad (4.5.5.3.1.4)$$

where

$$\Delta h_m = h_r - h_m \qquad (4.5.5.3.1.5)$$

L_{ori} is the street orientation correction factor, which takes into account the effect of rooftop-to-street diffraction into streets that are not perpendicular to the direction of propagation (see Fig. 4.5.5.3.1.1[b]); φ is the incidence angle, w is the width of the street, and h_r is the building height, as shown in Fig. 4.5.5.3.1.1(a).

Because the signal propagates passing over the rows of buildings, the multiple-screen diffraction loss L_{msd} depends on the BS antenna height relative to the building heights h_r and on the incidence angle φ. A criterion for grazing incidence is the "settled field distance," d_s:

$$d_s = \frac{\lambda d^2}{\Delta h_b^2} \qquad (4.5.5.3.1.6)$$

where

$$\Delta h_b = h_b - h_r \qquad (4.5.5.3.1.7)$$

For the calculation of L_{msd} for the distance $l > d_s$, and, more accurately, when $l \gg d_s$,

$$L_{msd} = L_{bsh} + K_a + K_d \log_{10}\left(\frac{d}{1000}\right) + K_f \log_{10}(f) - 9\log_{10}(b) \qquad (4.5.5.3.1.8)$$

where

$$L_{bsh} = \begin{cases} -18\log_{10}(1 + \Delta h_b) & \text{for } h_b > h_r \\ 0 & \text{for } h_b \leq h_r \end{cases} \qquad (4.5.5.3.1.9)$$

L_{bsh} is a loss term that depends on the BS height:

$$K_a = \begin{cases} 54 & \text{for } h_b > h_r \\ 54 - 0.8\Delta h_b & \text{for } h_b \leq h_r \text{ and } d \geq 500 \text{ m} \\ 54 - 1.6\Delta h_b \frac{d}{1000} & \text{for } h_b \leq h_r \text{ and } d < 500 \text{ m} \end{cases} \qquad (4.5.5.3.1.10)$$

$$K_d = \begin{cases} 18 & \text{for } h_b > h_r \\ 18 - 15\frac{\Delta h_b}{h_r} & \text{for } h_b \leq h_r \end{cases} \qquad (4.5.5.3.1.11)$$

$$K_f = \begin{cases} 0.7\left(\dfrac{f}{925}-1\right) & \text{for medium-size cities and suburban centers with medium tree density} \\[2em] 1.5\left(\dfrac{f}{925}-1\right) & \text{for metropolitan centers} \end{cases}$$

$$(4.5.5.3.1.12)$$

For the calculation of L_{msd} for $l < d_s$,

$$L_{msd} = 10\log_{10}\left(Q_M^2\right) \qquad (4.5.5.3.1.13)$$

$$Q_M = \begin{cases} 2.35\left[\dfrac{\Delta h_b}{d}\sqrt{\dfrac{b}{\lambda}}\right]^{0.9} & \text{for } h_b > h_r \\[1.5em] \dfrac{b}{d} & \text{for } h_b \approx h_r \\[1.5em] \dfrac{b}{2\pi d}\sqrt{\dfrac{\lambda}{p}}\left[\dfrac{1}{\theta}-\dfrac{1}{2\pi+\theta}\right] & \text{for } h_b < h_r \end{cases} \qquad (4.5.5.3.1.14)$$

$$\theta = \arctan\left[\dfrac{\Delta h_b}{b}\right] \qquad (4.5.5.3.1.15)$$

$$p = \sqrt{\Delta h_b^2 + b^2} \qquad (4.5.5.3.1.16)$$

We have described the model for the typical NLOS case 1 in this section. The following section will introduce different approaches.

4.5.5.3.2 Different Approaches for Rooftop Diffraction Calculations

For calculating the multiple diffraction loss over building rooftops, diffraction can occur around the sides of individual buildings. The diffraction angle over most rooftops is usually less than 1° for typical base station heights and distances, and the diffraction is largely unaffected by the particular shape of the obstacles. It is appropriate to treat the buildings by equivalent knife edges. The calculation of multiple diffraction loss will be separated into two parts. The first part considers the multiple diffractions across the first $(n-1)$ buildings, treated as knife edges. The second part treats a final building (see Fig. 4.5.5.3.2.1) either as a knife edge or as some more complex shape for which the diffraction coefficient is known.

For dealing with small diffraction angle much less than 1°, special methods have been developed to enable reasonably rapid calculation of the multiple diffraction integral for cases where accurate results are required and where the necessary data on the building positions and heights is available.[46]

4.5.5.3.3 Propagation along Street Canyons

Figure 4.5.5.3.3.1 depicts the situation of a BS_2 to MS_3 link in a typical microcellular NLOS case. This case is called NLOS2.

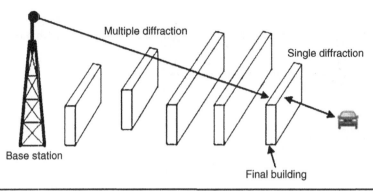

FIGURE 4.5.5.3.2.1 Multiple diffraction over building rooftops.

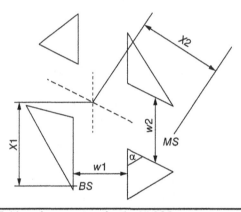

FIGURE 4.5.5.3.3.1 Definition of parameters for the NLOS2 case.

The relevant parameters shown in Fig. 4.5.5.3.3.1 for this situation are the following:

ω_1 = street width at the position of the BS (m),
ω_2 = street width at the position of the MS (m),
x_1 = distance BS to street crossing (m),
x_2 = distance MS to street crossing (m), and
α = is the corner angle (rad).

NLOS2 is the scenario for macrocells in urban high-rise environments and for micro- and picocells in urban low-rise environments.[47]

In NLOS2 situations, both antennas are below rooftop level, and diffracted and reflected waves at the corners of the street crossings must be considered (see Fig. 3.5.5.3.2.1):

$$L_{NLOS2} = -10 \log_{10} \left(10^{L_r/10} + 10^{Ld/10} \right) \qquad (4.5.5.3.3.1)$$

where L_r = The reflection path loss defined by

$$L_r = -20 \log_{10}(x_1 + x_2) + x_1 x_2 \frac{f(\infty)}{w_1 w_2} - 20 \log_{10}\left[\frac{4\pi}{\lambda}\right] \quad \text{dB} \qquad (4.5.5.3.3.2)$$

where

$$f(\infty) = \begin{cases} -41 + 110 \infty & \text{for } \infty \leq 0.33 \\ -13.94 + 28 \infty & \text{for } 0.33 < \infty \leq 0.42 \\ -5.33 + 7.51 \infty & \text{for } 0.42 < \infty \leq 0.71 \\ 0 & \text{for } \infty > 0.71 \end{cases} \qquad (4.5.5.3.3.3)$$

where L_d = The diffraction path loss defined by

$$L_d = -10 \log_{10}[x_2 x_1 (x_1 + x_2)] + 2D_a + 0.1\left[90 - \infty \frac{180}{\pi}\right] - 20 \log_{10}\left[\frac{4\pi}{\lambda}\right] \text{dB} \qquad (4.5.5.3.3.4)$$

$$D_a \approx -\left[\frac{40}{2\pi}\right]\left[\arctan\left(\frac{x_2}{w_2}\right) + \arctan\left(\frac{x_1}{w_1}\right) - \frac{\pi}{2}\right] \qquad (4.5.5.3.3.5)$$

4.6   Summary

In this chapter, microcell models for both empirical and deterministic methods are discussed and compared. The exponential growth of wireless system demands accurate and efficient propagation models. Although many researchers have been working hard during the past few decades in the area of field strength prediction, numerous problems remain to be solved. Many different prediction models have been proposed, and each of these has advantages and disadvantages and can be applied only in particular circumstances. Also, the models depend to a great extent on the accuracy of building, terrain, and morphology databases.

The trade-off between the accuracy of prediction and speed is critical for microcell prediction. Microcells often have to be deployed very quickly, and the environment is much more complicated as microcells are situated in dense urban areas. In the microcell system, signal coverage is usually not a problem. Therefore, the major goal of microcell prediction models is to solve the potential interference from other pico- (in-building), micro-, and macrocells, making prediction a challenging task. Even with a very high resolution databases, propagation prediction is still affected by many other factors, such as street furniture (signs, lampposts, and so on) and by details of the antenna siting and its interaction with neighboring cells. As we discussed earlier, terrain, tree, water, and other morphologies have a great impact on the accuracy of the model as well.

The Lee microcell prediction model is a statistical model based on street layout with buildings and empirical data to predict signal strength. The basic principles and algorithms of the model are simple and easy to implement. Using the statistical approach, it requires only 2D. The empirical data received on the streets include the loss due to the rooftop diffraction phenomenon. Therefore, in the Lee model, the heights of buildings need not be considered. The other physical models do consider the heights of the buildings, as shown in Sec. 4.5.3.

In Table 4.6.1, make comparisons in seven categories among selected models, some of which have been described in Chap. 3. The table gives us a general understanding of

Model Name	Suitable Environment	Complexity	Experimental Data	Details of Environment	Accuracy	Time	Other
Okumura Model	Macrocell	Simple	Based on experiments	No	Good	Little	Graphical path-loss data
Hata Model	Macrocell (early cellular)	Simple	No	No	Good	Little	
COST-231	Microcell (outdoor)	Simple	No	No	Good	Little	
Dual-slope	Microcell and picocell (LOS region)	Simple	No	No	Good	Little	
Ray-tracing	Outdoor and indoor	Complex	No	Yes	Very Good	Very Much	
FDTD	Indoor(small)	Complex	No	Every detail	Best	Very Much	Often combined with ray tracing
ANN	Outdoor and indoor	Complex	Yes	Detail	Very Good	Little	Takes time to learn from experimental data
TLM	Indoor	Complex	No	Yes	Good	Very Much	
Lee Microcell Model	Outdoor	Simple	For fine tuning	No	Very Good	Little	Integrated with measurement data

TABLE **4.6.1** Characteristics of the Different Microcell Models

the individual models. Of course, this is more or less a subjective comparison, and we have tried to be unbiased.

In this chapter, the Lee microcell prediction model has proven to be fairly accurate and not need a detailed database. It seeks balance among many different factors, such as speed, accuracy, and the ability to fine-tune measurement data. It also has the flexibility of quickly integrating with many other key parameters, such as terrain, trees, and water. As we will discuss in later chapters, the Lee macro-, micro-, and picocell models can be integrated and provide an effective solution for radio network planning and deployment.

References

1. Lee, W. C. Y. *Mobile Communications Design Fundamentals.* 2nd ed. New York: John Wiley & Sons, 1993: 88–94

2. Lee, W. C. Y. and Lee, D. J.Y. "Microcell Prediction in Dense Urban Area." *IEEE Transactions on Vehicular Technology* 47, no. 1 (1998): 246–53.

3. Lampard, G., and Dinh, V. D. "The Effect of Terrain on Radio Propagation in Urban Microcells." *IEEE Transactions on Vehicular Technology* 42, no. 3 (August 1993): 314–17.

4. Ikegami, F., Yoshida, S., Takeuchi, T., and Umehira, M. "Propagation Factors Controlling Mean Field Strength on Urban Street." *IEEE Transactions on Antennas and Propagation* 32, no. 8 (August 1984): 822–29.

5. Walfisch, J., and Bertoni, H. L. "A Theoretical Model of UHF Propagation in Urban Environments." *IEEE Transactions on Antennas and Propagation* 36, no. 12 (December 1988): 1788–96.

6. Rustako, A. J., Amitay, N., Owens, G. J., and Roman, R. S. "Radio Propagation at Microwave Frequencies of Line-of-Sight Microcellular Mobile and Personal Communication." *IEEE Transactions on Vehicular Technology* 40 (February 1991): 203–10.

7. Whitteker, J. H. "Measurements of Path Loss at 910 MHz for Proposed Microcell Urban Mobile Systems." *IEEE Transactions on Vehicular Technology* 37 (August 1988): 125–29.

8. Harley, P. "Short Distance Attenuation Measurement at 900 MHz and 1.8 GHz Using Low Antenna Height for Microcells." *IEEE Journal of Selected Areas in Communications* 7 (January 1989): 5–11.

9. Kostanic, Ivica, Guerra, Ivan, Faour, Nizar, Zec, Josko, and Susanj, Mladen. "Optimization and Application of W.C.Y Lee Micro-Cell Propagation Model in 850 MHZ Frequency Band." Wireless Center of Excellence (WiCE), Florida Institute of Technology, Melbourne, 2003. Also appeared in *Proceedings of the Wireless Networking Symposium*, Austin, Tx, October 22–4, 2003.

10. Lee, W. C. Y. "*Wireless and Cellular Telecommunications.*" 3rd ed., New York: McGraw-Hill, 2006. 370–372 (Near-in equation); pp. 664–74 (Lee's microcell system).

11. Lee, W. C. Y., *Mobile Communications Design Fundamentals*, Ibid. p. 92.

12. Lee, W. C. Y., and Lee, D. J. Y. "The Propagation Characteristics in Cell Coverage Area." VTC'97, Phoenix, AZ, May 5, 1997, 2238–42.

13. Lee, W. C. Y., and Lee, D. J. Y. "Microcell Prediction Enhancement for Terrain." 7th IEEE International Symposium on Personal, Indoor and Mobile Radio Communications, vol. 2, PIMRC'96, Taipei, 1996, 286–90.

14. Lee, W. C. Y. 8 patents granted in Microcell systems and implemented in Los Angeles and San Diego in 1991 by Pactel Corp. and listed in Ref. 27 of Chap. 1.

15. Lee, W. C. Y. "Lee's Essentials of Wireless Communications" McGraw-Hill, 2001, pp. 75–80.

16. Lee, W. C. Y. "Smaller Cells for Greater Performance." *IEEE Communications Magazine* (November 1991): 19–23.

17. Lee, W. C. Y. and Lee, D. J. Y. "Pathloss Prediction from Microcell to Macrocell." IEEE 51st VTC2000'S, vol. 3, Tokyo, 2000, 1988–92.

18. Lee, W. C. Y. and Lee, D. J. Y. "Microcell Prediction in Dense Urban Area." *IEEE Transactions on Vehicular Technology* 47, no. 1 (1998): 246–53.

19. Agilent Technologies. *WIZARD Users Manual*. Santa Clara, CA: Agilent Technologies, 2003.

20. Green, E. "Radio Link Design for Microcellular Systems." *BT Technology Journal* 8, no. 1 (1990): 85–96.

21. Chia, S. T. S. "Radiowave Propagation and Handover Criteria for Microcells." *BT Technology Journal* 8, no. 4 (1990): 50–61.

22. Bultitude, R. J. C., and Hughes, D. A. "Propagation Loss at 1.8 GHz on Microcellular Mobile Radio Channels." Proceedings IEEE International Symposium on Personal, Indoor and Mobile Radio Communications (PIMRC96), October 1996, 786–90.

23. Kaji, M., and Akeyama, A. "UHF-Band Radio Propagation Characteristics for Land Mobile Radio System Using Low Antenna Height Base Station." *AP-S International Symposium Digest* 2 (1985) 835–38.

24. Bultitude, R. J. C., and Bedal, G. K. "Propagation Characteristics on Microcellular Urban Mobile Radio Channels at 910 MHz." IEEE Journal of Selected Areas in Communications 7 (January 1989): 31–39.

25. Greenstein, L. J., Armitay, N., Chu, T. S., et al. "Microcells in Personal Communication Systems." *IEEE Communications Magazine* 30, no. 12 (December 1992): 76–88.

26 Jakes, W. C., Jr. Microwave Mobile Communications, John Wiley & Sons, 1974, Chap. 4.

27. Lee, W. C. Y. Mobile Communications Engineering. 2nd ed. New York: McGraw-Hill, 1998: 289–90.

28. Blaunstein, N. "Average Field Attenuation in the Nonregular Impedance Street Waveguide." *IEEE Transactions on Antenna and Propagation* 46, no. 12 (December 1998): 1782–89.

29. Mazar, R., and Bronshtein, A. "Propagation Model of a City Street for Personal and Microcellular Communications." *Electronics Letters* 33, no. 1 (1997): 91–2.

30. Blaunstein, N., and Levin, M. "Prediction of UHF-Wave Propagation in Suburban and Rural Environments." Proceedings URSI Symposium Comm-Sphere '95, 1995, 191–200.

31. Blaunstein, N., Giladi, R., and Levin, M. "Characteristics Prediction in Urban and Suburban Environments." *IEEE Transactions on Vehicular Technology* 47, no. 1 (January 1998): 225–34.

32. Dersch, U., and Zollinger, E. "Propagation Mechanisms in Microcell and Indoor Environments." *IEEE Transactions on Vehicular Technology* 43, no. 4 (April 1994): 1058–66.

33. Bertoni, H. L., Honcharenko, W., Macie 1, L. R., and Xia, H. H. "UHF Propagation Prediction for Wireless Personal Communications." *Proceedings of the IEEE* 82 (1994): 1333–59.

34. Erceg, V., Rustako, A. J., and Roman, P. S. "Diffraction around Corners and Its Effects on the Microcell Coverage Area in Urban and Suburban Environments at 900 MHz, 2 GHz, and 6 GHz." *IEEE Transactions on Vehicular Technology* 43, no. 3 (1994): 762–66.

35. Sarnecki, J., Vinodrai, C., Javed, A., O'Kelly, P., and Dick, K. "Microcell Design Principles." *IEEE Communication Magazine* 31, no. 4 (1993): 76–82.

36. Feuerstein, M. J., Blackard, K. L., Rappaport, T. S., and Seidel, S. Y. "Path Loss, Delay Spread and Outage Models as Functions of Antenna Height for Microcellular System Design." *IEEE Transactions on Vehicular Technology* 43, no. 3 (1994): 487–98.

37. Arnold, H. W., Cox, D. C., and Murray, R. R. "Macroscopic Diversity Performance Measured in the 800 MHz Portable Radio Communications Environment." *IEEE Transactions on Antennas and Propagation* 36, no. 2 (1988): 277–80.

38. Zhang, Y. P., Hwang, Y., and Parsons, J. D. "UHF Radio Propagation Characteristics in Straight Open-Groove Structures." *IEEE Transactions on Vehicular Technology* 48, no. 1 (1999): 249–54.

39. Goldsmith, A. J., and Goldsmith, L. J. "A Measurement-Based Model for Predicting Coverage Areas of Urban Microcells." *IEEE Journal on Selected Areas in Communication* 11, no. 7 (1993): 1013–23.

40. Dehghan, S., and Steele, R. "Small Cell City." *IEEE Communications Magazine* 35, no. 8 (1997): 52–9.

41. Wang, L. C., Stubea, G. L., and Lea, C. T. "Architecture Design Frequency Planning and Performance Analysis for a Microcell/Macrocell Overlaying System." *IEEE Transactions on Vehicular Technology* 46, no. 4 (1997): 836–48.

42. Saunders, S. R., and Boner, F. "Prediction of Mobile Radio Propagation over Buildings of Irregular Heights and Spacing." *IEEE Transactions on Antennas and Propagation* 42, no. 2 (1994): 137–44.

43. Berg, J. E. "A Recursive Method for Street Microcell Path Loss Calculations." Proceedings of the IEEE International Symposium on Personal, Indoor and Mobile Radio Communications, (PIMRC '95), vol. 1, 1995, 140–3.

44. Recommendation ITU-R P.1411, "Propagation Data and Prediction Methods for the Planning of Short-Range Outdoor Radio Communication Systems and Radio Local Area Networks in the Frequency Range 300 MHz to 100 GHz." 2012.

45. Recommendation ITU-R P.341-5, "The Concept of Transmission Loss for Radio Links." 1999.

46. Maciel, L. R., and Bertoni, H. L. "Cell Shape for Microcellular Systems in Residential and Commercial Environments." *IEEE Transactions on Vehicular Technology* 43, no. 2 (1994): 270–8.

47. Wiart, W. "Microcellular Modeling When Base Station Is Below Rooftops." *Proceedings of IEEE VTC'94 Conference*, Stockholm, Sweden, 1994, 200–4.

CHAPTER 5

In-Building (Picocell) Prediction Models

5.1 Introduction

5.1.1 Differences from Other Models

The principal characteristics of an indoor RF propagation environment that distinguish it from an outdoor environment are that the multipath is usually severe, a line-of-sight (LOS) path may not exist, and the characteristics of the environment can change drastically over a very short time or distance.[1] The ranges involved tend to be rather short, on the order of 100 m or less. Walls, doors, furniture, and people can cause significant signal loss. Indoor path loss can change dramatically with either time or position because of the amount of multipath present and the movement of people, equipment, and/or doors.[2] As discussed in the previous chapter, multiple reflections can produce signal smearing (delay spread) and may cause signal fading.

Propagation prediction for indoor radio systems has some unique considerations, requirements, and challenges. As we have mentioned earlier, the in-building prediction has become more and more crucial to the performance of many wireless systems, especially as WLAN, WiMax, 3G, and 4G become more and more popular. There are several ways of implementing an in-building prediction model. From the empirical model based on the earlier measurement data, we include building layout to characterize the radio behavior in a specific building. With more building data available and the improvement of computer speed, site-specific propagation models can be created, including building materials of walls, ceilings, and partitions into its calculation. Some radio propagation theories can now be applied to calculate the propagation for inside and outside buildings.[3–16] For example, ray tracing combined with GTD/UTD, FDTD, and TLM models can all be used to predict the coverage of in-building systems. The ray-tracing technique will be described in Sec. 5.6.1. Once the required coverage is defined, all-critical parameters can be used to make a suitable prediction for the signal coverage. Furthermore, taking some specific parameters due to a particular area under test into the model so that the deduction of interference, the capacity, the system performance, and handoffs can be optimized.

The ultimate purposes, just as when dealing with outdoor systems, are to ensure sufficient signal coverage in a required area (or to ensure a reliable path in the point-to-point systems) and to avoid interference both within its own system and to other systems.

However, in the indoor environment, the boundary of coverage is well defined by the geometry of the building, and the walls of the building itself will affect the propagation. In addition to the frequency reuse on the same floor of a building, there is often a desire to have frequency reuse between floors of the same building, adding a third dimension to managing interference issues. Finally, in a very short range, particularly where millimeter wave frequencies are used, small changes in the immediate environment of a radio path may have substantial effects on the propagation characteristics. Because of the complex nature of these factors, if the specific planning of an indoor radio system were undertaken, the detailed knowledge of a particular site would be required, including geometry, materials, furniture, and expected usage patterns. However, for an initial system planning, it is necessary to estimate the number of base stations with their proper locations to provide coverage to the scattered mobile stations within the area and also to estimate potential interference to other wireless services or between systems. At the same time, the model should not require a lot of input information by the user in order to carry out the calculations.

5.1.2   Propagation Impairments and Measure of Quality in Indoor Radio Systems

Propagation impairments in an indoor radio channel are caused mainly by the following:

- Reflection from and diffraction around objects (including walls and floors) within the rooms
- Transmission loss through walls, floors, and other obstacles
- Channeling of energy, especially in corridors at high frequencies
- Motion of persons and objects in the room, including possibly one or both ends of the radio link, giving rise to impairments such as the following:
  - Path loss—not only the free space loss but also additional loss due to obstacles and transmission through building materials and possible mitigation of free space loss by channeling
  - Temporal and spatial variation of path loss
  - Multipath effects from reflected and diffracted components of the wave
  - Polarization of the radio wave mismatches due to random alignment of the mobile terminal

This chapter presents mainly general site-independent models and the description of propagation impairments encountered in the indoor radio environment. In many cases, the available data on basic models were limited in either frequency ranges or test environments; it is hoped that the accuracy of the models will be improved with more available data and experience in their application.

It is useful to define which propagation characteristics of a channel are most appropriate in describing their significance for different applications, such as voice communications, data transfer at different speeds, and video services.

5.1.3   The Highlights of the Lee In-Building Model[1,3]

This section explains the features of newly constructed office buildings that influence the radio-wave propagation between a transmitter and a receiver located on the same floor.

Many in-building propagation research material and predictions have never been published before and are disclosed in this chapter. The first feature that impacts communication is assuming a clear space loss between the transmitter and receiver. The second feature is the loss due to reflections from interior and exterior walls. The third feature is using different propagation loss formulas for regular rooms (such as conference rooms or offices) and special rooms (such as elevators and utility rooms). This model provides an improved prediction by allowing one to understand the effect of the building environment on the propagation characteristics of a mobile receiver. The prediction is done basically along a radial by applying different formulas while passing through different environments. Because there are many different propagation paths from the transmitting antenna to the receiving antenna, the paths exhibit many different starting directions at the transmitting antenna and many different directions of arrival at the receiving antenna. At any point, the signal can be higher or lower than the free space loss due to the multipath. However, the average signal loss is always higher than free space. The measured data were collected by moving the transmitter or the receiver over a spatial area (often running in a circular path) and then averaged by following the traditional method to find the local mean, as stated in Sec. 1.6.3.1. The area had linear dimensions of 10 to 20 wavelengths for obtaining the local mean signals. These local means can remove the rapid variation of the received signal.[17,18]

Once the room penetration loss characteristics are derived, the optimized placement of antenna can be determined by moving the mobile around at each of different antenna locations. In general, the mobile communication channel often consists of a few strong paths combined with a number of weaker paths; the indoor channel exhibits similar characteristics.[19] However, the prediction from the model is based only on the average of the signal strength received by the mobile receiver.

5.2   The Lee In-Building Prediction Model

The potential implementation of an in-building prediction model for the wireless local area networks (LAN) and personal communication services (PCS) inside buildings requires a thorough understanding of signal propagation in a building. In this section, the Lee in-building (Picocell) prediction model is presented. This prediction model focuses on a single floor of a building but is also applicable to different floors of the building. This model is also applicable to the through-building propagation loss by applying the same principle. The validation of this model was done in two different buildings of similar construction in the 900-MHz band. A special feature of this model is its ability to handle different types of obstructions. The model is validated by gathering the measured data for a specific floor of a building and comparing them with the predicted values. The standard deviation between measured data and predicted values is within 5 dB.

5.2.1   Derivation of Close-In Distance for the In-Building Model

In this chapter, the near-in distance is derived for the microcell model. Here, we are dealing with, a different signal propagation environment. It is an in-building environment, so it is a close-in environment. Since the microcell environment is an open environment, the near-in distance used in the microcell environment can not be applied to this close-in environment. Therefore, we have derived another propagation distance that is from the base station to a short distance at which the signal still maintains strong can be treated as a free space path-loss signal. We name the distance the *close-in distance*.

Also, in a close-in environment, there are four planes: one ground, one ceiling, and two side planes. These four planes will generate four reflected waves received by the receiver besides the direct wave. Because the transmitting antenna is usually set up high and close to the ceiling, among the four reflected waves, only the ground-reflected wave is a dominant wave because the reflected point on the ground from the ground-reflected wave is closer to the receiver than the other three reflected points on the other three planes from the other three reflected waves. The three reflected points from three reflected waves occurred further away from the receiver would scatter the energy after reflected and cause the three reflected waves to carry the weak energy to the receiver. Therefore, we may ignore the other three reflected waves and consider only the direct wave and the ground-reflected wave as the two main components.

5.2.1.1   Based on the Two-Ray Model

The close-in distance can be derived from the two-ray model, which was described in Sec. 1.9.1.3.[20] The received signal power is expressed as

$$P_r = P_0 \left(\frac{1}{4\pi d/\lambda} \right)^2 \left| 1 + a_v e^{-j\Delta\phi} \right|^2 \tag{1.9.1.3.1}$$

where the reflection coefficient a_v appeared in Eq. (1.9.1.1.2) as

$$a_v = \frac{\varepsilon_c \sin\theta_1 - (\varepsilon_c - \cos^2\theta_1)^{1/2}}{\varepsilon_c \sin\theta_1 + (\varepsilon_c - \cos^2\theta_1)^{1/2}} \qquad \text{(vertical)} \tag{1.9.1.1.2}$$

and $\Delta\phi$ is the phase difference between the direct wave and reflected wave at the point of reception, θ_1 is the incident angle, and ε_c is listed in Eq. (1.9.1.1.3) as

$$\varepsilon_c = \varepsilon_r - j60\sigma\lambda \tag{1.9.1.1.3}$$

where the permittivity ε_r is the principal component of the dielectric constant ε_c, σ is the conductivity of the dielectric medium in Siemens per meter, and λ is the wavelength.[21]

Some typical values of permittivity and conductivity for various common types of media are shown in Table 5.2.1.1a.

Medium	Permittivity ε_r	Conductivity σ, S/m
Copper	1	5.8×10^7
Seawater	80	4
Rural ground (Ohio)	14	10^{-2}
Urban ground	3	10^{-4}
Fresh water	80	10^{-3}
Turf with short, dry grass	3	5×10^{-2}
Turf with short, wet grass	6	1×10^{-1}
Bare, dry, sandy loam	2	3×10^{-2}
Bare, sandy loam saturated with water	24	6×10^{-1}

TABLE 5.2.1.1a   Data of ε_c on Different Surfaces[22]

5.2.1.1.1. Estimate ε_r[23-27]

Since ε_r is the dominated factor in Eq. (1.9.1.1.3), we should look into the values of ε_r from the building material.

Stone wall: Inside building $\varepsilon_r = 4.5$
Outside building $\varepsilon_r = 7.9$
Concrete wall: Inside building $\varepsilon_r = 5.4$
Glass wall: Inside building $\varepsilon_r = 2.3$

5.2.1.2   The Close-In Distance Is Determined When the Reflection Coefficient $a_v = 0$

We may find the condition of the reflection coefficient $a_v = 0$ from Eq. (1.9.1.1.2) as

$$\varepsilon_c \sin\theta_1 = (\varepsilon_c - \cos^2 \theta_1)^{1/2} \tag{5.2.1.2.1}$$

Solving Eq. (5.2.1.1), we get

$$\sin\theta_1 = \frac{1}{\sqrt{\varepsilon_c + 1}} \tag{5.2.1.2.2}$$

$$\tan\theta_1 = \frac{1}{\sqrt{\varepsilon_c}} \tag{5.2.1.2.3}$$

where the dielectric contant ε_c consists of two parts—the permittivity ε_r and the conductivity σ, expressed in Eq. (1.9.1.1.3).

We select two from Table 5.2.1.1b, the permittivities and the conductivities are listed in the range 900 MHz to 2.44 GHz.

For wooden floor, $\varepsilon_r = 3$, and $\sigma = 0.001$, them $\varepsilon_c = 3 - j\,60 \times 0.001\,\lambda = 3 - j\,0.06\,\lambda$
For concrete ceiling/floor, $\varepsilon_r = 7$, and $\sigma = 0.05$, then $\varepsilon_c = 7 - j\,60 \times 0.05\,\lambda = 7 - j\,3.0\,\lambda$
We also select one from Table 5.2.1.1c at 5.24 GHz:

	Ext. Wall (brick)	Int. Wall (brick)	Wooden Door	Glass Window	Ceiling/ Floor
ε_r	6	5	3	4	7
σ (S/m)	0.05	0.02	0.001	0.001	0.05

TABLE 5.2.1.1b   Reference Dielectric Parameters for Different Building Materials at the Frequency 900 MHz to 2.44 GHz[23]

Wall Type	ε_r	σ
10-cm-thick plasterboard wall	8.37	0.0183
5-cm-thick soft partition	6.87	0.164
6-mm-thick window	3.41	0.733
3-cm-thick desktop	6	0.03
25-cm-thick concrete floor/ceiling backed by perfect conductor	6	0.03

TABLE 5.2.1.1c   The Data of ε_c at 5.24 GHz[24]

Wall Type		Estimated ε_r
Concrete		5.4
Stone	Indoor	4.5
	Outdoor	7.9
Glass D 4		2.3

TABLE 5.2.1.1d   Data of ε_r at 2.4 GHz[25]

For 10-cm-thick plasterboard wall : $\varepsilon_r = 8.37$, and $\sigma = 0.0183$, then

$$\varepsilon_c = 8.37 - j\,60 \times 0.0183\,\lambda = 8.37 - j\,1.098\,\lambda$$

For 25-cm-thick concrete floor/ceiling backed by perfect conductor, $\varepsilon_r = 6$, and $\sigma = 0.03$, then

$$\varepsilon_c = 6 - j\,60 \times 0.03\,\lambda = 6 - j\,1.8\,\lambda$$

Because the conductivity σ is measured in S/m, the wavelength λ should be in meters. At 900 MHz, $\lambda = 0.333$ m, at 2.4 GHz $\lambda = 0.125$ m , and at 5.24 GHz $\lambda = 0.057$ m. Therefore, for the frequency greater than 900 MHz, the permittivity ε_r is the dominated factor for ε_c as shown in the above three cases. Then Eq. (5.2.1.2) and Eq. (5.2.1.3) can be stated as,

$$\sin\theta_1 \frac{1}{\sqrt{\varepsilon_r + 1}} \qquad (5.2.1.2.4)$$

$$\tan\theta_1 = \frac{1}{\sqrt{\varepsilon_r}} \qquad (5.2.1.2.5)$$

As shown from Fig. 5.2.1.2.1, d_1 is the distance from the base station to the reflection point, and d_2 is the distance from the reflection point to the mobile. h_1 and h_2 are the antenna heights of the base station and the mobile, respectively. The relationship among h_1, h_2, and d_1, d_2 based on the Snell's law can be expressed as:

$$\tan\theta_1 = \frac{h_1}{d_1} = \frac{h_2}{d_2} \qquad (5.2.1.2.6)$$

and the close-in distance D_c is

$$D_c = d_1 + d_2 \qquad (5.2.1.2.7)$$

Solving Eqs. (5.2.1.2.5), (5.2.1.2.6), and (5.2.1.2.7), we get the formula for the close-in distance d_c as

$$D_c = (h_1 + h_2)\sqrt{\varepsilon_r} \qquad (5.2.1.2.8)$$

In Eq. (5.2.1.2.8), we can visualize that the higher the value of the permittivity ε_r the larger the close-in distance. Also, from Eq. (5.2.1.2.8), we observe that close-in distance D_c is independent of frequency.

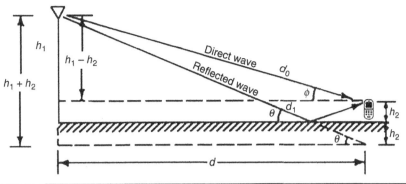

FIGURE 5.2.1.2.1   The two-ray model with large incident angle θ_1

Then we assume that on the same floor of the building, $h_1 = 8$ ft (2.44 m) and $h_2 = 4$ ft (1.22 m), so the close-in distance D_c can be found:

For wooden floor at 900 MHz to 2.4 GHz, $\varepsilon_r = 3$, $D_c = 20.78$ ft (6.34 m)

For concrete ceiling/floor at 900 MHz to 2.4 GHz, $\varepsilon_r = 7$, $D_c = 12 \times \sqrt{7} = 31.75$ ft (9.68 m)

For 10-cm-thick plasterboard wall at 5.24 GHz, $\varepsilon_r = 8.37$, $D_c = 12 \times \sqrt{8.37} = 34.72$ft (10.58 m)

For 25-cm-thick concrete floor/ceiling backed by perfect conductor, $\varepsilon_r = 6$, $D_c = 12 \times \sqrt{6} = 29.39$ ft (8.96 m)

In the office building, the floors are made of concrete, so the close-in distance can be roughly guessed at around 30 to 35 ft at a frequency range of 900 to 5.24 MHz. We may set $\varepsilon_r = 7$ in Eq. (5.2.1.2.8) for our default value. Then the nominal close-in distance D_c will be

$$D_c = (h_1 + h_2)\sqrt{7} = 2.646\,(h_1 + h_2) \qquad \text{for concrete ceiling/floor} \qquad (5.2.1.2.9)$$

and the incident angle θ_1 is

$$\theta_1 = \sin^{-1}\frac{1}{\sqrt{\varepsilon_r + 1}} = \sin^{-1}\frac{1}{\sqrt{8}} = 20.7° \qquad (5.2.1.2.10)$$

From Eq. (5.2.1.2.10), the incident angle from the concrete floor is about 20.7°.

5.2.1.3   Verification with Measured Data

The indoor measured data were collected at three different frequencies: 2.4, 4.75, and 11.5 GHz.[27] Measured data were taken at the hallway at the TNO Laboratory. The size of the hallway was 35 m long, 3 m wide, and 3.5 m high. The transmitter antenna height was $h_1 = 1.5$ m and 3 m. The receiving antenna height was $h_2 = 1.5$ m. We can calculate the close-in distance D_c from Eq. (5.2.1.2.9) as

$$D_{c1} = 2.646\,(1.5 + 1.5) = 7.9 \text{ m} \qquad \text{for } h_1 = 1.5 \text{ m and } h_2 = 1.5 \text{ m}$$
$$D_{c2} = 2.646\,(1.5 + 3) = 11.9 \text{ m} \qquad \text{for } h_1 = 3 \text{ m and } h_2 = 1.5 \text{ m}$$

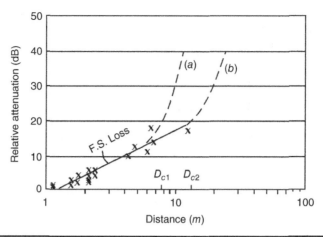

FIGURE 5.2.1.3.1 Attenuation-distance profile: Curve (a) for $h_1 = 1.5$ m; curve (b) for $h_1 = 3$ m; under LOS condition at 2.4 GHz (measured data reprinted from reference[27]).

Figure 5.2.1.3.1 shows the measured data at a LOS condition at 2.4 GHz. Some data were collected at $h_1 = 1.5$ m and some at $h_1 = 3$ m. We can see from the measured data that some data points were changing their attenuation from free space loss to higher loss after roughly 8 m and that some were after 12 m. Since we do not know the actual material of the hallway floor, the permittivity $\varepsilon_r = 7$ that we used in Eq. (5.2.1.2.9) may be a little bit too high. Still, the data agree fairly well with the prediction of the close-in distances D_{c1} and D_{c2}. Figures 5.2.1.3.2 and 5.2.1.3.3 show the measured data at the LOS conditions at 4.75, and 11.5 GHz, respectively. At two different frequencies, the measured data show fairly good agreement with the predicted close-in distance. Also, comparing all three figures, we show that the close-in distance frequency is independent of frequency.

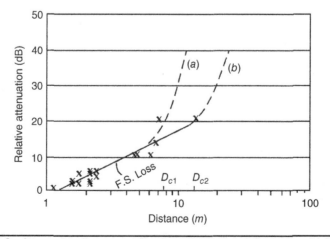

FIGURE 5.2.1.3.2 Attenuation-distance profile: Curve (a) for $h_1 = 1.5$ m; curve (b) for $h_1 = 3$ m; under LOS condition at 4.75 GHz (measured data reprinted from reference[27]).

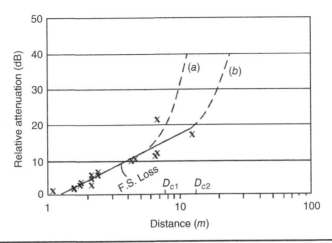

FIGURE 5.2.1.3.3 Attenuation-distance profile: Curve (*a*) for $h_1 = 1.5$ m; curve (*b*) for $h_1 = 3$ m; under LOS condition at 11.5 GHz (measured data reprinted from reference[27]).

5.2.2 The Single-Floor (Same Floor) Model

In this section, we concentrate on making the prediction of the signal coverage on just one floor of the office building since the same principle for computing transmission losses is applicable to all floors and buildings. The input information for this model is in an AutoCAD file format (AutoCAD file format: dwg; http://en.wikipedia.org/wiki/.dwg). The different types of information in this file are in AutoCAD file format:

The building boundary

Rooms

Special rooms (elevators, utility rooms, and so on)

The model is capable of computing losses from the above details of the building. This is explained in the following section. Figure 5.2.2.1 shows a sample building layout and the different distances along where path loss occurs.

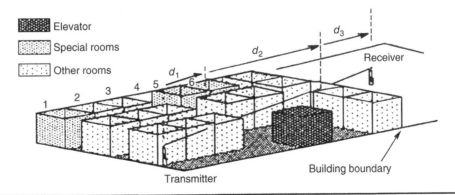

FIGURE 5.2.2.1 Sample in-building layout.

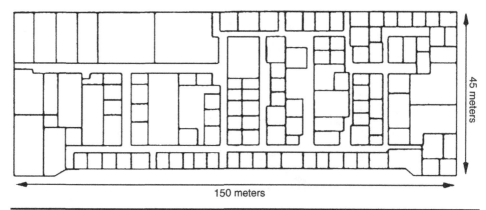

Figure 5.2.2.2 Qualcomm building A (first floor) used for validation.

The parameters P_t, d_1, d_2, d_3, and P_r are shown in Fig. 5.2.2 1, where

P_t = power transmitted,

d_1 = distance from the transmitter in the fifth room to the sixth-room intersection,

d_2 = distance from the sixth-room intersection to the building boundary, and

d_3 = distance from the building boundary to the receiver.

Note that d_3 will be zero if the receiver is in the building boundary, in which case d_2 will be the distance from the first wall intersection to the receiver.

Figure 5.2.2.2 shows the first floor of the Qualcomm building. This building is 45 by 150 m. There are about 50 rooms of different sizes, but most of them are made of the same 5/8-in sheetrock material with metal studs. The Lee in-building prediction model was tested on this floor, and the results are compared in Sec. 5.2.6.

When considering the path loss between transmitter and receiver, the losses could be of three types. One is the standard LOS loss. The LOS condition also should be found from the building layout to ensure that the LOS signal is in the close-in zone between the antenna and any obstruction. The other two losses are due to the signal passing through rooms when the receiver is either inside or outside the close-in zone. With these facts in mind, the first step is to calculate the loss caused by walls of regular rooms and special rooms. A special room is defined as a room that is built with different materials than most of the rooms of in the same building/floor and usually includes elevators and a utility room.

5.2.2.1 LOS Condition

In the LOS case, the receiver is in a direct LOS of the transmitter. Figure 5.2.2.1.1 shows the top view of the building layout in Fig. 5.2.2.1. In this figure, the radio path from the transmitter to the receiver is not obstructed.

Since there are no intersecting rooms blocking the radio path, both d_2 and d_3 shown in Fig. 5.2.2.1 are equal to zero in this setting. Thus, the only distance that contributes to the path loss is d_1. This path loss L_{LOS} is then given:

$$L_{LOS} = 20 \log \frac{4\pi d_1}{\lambda} \qquad d_1 < D_c \tag{5.2.2.1.1}$$

where λ is the signal wavelength and d_1 is the distance within the close-in distance D_c.

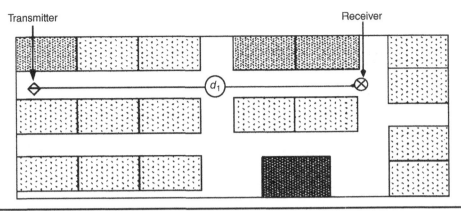

Figure 5.2.2.1.1   Top view, LOS.

The close-in distance D_c is defined in Eq. (5.2.1.2.8) as

$$D_c = (h_a + h_m) \sqrt{\varepsilon_r}$$ (5.2.2.1.2)

or the nominal equation of close-in distance from Eq. (5.2.1.2.9) as

$$D_c = 2.646 \, (h_a + h_m) \qquad \text{for concrete ceiling/floor}$$ (5.2.2.1.3)

where in Eq. (5.2.2.1.3), h_a (the antenna is mounted on the ceiling, h_a = 8 ft as a default number) and h_a are the antenna heights at the base and at the mobile, respectively.

P_t is the power transmitted, and P_r is the total power at the receiver:

$$P_r = P_t + G_t - L_{LOS} + G_r$$ (5.2.2.1.4)

where G_t is the transmitter antenna gain and G_r is the receiver antenna gain.

5.2.2.2   Non-LOS (NLOS) Condition

When the receiver is situated so that the LOS signal is not received from the transmitter, the path losses could be classified as two kinds. The first path-loss component yields when the receiver is in the close-in zone, and the second path-loss component yields when the receiver is not in the close-in zone. These cases are illustrated below.

5.2.2.2.1   Receiver in Close-In Region   As shown in Fig. 5.2.2.2.1, a wall obstructs the transmitter and receiver. The receiver is close enough to the transmitter to be in its close-in region:

$$L_{LOS} = 20 \log \frac{4\pi d_1}{\lambda} + F_{LOS} \qquad d_1 < D_c$$ (5.2.2.2.1)

where λ is the signal wavelength, d_1 is the distance within the close-in distance D_c, and F_{LOS} is the loss because of lacking of close-in clearance from obstruction between the antenna and the close-in distance D_c as:

$$F_{LOS} = 12.5 \log \left(\frac{B + D_c}{D_c} \right)$$ (5.2.2.2.2)

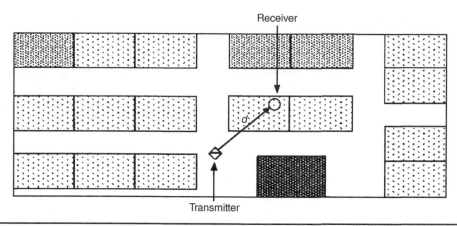

FIGURE 5.2.2.2.1 Top view—receiver in close-in region.

where the thickness of the obstruction B is in feet. Equation (5.2.2.2.2) is from the empirical data. The close-in distance D_c is defined in Eq. (5.2.1.2.8).

This would give us a path loss L_{LOS} shown in Eq. (5.2.2.2.1) and the power P_r at the receiver shown in Eq. (5.2.2.1.4). The measurement shows that when the receiver is in the close-in region, the signal will be affected by penetrating through the wall. The sum of thickness of number of walls B is shown in Eq. (5.2.2.2.2) for calculating the excessive loss due to the obstruction.

5.2.2.2.2 Receiver beyond the Close-in Region When the receiver is located beyond the close-in region and obstructed by a wall or walls, a new component is added to the path loss. This loss is related to the wall thickness and the material that the wall is made of. In a building, walls that separate rooms are mostly made of the same material. The signal loss characteristics can be easily derived through a linear regression. This will be introduced in Sec. 5.3.4. Without measuring the actual wall thickness, which we do within the close-in distance, the measured data extrapolate an additional path-loss slope of the signal for penetrating through rooms, depending the material of the walls.

In Fig. 5.2.2.2.2, we see that d_1 is the distance from the transmitter to the intersection (wall) of the first room, while d_2 is the distance from the intersection of the first room to the receiver. The radio path along d_1 is in the direct LOS path from the transmitter. Hence, the path loss can have two components. The first one, L_{LOS} due to $d_1 < d$, is given as

$$L_{LOS} = 20 \log \frac{4\pi d_1}{\lambda} + F_{LOS} \qquad \text{for } d_1 < D_c \qquad (5.2.2.2.1)$$

The additional path-loss component (L_{room}) due to the distance d_2 from the first-room intersection to the receiver is given by

$$L_{room} = m_{room} \log\left(1 + \frac{d_2}{d_1}\right) \qquad \text{for } D_c < d_2 \qquad (5.2.2.2.3)$$

where m_{room} is the an additional path-loss slope of the signal for penetrating through rooms, depending the material of the walls; m_{room} is usually around 27 dB/dec.

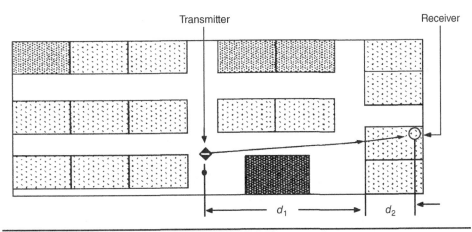

FIGURE 5.2.2.2.2 Top view—receiver not in close-in region.

The received signal strength is calculated on a grid map based on a building layout, and d_2 and d_1 are scaled on the grid map from the user's input. Typically, the room slope m_{room} in Eq. (5.2.2.2.3) is 40, which is a surprising value from the measurement data. This received signal in this case can be derived from the two path-loss components and is the same as used in the mobile environment.

Thus, the predicted power P_r at the receiver is

$$P_r = P_t + G_t - L_{LOS} - L_{room} + G_r \qquad (5.2.2.2.4)$$

5.2.2.3 Receiver in a Special Room

A special room is usually a room that is built with different materials than most of the other rooms in a building. For example, a utility room and elevators are built with different materials than offices. In Fig. 5.2.2.2.2, if the receiver is in a special room, then the following two path-loss components L_{LOS} and $L_{special\ room}$ hold true:

$$L_{LOS} = 20\ \log\ \frac{4\pi d_1}{\lambda} + F_{LOS} \qquad \text{for } d_1 < D_c \qquad (5.2.2.2.1)$$

$$L_{special\ room} = m_{special\ room}\ \log\left(1 + \frac{d_2}{d_1}\right) \qquad \text{for } D_c < d_2 \qquad (5.2.2.3.1)$$

where L_{LOS} is calculated along distance d_1 using the same equation of Eq. (5.2.2.1.1) and $L_{special\ room}$ is along distance d_2. The value of the path-loss slope for the special room $m_{special\ room}$ is typically greater than 40 db/dec.

The received power P_r is then given as

$$P_r = P_t + G_t - L_{LOS} - L_{special\ room} + G_r \qquad (5.2.2.3.2)$$

5.2.2.4 Receiver Outside the Building

Finally, we look at the path loss at a receiver that is located outside the building. In this case, as shown in Fig. 5.2.2.4.1, the total path loss will be affected by three distance components.

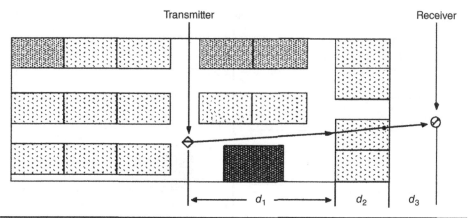

Transmitter   Receiver

FIGURE 5.2.2.4.1   Top view—receiver outside the building.

As explained before, the path loss of L_{LOS} caused by the d_1 component is the same as shown in Eq. (5.2.2.2.1), and the path loss of L_{room} caused by the d_2 component is the same as shown in Eq. (5.2.2.2.3). A new path loss $L_{outside}$ is introduced because of d_3. The equations for these three path-loss components are given below:

$$L_{LOS} = 20 \log \frac{4\pi d_1}{\lambda} + F_{LOS} \qquad \text{for } d_1 < D_c \qquad (5.2.2.2.1)$$

$$L_{room} = 40 \log\left(1 + \frac{d_2}{d_1}\right) \qquad \text{for } D_c < d_2 \qquad (5.2.2.2.3)$$

$$L_{outside} = L_{external\ wall}\ 20 \log\left(1 + \frac{d_3}{d_1 + d_2}\right) \qquad \text{for } d_2 < d_3 \qquad (5.2.2.4.1)$$

The three formulas are specified here. The measured data has been collected to validate the loss due to the signal penetrating through the external wall of the building. The wall penetration loss $L_{external\ wall}$ shown in Eq. (5.2.2.4.1) is typically between 15 and 20 dB.

The power received at the receiver P_r is then given by

$$P_r = P_t + G_t - L_{LOS} - L_{room} - L_{outside} + G_r \qquad (5.2.2.4.2)$$

Equation (5.2.2.4.2) has been introduced to calculate the loss through buildings. However, the measured data are collected to validate the accuracy of the formulas depicted in the later sections.

5.2.3   Determining Path-Loss Slope in a Room

The path-loss slopes in a regular room and a special room are determined empirically. It is based on a close approximation of the measured data to the predicted values. How the slope is derived is discussed in this section. First, a set of measured signal strength data at various locations in the building is collected and processed to derive the slope for the wall penetration loss. Once the penetration loss for wall is derived, the received signal strength can be calculated based on the propagation characteristics in different locations. The conditions that need to be checked include humidity and, LOS, it is the number of rooms blocking the signal, and the distance the receiver is from the transmitter. All this information is stored in the AutoCAD file format. This information makes the model more capable of handling the positioning of different antennas and easier to optimize the design. Also, this set of slopes is used for prediction in the model. Every predicted point is calculated individually based on the multiple-distance component model to get the signal path-loss value because of its unique position in the building. The slope in each equation is derived by averaging the loss value from all measured points. Different sets of slopes will be derived based on the type of rooms (regular rooms and special rooms).

This empirical slope is used by the Lee in-building (Picocell) model and the Lee method.[3] This is because each building is built differently. But the structure of the general formula used in the model is the same. The prediction is made easier by inserting the empirical slopes. This is the same approach that was used in both the macrocell prediction and the microcell prediction.

5.2.4   Applications of the Lee Model

The model provides two convenient and easy means for designing in-building systems. First, buildings are classified into different categories based on the construction materials. These building loss curves derived from the measurements in different buildings can be put into different look-up tables. As more and more measurement data are collected, the tables will cover more building loss curves and can be applied to predict the losses from different buildings accordingly. Second, once the building loss curve is identified based on the building type, the model can be used to design an in-building system by placing the transmitter at different locations and at different heights and with different antennas and different ERPs to ensure that the optimized coverage is achieved with minimum interference.

5.2.5   Characteristics of the Measured Data

Two sets of measured data were collected in two different office buildings. As more measured data are collected, the model will be fine-tuned. The first set of building characteristics was studied by covering an area from three different transmitter locations (nodes 1, 2, and 3) in the Qualcomm building, as shown in Fig. 5.2.5.1. Some special cases were examined from this set of tests in the building, such as the different signal attenuations that from the elevators. The path loss from LOS and non-LOS conditions crossing a single room and multiple rooms are also considered.

The second set of building characteristics was collected from the fifth floor of the AirTouch building located in Walnut Creek, California, shown in Fig. 5.2.5.2.

The size of the building is shown in the figure. All walls are made of business building material with standard sheetrock and metal studs. In Fig. 5.2.5.2, the clear circles are three transmitter locations: 1, 2, and 3. The shaded circles marked "1" are the receiver spots. The solid dark squares and circles are building beams. Each time, only a transmitter was used, and the measurements were taken the same way as in the Qualcomm building. The position of the transmitter and the routes of measurements taken are shown in the figures. The comparison of the measured data with the predicted results will be shown later.

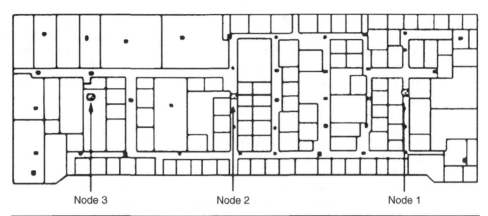

Node 3   Node 2   Node 1

FIGURE 5.2.5.1   Qualcomm building A (first floor) showing the three sectors.

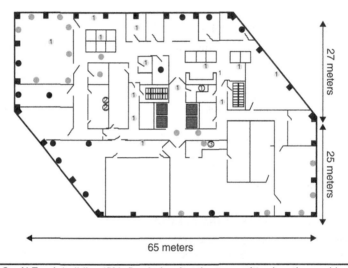

27 meters

25 meters

65 meters

FIGURE 5.2.5.2   AirTouch building (fifth floor) showing the transmitter location and locations of measurement.

The layouts of the building floors, shown in Figs. 5.2.5.3 and 5.2.5.4, were created from virtual buildings for the purpose of exercising the prediction tool. They were rectangular shapes about 100×35 m. The wall was assumed to be made of typical business building material with standard sheetrock and metal studs. Both figures show the coverage plots.

Figure 5.2.5.3 shows the contour plot of a single cell. The signal strength behind the elevator (a special room), indicated in Fig. 5.2.5.3, was attenuated about 10 dB more than other locations in a regular room.

Figure 5.2.5.4 shows the best server plot with different colors indicating different signal strength thresholds. There are two cell sites on the same floor: C-481 and C-485.

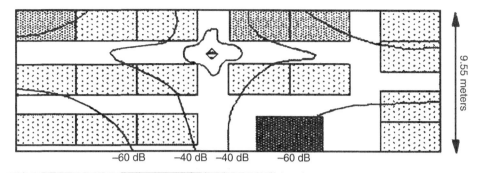

FIGURE 5.2.5.3 AirTouch building contour coverage plot.

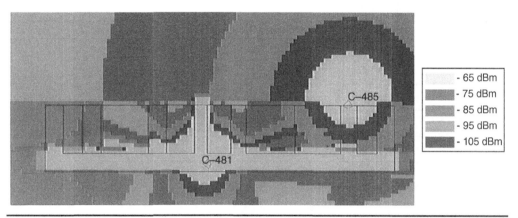

FIGURE 5.2.5.4 Sample best server plot with different signal strength thresholds. (A color version of this figure is available at www.mhprofessional.com/iwpm.)

These two figures show the signal coverage distributions in the different floor layouts based on the path-loss prediction model. The hallways that act like waveguides have stronger coverage.

5.2.6 Validation of the Model (Measured versus Predicted)

Figure 5.2.6.1 shows the measured versus predicted chart on the first floor of the Qualcomm building A. Both curves are matched fairly well. We also follow the process specified in Sec. 5.2.5 to obtain the slopes of the regular rooms and special rooms applied to the in-building model.

Figure 5.2.6.2 shows the CDF of the deviation (difference between measured and predicted) chart for the Qualcomm building. The mean deviation between measured and predicted signal strength was found to be 2.13 dB, and more than 85 percent of the predicted values were within a deviation value of 4 dB. The prediction model is fairly good based on the low deviation shown in the figure.

Figure 5.2.6.3 shows the comparison between predictions and measurements on the fifth floor of the AirTouch building (see also Fig. 5.2.5.2). The mean deviation is about 3 dB, and more than 80 percent of the predicted values were within a deviation value of 4 dB. It shows that the prediction model is fairly good.

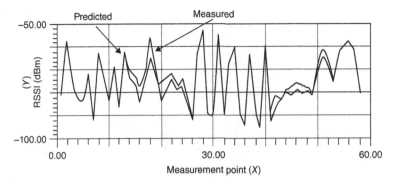

X: Measurement points are locations of where signals were measured
Y: Received signal strength is the measured signal strength (dBm) at
   the measured location

Figure 5.2.6.1   Measured versus predicted (Qualcomm building).

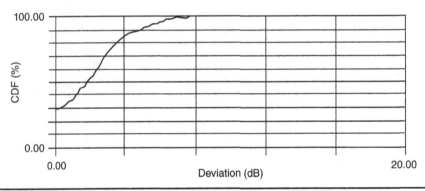

Figure 5.2.6.2   Delta chart (Qualcomm building).

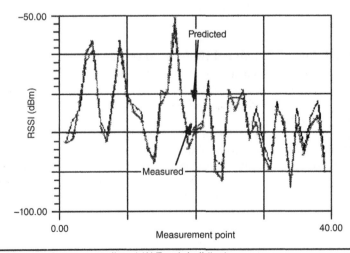

Figure 5.2.6.3   Measured versus predicted (AirTouch building).

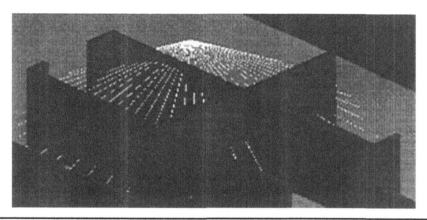

FIGURE 5.2.7.1 3D capability of in-building coverage.

5.2.7 Balance Between Coverage and Interference

The Lee model was created to predict propagation on a single floor in different buildings based on the set of measured data. We have also shown that the predicted values are a close approximation of the measured ones. However, more measured data need to be collected in order to fine-tune this model, shown in Sec. 5.3. Further work should concentrate on making this model applicable to intrafloor coverage/interference of a building and fine-tuning the model based on more measured data.

Balancing the coverage and interference is extremely important in the application of in-building system design. Figure 5.2.7.1 shows a 3D capability of in-building coverage. Figure 5.2.7.2 shows interference that especially can come from macro-, micro-, and picocells. Minimizing interference in managing a desired outcome between coverage and capacity in a dense urban area is challenging.

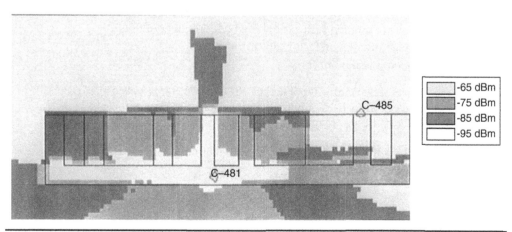

FIGURE 5.2.7.2 In-building interference plot. (A color version of this figure is available at www .mhprofessional.com/iwpm.)

5.2.8   Analyzing the Lee In-Building Prediction Model

It is necessary to evaluate the prediction model by estimating the link budget for a typical indoor wireless system. Both the Lee and the Keenan–Motley in-building propagation models are analyzed in two situations—LOS and NLOS—and from the estimation of the link budget for the coverage boundary in the in-building applications. Some issues regarding the Lee and modified Keenan–Motley models are identified and discussed. Finally, an enhancement of the Lee model is presented that provides better flexibility and accuracy.

5.2.8.1   Description of the Keenan–Motley Model and the Lee In-Building Model

Depending on the particular values used for different parameters, we set up the link budget to be 110 dB. Then, from the measurement data, we found that the coverage distance under the LOS situation varies from 200 to 300 m. Under the NLOS situation and with one room at the end of building, the coverage distance varies from 160 to 200 m. This demonstrates largely that the coverage is limited by the NLOS situation. For example, in the LOS situation, 3 dB can extend coverage distance from 15 to 30 m. In the NLOS situation, 3 dB does not provide much extension on coverage. Nevertheless, it does provide some enhancement of the RF signal quality, especially in the C/I calculation.

5.2.8.1.1   The Keenan–Motley Model[12]

$$P_r = P_t - L \ (\text{dB}) = P_t - 32.5 + 20 \ \log(f) + 20 \ \log(d) + K \cdot F(K)$$
$$+ \ P \cdot W(K) + D(d - D_b) \tag{5.2.8.1.1}$$

where

Free space formula $= -32.5 + 20 \ \log(f) + 20 \ \log(d)$

$\quad L = $ path loss (dB),

$\quad f = $ frequency in MHz,

$\quad d = $ transmitter to receiver separation in km,

$\quad K = $ number of floors traversed by the direct wave,

$\quad F(K) = $ floor attenuation factor (dB) based on K number of floors,

$\quad P = $ number of walls traversed by the direct wave,

$\quad W(K) = $ wall attenuation factor (dB) based on K number of floors,

$\quad D = $ linear attenuation factor (dB/m), and

$\quad D_b = $ indoor breakpoint (m).

For distances above the breakpoint, typically add 0.2 dB/m (typical breakpoint = 65 m). Since this section focuses only on the same floor prediction, those parameters that deal with different floors are not discussed.

5.2.8.1.2   The Lee Model   The Lee model formula can be expressed as

$$P_r = P_t + G_t - L_{LOS} - L_{(B)} - L_{(c)} + G_r \tag{5.2.8.1.2}$$

The symbols P_t, G_t, L_{LOS}, and G_r, are shown in Sec. 5.2.2.1.

- When in the LOS condition

$\quad L_{(B)} = 0$

$\quad L_{(C)} = 0$

- When in the NLOS condition

  (a) Receiver is in a regular room

  $$L_{(B)} = L_{room} \qquad \text{(see Eq. (5.2.2.2.3))}$$
  $$L_{(C)} = 0$$

  (b) Receiver is in a special room

  $$L_{(B)} = L_{Special\ room} \qquad \text{(see Eq. (5.2.2.3.1))}$$
  $$L_{(C)} = 0$$

  (c) Receiver is outside the building

  $$L_{(B)} = L_{room} \qquad \text{(see Eq. (5.2.2.2.3))}$$
  $$L_{(C)} = L_{outside} \qquad \text{(see Eq. (5.2.2.4.1))}$$

5.2.8.2 Sensitivity of Setting the Attenuation Factors in the Keenan–Motley Model

The Keenan–Motley model is analyzed through its attenuation factors, which provide benchmark data in different scenarios for in-building coverage and the sensitivity of the attenuation factors. The fundamental three cases are LOS, NLOS coverage at the building edge, and NLOS coverage through multiple rooms.

5.2.8.2.1 LOS Situation Figure 5.2.8.2.1 shows an assumed scenario that the client could see the antenna directly in the outdoor environment where $h_1 = 30$ m and $h_2 = 3$ m on a narrow street. There are two propagation characteristics in this situation. Within the close-in distance, the propagation loss from the antenna to mobile experiences the free space loss. The free space formula is used to calculate the path loss. Beyond the close-in distance, additional LOS loss will be applied to the free space formula.

Figure 5.2.8.2.2 shows the difference between free space path loss and LOS path loss according to the empirical formula for illustration. The transition point D_b can be clearly seen at about 74 m, as described in Sec. 5.2.1.3. The parameter D in Keenan–Motley's Model for LOS after free space loss (i.e., after the breakpoint D_b) is suggested to be 0.2 dB/m based on the measured field data. However, it usually varies, depending on the kind of building and antenna location. Figure 5.2.8.2.3 shows the sensitivity of the path loss when the attenuation parameter D is less than 0.2 dB/m.

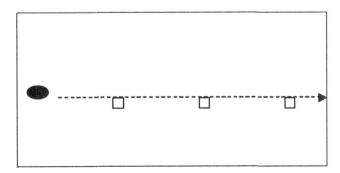

FIGURE 5.2.8.2.1 LOS situation.

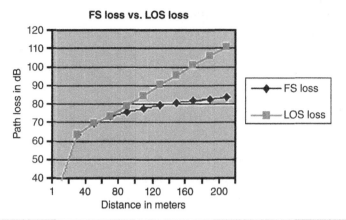

FIGURE 5.2.8.2.2   Free space loss and LOS loss.

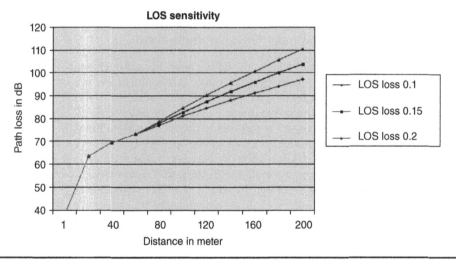

FIGURE 5.2.8.2.3   LOS loss sensitivity analysis at 1800 MHz. (A color version of this figure is available at www.mhprofessional.com/iwpm.)

As shown in Fig. 5.2.8.2.3, the attenuation parameter D plays a critical role in the limitation of coverage area in the LOS scenario. Taking a 110-dB link budget as a guideline, then, when the 0.1-dB/m value is applied, the LOS coverage for 1800 MHz can be up to 300 m. When the 0.2-dB/m value is used, the LOS coverage for 1800 MHz can be only up to 200 m.

Usually, there is a loss difference, $20 \log(f_1/f_2)$, between two different frequency bands. For example, the RF signal at 900 MHz is about 6 dB stronger than 1800 MHz in the same location. However, it is highly correlated to the propagation environment. In dense urban areas, the difference can be up to 10 dB. As shown in Fig. 5.2.8.2.4, at 900 MHz, an attenuation factor of 0.1 dB/m provides better coverage than all other situations. However, if the coverage for different frequencies is considered, path-loss differences among the different frequencies are not as critical. If we apply $D = 0.2\text{dB/m}$

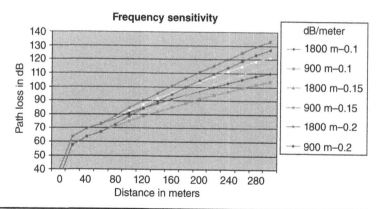

FIGURE 5.2.8.2.4 Frequency sensitivity analysis for LOS situations. (A color version of this figure is available at www.mhprofessional.com/iwpm.)

to calculate the coverage difference between 900 and 1800 MHz, the frequency of 900 MHz takes only 30 m to compensate for a 6-dB gain over 1800 MHz, as seen from Fig. 5.2.8.2.4.

5.2.8.2.2 NLOS Situation In general, the NLOS situation mostly decides the ultimate coverage area for the RF signal. As shown in the Fig. 5.2.8.2.5, the scenario of the impact on coverage in a multiple-walls situation is demonstrated with the Keenan–Motley model. Assume that the transmitter was about 60 m away from the antenna, that there are three rooms lined up straight, and that each room is about 10 m in width.

The Keenen–Motley model uses a direct wall room loss formula in the scenario that the direct wave intersects one or more rooms. The wall loss for each room, again, is building material dependent. Usually, 10 dB is applied to per room loss.

As shown in Fig. 5.2.8.2.6, the wall loss factor is the dominant factor in determining how far the coverage is. Depending on which propagation environment formula is used, the coverage area can be from 75 m (15 dB/wall—two walls) to more than 100 m (5 dB/wall—three walls). Again, the coverage area is calculated based on a 110-dB link budget.

In most cases, one scenario is used by taking a straight LOS to the room at the other end of the building, as shown in Fig. 5.2.8.2.7.

In Fig. 5.2.8.2.7, the scenario shows a direct LOS wave received in the room at the end of the building. However, under certain sensitivity analyses, a rough estimation of the coverage area can be obtained. This estimation can be used as a baseline for any specific building coverage.

As shown in Fig. 5.2.8.2.8, the coverage can vary from 160 m based on an attenuation factors of 0.2 dB/m. to 220 m. based on 0.1 dB/m, with a 10-dB/wall loss at a link budget of 110 dB.

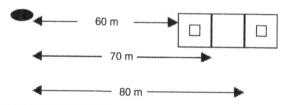

FIGURE 5.2.8.2.5 Multiple-room scenario.

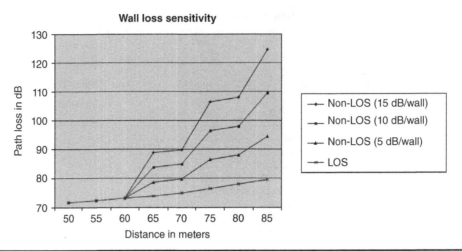

FIGURE 5.2.8.2.6 Wall-loss sensitivity analysis. (A color version of this figure is available at www.mhprofessional.com/iwpm.)

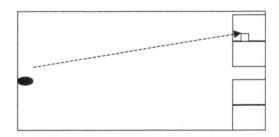

FIGURE 5.2.8.2.7 A scenario that the receiver is in a room at the other end of the floor.

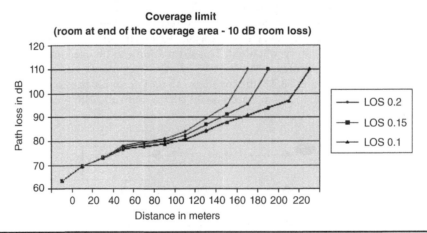

FIGURE 5.2.8.2.8 Coverage distance sensitivity (from the scenario of Fig. 5.2.8.2.7). (A color version of this figure is available at www.mhprofessional.com/iwpm.)

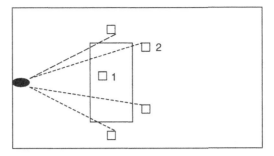

FIGURE 5.2.8.3.1   Mobile outside the room but blocked.

5.2.8.3   Potential Areas for Lee Model Enhancements

To make the model effective, simple formulas are provided by the Lee model to cover general cases. However, there are some special cases need to be addressed.

Potential issue with the Keenan–Motley model are the following:[12]

1. The LOS loss applied after the breakpoint can be in question. This number can be potentially huge and is not practical. For example, the 0.2-dB/m attenuation factor after the breakpoint can potentially run to a 40-dB loss within a 200-m span. Also, the loss within the LOS category is difficult to justify and apply. This issue was discussed in the Lee model.

2. The NLOS loss right behind an office in a corridor is not properly addressed. As shown in the Fig. 5.2.8.3.1, the situation when the mobile is behind the room (not inside the room) is not properly addressed. For example, in Fig. 5.2.8.3.1, the mobile at locations 1 and 2 will have the same predicted signal strengths, while in reality, the signal strength at point 2 will most likely be much higher because of the deflection of the radio wave. Applying a flat wall loss will not be general enough to address different locations behind and around a room. This issue has been better addressed in the Lee model.

5.2.8.3.1   Enhancements of the Lee Model

The following areas are properly taken care in the enhancement of Lee model in Sec. 5.3:

1. LOS loss is represented by the L_{LOS} parameter, in which the RF characteristics are properly addressed and used after the close-in distance.

2. Loss inside the room is properly addressed with its RF characteristics inside the room.

3. Aggregate calculation of the model takes care of not only the thickness of wall but also the number of rooms where the radio wave actually travels.

Beside the areas of concern stated above, there are two additional enhancements. The layering concept is used to include the quality of service (QoS) requirement and the characteristics of room/area. A special area can be identified by marking that area with a specific value. The model has the intelligence to handle the propagation in these two areas. These two enhancements can ensure that the deployment of WLAN in picocells

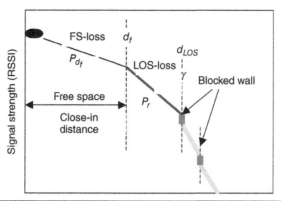

Figure 5.2.8.3.2   Illustration of the enhancement by adding losses blocked by walls.

can meet the QoS requirement and is more cost effective. The enhancement gained by adding the loss due to the blocks of waves by walls is illustrated in Fig. 5.2.8.3.2.

As illustrated in Fig. 5.2.8.3.3, the enhancement is taking care of the one LOS and the two NLOS scenarios. The enhancements shown in dotted lines are affected by the number of walls.

The prediction of signal strength in four different scenarios can be described by four different propagation curves. The first curve illustrates the LOS curve that is composed of the free space curve, the breakpoint, and the LOS curve after the breakpoint. The second curve illustrates the received signal strength indicator (RSSI) curve with the wall loss that is directly related to the type of the wall and total aggregated thickness of the wall. The third curve illustrates the loss due to the thickness of walls and beyond the

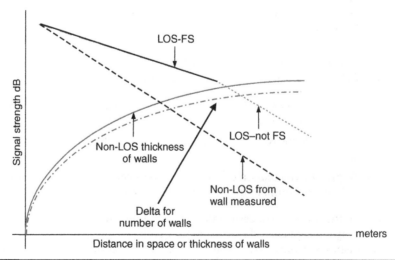

Figure 5.2.8.3.3   Illustration of the enhancements in dotted lines shown the effect by considering the number of walls.

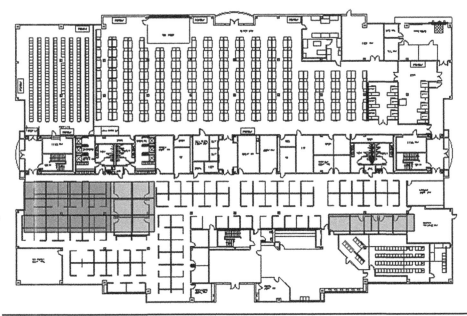

Figure 5.2.8.3.4 The layout of building.

last wall. The fourth curve illustrates the correction from the third curve by considering the number of the walls. At any specific point in the building, the received signal strength can be calculated based on the enhancement from two dotted curves.

Another enhancement on special areas with their QoS is shown in Fig. 5.2.8.3.4. Here, the QoS and area of interest can be marked. These marked data will be used for the enhanced Lee in-building model to calculate associated parameters. The offices for CEO/VPs can be marked with high QoS. During the calculation, these areas can be given more attention by an algorithm to provide better coverage without creating interference, or the interference can be tolerated by lowering the special QoS in some areas. The same principle is applicable to other special concerns. Thus, a good wireless network actually handles coverage, interference, QoS requirement, and the treatment of special areas. Figure 5.2.8.3.5 indicates that the signal path passes each bin in the building. Each bin has the information data stored, as shown in the figure.

As shown in Fig. 5.2.8.3.5, there are many factors that show clearly what it means to deploy a good in-building wireless system. Along the radial, many characteristics are identified in the factors so that the final design of the network is the combination of all these factors. Each bin along the radio path can potentially have an impact on the signal strength. As shown in Fig. 5.2.8.3.5, the highlighted demographic bin might have special characteristics that reduce the signal loss. The dotted line represents the signal strength without considering the characteristic of the bin along the radial path. The solid line represents the actual signal strength with the characteristic of the bin considered.

The received signal, after passing through close-in distance D_c to reach the room, is

$$P_r = P_{LOS} = P_{D_c} - \gamma \log \frac{d}{D_c}$$
(5.2.8.3.1)

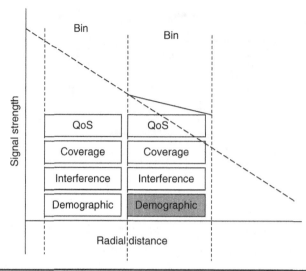

FIGURE **5.2.8.3.5** Specific data stored in radial array.

where P_r = the received signal strength in dBm in the LOS but in the non–free-space area,
  P_D' = the received signal in dBm is at the close-in distance where is the intercept of the free space slope line,
  γ = the slope in range from the location at D_c to the wall (the transition slope),
  D_c = close-in distance in meters,
  FS_{LOS} = free space path loss, and
  d = the mobile radial distance from the base transmitter to the mobile receiver in meters.

The line γ is drawn so that it intersects at a point within the close-in distance D_c to the first wall. Thus, the received signal strength from D_c to the room beyond the wall is obtained.

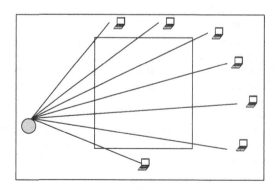

FIGURE **5.2.8.3.6** Client blocked by one room.

The received signal inside the room is

$$P_r = P_{LOS} - L_{in\ wall} - L_{room}$$

$$= P_{LOS} - \underbrace{\sum_{i=1}^{n} m_{in\ wall_i}\ Wall_{loss_i}}_{L_{in\ wall}} - \underbrace{\sum_{i=1}^{n} m_{room_i} \log\left[(r + d_{LOS})/d_{LOS}\right]}_{L_{room}} \qquad (5.2.8.3.2)$$

where P_r = received power in dBm in the room,

P_{LOS} = the power at the interception point of LOS and the first wall in dBm,

n = number of walls,

$m_{in\ wall}$ = the path-loss slope for each wall in dB/dec (obtained in Sec. 5.3.2.1),

$Wall_{loss_i}$ = loss due to the ith wall,

r = the mobile radial distance from the base transmitter to the mobile receiver in meters,

m_{room} = the path-loss slope for inside the room in dB/dec, and

d_{LOS} = LOS distance from the transmitter to the first wall in meters.

The general formula of the Lee model is shown in Sec. 5.3.5, where $L_{in\ wall} = L_{(D)}$ is expressed in Eq. (5.2.8.3.2).

5.2.8.3.2   Measurement Integration   The success of any prediction model relies on taking advantage of field measurement data and fine-tuning the model formula for different environments. Due to the enhancement on the Lee model, the four curves shown in Fig. 5.2.8.3.3 can be adjusted through measurement integration.

As shown in Fig. 5.2.8.3.7, different setups are used in collecting measured data to tune the Lee model.

The first one will be in the LOS situation. In the LOS situation, two different cases need to be specified. One is within the Fresnel zone, and the other is after the Fresnel zone area. This needs to be distinguished so that behavior after the close-in distance can be further fine-tuned. Based on the measured data for handling this category, the path-loss curve can be derived. The various path-loss setups are highlighted in Fig. 5.2.8.3.8.

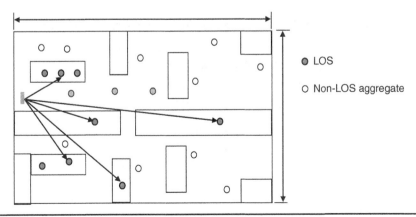

Figure 5.2.8.3.7   Measurement data collection.

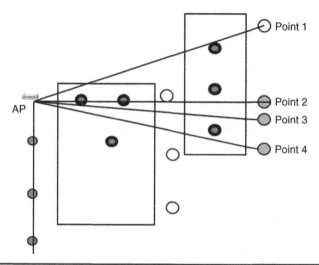

FIGURE 5.2.8.3.8 Measurement points collection.

The next category is for handling the NLOS situation when the client is inside or outside the room. This category needs handle two different cases since the behaviors are different in these two cases. The first case is for those data points that are inside the room but in the NLOS situation. The second case is for those data points that are outside the room but in the NLOS situation. Based on the location of the client, the accumulated thickness of the wall can be extracted. Thus, the path-loss curve against the blockage thickness can be derived.

In circumstance when the client is in the room, the method of calculating wall thickness needs to be enhanced to consider the distance from the wall. This issue is not addressed in the Keenan–Motley models, as shown in its formula:

$$L \text{ (dB)} = 32.5 + 20 \log(F) + 20 \log(d) + K \cdot F(k) + P \cdot W(k)$$
$$+ D(D - Db) + \text{Flag (in room)} \tag{5.2.8.3.5}$$

where Flag (in room) is the value of the loss, depending on the client is whether inside or outside the room. The parameters of Eq. (5.2.8.3.5) are shown in Sec. 5.2.8.1.

5.3 Enhanced Lee In-Building Model

At Republic Polytechnic (RP) in Singapore, measurement data were collected from a 2.4-GHz WLAN-developed system installed in many different floors in two different wings (towers) that are connected by a corridor. The measured data were collected from single-floor, interfloor, and interbuilding measurements. This section provides a complete solution for in-building-related propagation by validating the Lee in-building model[1,3,28] for a single-floor scenario first. Then the performance of the Lee in-building model for interfloor and interbuilding scenarios are examined. Also, the FDTD model[29] was used to validate the measured data with the Lee model to provide another reference point. The results show that the Lee model outperforms the FDTD model in calculating both speed and accuracy.

The in-building propagation environment is unique and challenging for system deployment. The Lee model was used to optimize the coverage and interference while using fine-tuning radio parameters (e.g., antenna types, ERP, and downtilt) to improve the system performance and to enhance the user's experience. The networks maintained continuous connectivity with smooth roaming and high data throughput, providing a great experience for users.

5.3.1   Highlight of the Enhanced Lee Model

In-building wireless system deployment and performance have their unique challenges, especially in meeting the high traffic density, high data throughput, and large number of users in a vertically concentrated system. This is particularly common and challenging in a skyscraper type of office building in dense urban areas. Due to the complications of the indoor environment, there is much research literature related to in-building propagation besides the Lee in-building model.[4,6,14,29,30,31]

This section provides a complete solution for in-building-related propagation. We first present the measured signal results received inside and between the buildings, examine the accuracy of the Lee model for the same-floor prediction, and then extend the Lee model's principle to derive the formulas for interfloor and interbuilding scenarios.

Because Finite-difference time-domain (FDTD) is a popular computational technique for the electrodynamics modeling. It is considerable easy to understand and easy to implement in software. The commercial FDTD model and the in-house developed FDTD models are used to validate the Lee model with the measured data and to make sure all data and results are valid.

5.3.2   Studying Measured Data in various Cases

Measurements were taken from the first to fifth floors in the north and south wings of a building in the Republic Polytechnic (RP) in Singapore. The RP building is a 12 floors high. The north and south wings are of the same shape in both buildings. The dimension of each floor is 31, 8.4, and 2.5 m in length, width, and height, respectively. Figure 5.3.2.0.1 shows a picture of the buildings.

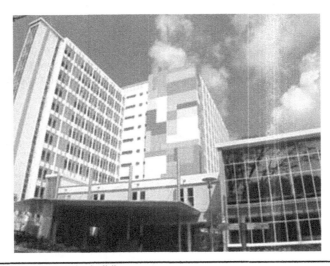

Figure 5.3.2.0.1   Picture of the buildings.

Between these two wings, there is a corridor junction, which is 8×10 m in length and width, as shown in Fig. 5.3.2.0.2. On each floor, the brick walls were used to divide several classrooms (three classrooms in the north wing and three classrooms in the south wing) and an aisle. The furniture in the classroom includes school desks and chairs, which were made of plastic and reinforced metal brackets. Both the north and the south sidewalls were full of windows. There were three access points (AP) in the north and south wings on each floor. The three APs were operational on three signal channels: channels 1, 6, and 11. The APs were mounted fixed at a height of 1.5 m.

5.3.2.1   A Same-Floor Case

Figure 5.3.2.1.1 shows a top view of south wing and the locations where the signal strengths were measured and the locations of APs. On the same floor, we took the measurements on each floor from the second to the fourth floor. On every floor, nine APs are shown, and 18 spots were picked up to measure the signal strength. The positions of these spots are the same for each floor.

In Figs. 5.3.2.1.1 and 5.3.2.1.2, the green circles represent the desks and the surrounding blue circles are chairs. The red rectangles represent the measurement points. The yellow bars on each side of the building are windows, and the black bars are the doors. The frequencies are transmitted on channels 1, 6, and 11 on 802.11b. Some pictures were taken to correlate the measured data with the environment. The red arrows indicate the directions of the camera.

Figure 5.3.2.1.2 shows a top view of north wing and the locations where the signal strengths were measured as well as locations of APs. The layouts for both south and north towers are almost the same. Both have nine APs and many windows on each floor. The divider in each floor is almost the same as well. Two towers are connected with a corridor. This makes radio planning extremely difficult.

In Fig. 5.3.2.1.2, we divided these points into two groups: one belongs to the LOS, and the other belongs to the NLOS. P2, P4, and P6 were transmitted into NLOS areas. The rest APs were transmitted in the LOS areas. As there are APs on both sides of the building inside and outside the room, it is difficult to mark these measurement points as LOS and NLOS in Figs. 5.3.2.1.1 and 5.3.2.1.2. The Lee model uses these points to calculate the prediction of path loss. For the LOS case, the loss can be calculated from Sec. 5.2.2.1.

The slope of the interior wall loss is derived through the standard integration process of measurement, as shown in Fig. 5.3.2.1.3. The slopes of wall loss from both wings of the building were obtained from the best fit of the measured data. It replaced the old value of $m_{in\,wall}$ in Eq. (5.2.8.3.2). Then the new value was used to calculate the prediction values.

5.3.2.2   An Interfloor Case

Measurement data were also collected from the nine APs at different floors and different wings.

The locations of the measurement spots and the APs are exactly the same as we used in the same-floor case. The APs were fixed at one side of the ceilings (south side of north wing and north side of south wing), while the test spots were spread all around on each floor, as indicated in Fig. 5.3.2.1.2. In general, we believe that the interfloor propagation is a combination of diffraction and penetration.

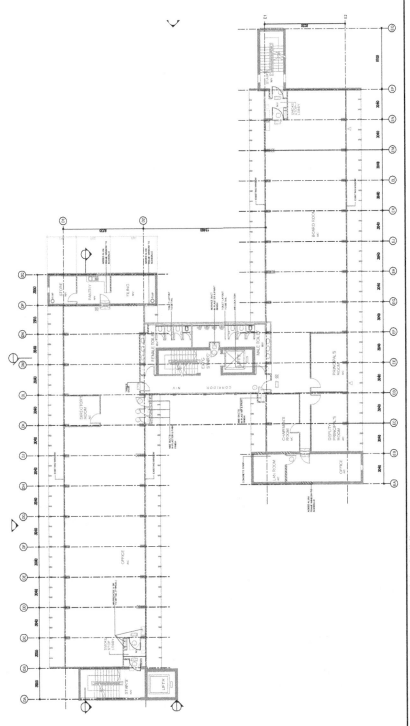

Figure 5.3.2.0.2 Top view of the buildings.

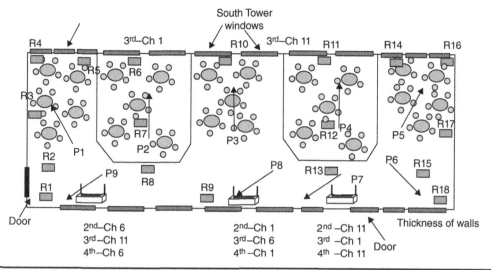

FIGURE 5.3.2.1.1 Layout of the floor and measurement locations of the south tower. (A color version of this figure is available at www.mhprofessional.com/iwpm.)

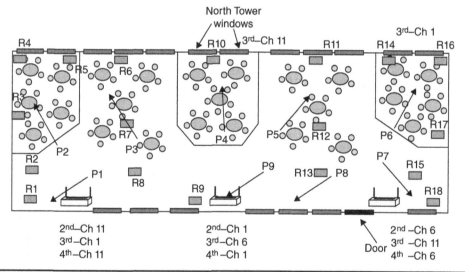

FIGURE 5.3.2.1.2 Layout of the floor and measurement locations of the north tower. (A color version of this figure is available at www.mhprofessional.com/iwpm.)

5.3.2.2.1 Interfloor Measured Data

The transmitting power is 1 mw at a frequency of 2.4 GHz. Figure 5.3.2.2.1.1 shows the path loss in relative decibels versus the distance between T_x and R_x in meters plotted on a decade scale x, or 10^x meters. There are four scenarios of the interfloor propagation: two in the north wing and two in the south wing. The scenarios were propagating such as that two of them are the loss from one-floor separation and another two of them are loss from two-floor separation, as indicated in the figure.

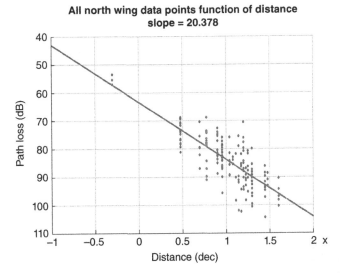

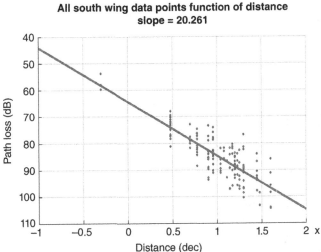

FIGURE 5.3.2.1.3   Slope of the wall loss (10^x in meters).

The red data points show the test spots, which had the clear diffraction path to the APs, while the blue points show the test spots, which had only the penetration path to the APs. We noticed that the penetration propagation suffered more loss than the diffraction propagation. Similarly, in the same-floor case, depending on whether there is a diffraction path, we divided the data points into two groups—the diffraction group and the penetration group—based on the locations of the spots.

From the measurement plots, we cannot really see the effects clearly caused by either the diffraction loss or the penetration loss. It is caused by a combination of kinds of loss.

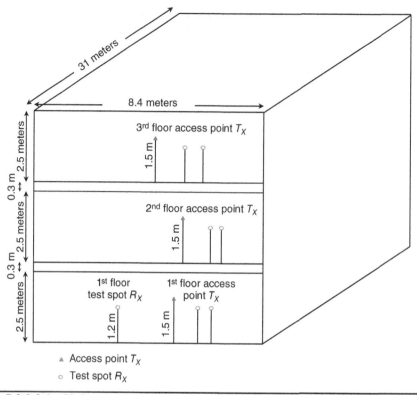

FIGURE 5.3.2.2.1   3D Cross-sectional view of building.

One-Floor Separation Case: At 1-m distance (0 dec), which is from the transmitter to the ceiling, the path loss is 65 dB, as shown in Fig. 5.3.2.1.3 (same floor). In Fig. 5.3.2.2.1.1(*a*) and (*b*), the minimum distance between the transmitter at one floor and the receiver at the second floor is 2.5 m (0.4 dec). From the plot (*a*) and (*b*), the losses at 2.5 m are 76 dB. The difference in dB is 11. Hence, the one-floor separation path loss from the measured data is 11 dB/floor.

Two-Floor Separation Case: From Fig. 5.3.2.2.1.1(*c*) and (*d*), the minimum distance from transmitter at one floor and receiver at the third floor is 5 m (0.7 dec). We have already found from Fig. 5.3.2.1.3 that the path loss is 65 dB from the transmitter to the ceiling, which is 1 m. From Fig. 5.3.2.2.1.1(*c*) and (*d*), the losses at 5 m are about 84.5 dB. The difference in dB is 19.5 dB from two-floor separation loss. Hence, the loss/floor from the measured data is 9.75 dB.

These measured data were collected at 2.4 GHz.

5.3.2.2.2   Implementation of the Interfloor Model

As explained above, the penetration propagation will be included in our current scenario. The general signal strength prediction of Eq. (5.2.2.2.4) for the same-floor case will be modified.

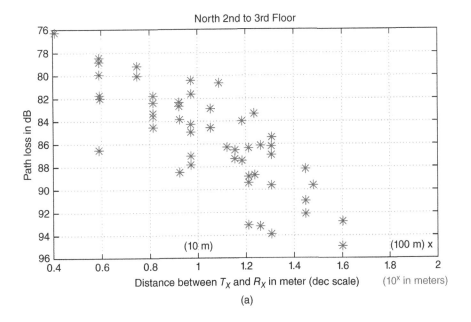

(a)

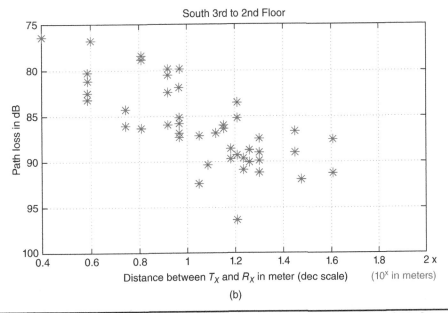

(b)

FIGURE 5.3.2.2.1.1 Interfloor measurement data characteristics. (A color version of this figure is available at www.mhprofessional.com/iwpm.)

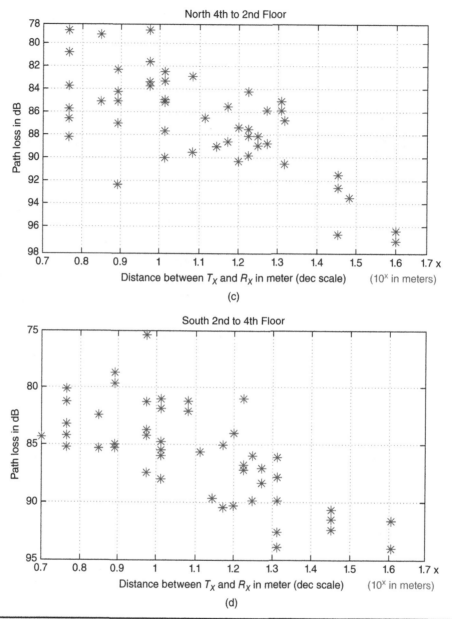

FIGURE 5.3.2.2.1.1   *(Continued)*

When the transmitter and the receiver are separated on different floors, the path loss is due to the two components: one from the diffraction loss, and the other from the penetration loss. However, it is very hard to distinguish which component contributes the most at a certain location. Therefore, we just use the measured data from many sources,[4,32–36] including our measured data, to come up with an empirical model.

This model applies only to conventional office buildings, which are not odd shaped or attached to other unmatched buildings. We have shown that the RP building in Singapore (Sec. 5.3.2.2.1) is not a conventional building; thus, the data collected from the RP building have different characteristics.

A. The Lee Empirical Model for Interfloors   The path loss $L_{i\text{-}f}$ due to the interfloor case when the transmitter and the receiver are in the different floors can be expressed as follows:

For the path loss between any two floors at first seven floors (six-floor separation), the interfloor loss $L'_{i\text{-}f}$ goes linearly as the loss per floor for every floor.

$$L'_{i\text{-}f} = m_{f1} \cdot n_{\Delta f} + L_{s\text{-}f} \qquad (5.3.2.2.2.1)$$

where m_{f1} is the loss per floor, $n_{\Delta f}$ is the number of floor separations, $L_{s\text{-}f}$ is the loss received at the receiver when it is on the same floor as the transmitter and is located either-lose to the floor/ceiling or to the floor/ground, and m_{f1} is a function of frequency. From the empirical data, we can express it as

$$m_{f1} = 1.93 f_{\text{GHz}} + 6.36 \qquad (5.3.2.2.2.2)$$

At 850 MHz, $m_{f1} = 8$ dB/floor, and $L_{s\text{-}f} = 10$ dB (from Fig. 5.3.2.2.2.1).

$$L'_{i\text{-}f} = 8 \cdot n_{\Delta f} + 10 \qquad (5.3.2.2.2.3)$$

At 2.3 GHz, $m_{f1} = 10.8$ dB/floor, and $L_{s\text{-}f} = 10$ dB (from Fig. 5.3.2.2.2.1)

$$L'_{i\text{-}f} = 10.8 \cdot n_{\Delta f} + 10 \qquad \text{in dB} \qquad (5.3.2.2.2.4)$$

From the measured data, we discovered that after the first six-floor separations, the interfloor loss $L''_{i\text{-}f}$ is no longer linear but follows the logarithm scale:

$$L''_{i\text{-}f} = m_{f2} \cdot \log[n_{\Delta f}] + A \qquad \text{in dB} \qquad (5.3.2.2.2.5)$$

where m_{f2} is the loss slope per six floors on a logarithm scale; that is, the loss of the first six-floor separations is 10 dB, the loss of next 12 floor separations is 10 dB, the loss of the next 20 four-floor separations is 10 dB, and so on. The general path-loss slope of m_{f2} is the exponent doubled over a given number of floor separations and can be expressed as

$$m_{f2} = \ell_2 / \log 2 \qquad \text{in dB} \qquad (5.3.2.2.2.6)$$

where ℓ_2 is the path loss over a given number of floor separations as a transition at this separation for different loss slopes. In our case, the number of floor separations is six, and ℓ_2 is 10 db. Thus, the path-loss slope from the first six-floor separations to that after the six-floor separation are different, and the transition is at the six-floor separations.

A is the constant that makes two interfloor loss curves, Eq. (5.3.2.2.2.1) and Eq. (5.3.2.2.2.5), to be connected at a given floor separation as shown in Fig. 5.3.2.2.2.1.

The constant *A* can be found by setting two equations equal to $n_{\Delta f} = 6$, that is, six-floor separation.

$$A = 6 m_{f1} + L_{s\text{-}f} - m_{f2} \cdot \log 6 - \alpha \qquad (5.3.2.2.2.7)$$

where α is the correction factor in dB value added when the measured data show the difference between two slopes at the transition.

From the measured data collected at 900 MHz or lower, the two path-loss slopes are not continuous at the transition, which is six-floor separations. There is a 5-dB-less loss being found at the six-floor separations. At 1800 MHz and greater, the two path-loss slopes are continuous at the transition. Therefore, a correction factor α is

$$\alpha = \begin{cases} 5\,dB & \text{for 900 MHz and lower} \\ 0 & \text{for 1800 MHz and higher} \end{cases}$$

At 850 MHz, $m_{f_2} = 10/\log 2 = 33.33$ is from Eq. (5.3.2.2.2.6), and $\alpha = 5$, $A = 32 - 5 = 27$, is from Eq. (5.3.2.2.2.7): Then Eq. (5.3.2.2.2.5) becomes

$$L''_{i\text{-}f} = 33.33 \cdot \log [n_{\Delta f}] + 27 \tag{5.3.2.2.2.8}$$

At 2.3 GHz, $m_{f_2} = 10/\log 2 = 33.33$, $\alpha = 0$, and $A = 49$ are found from Eq. (5.3.2.2.2.7). Then

$$L''_{i\text{-}f} = 33.33 \cdot \log [n_{\Delta f}] + 27 \tag{5.3.2.2.2.9}$$

From Fig. 5.3.2.2.2.1, we have noticed that most of the path loss is from a one-floor separation to six-floor separations. After six-floor separations, the interfloor path loss becomes weaker as observed from the measured data. All the empirical formulas of the interfloor model are applied only to the conventional office buildings.

B. Moving on the Separated floor   The receiver was moving on the separated floor as the path-loss distance from the transmitter to the receiver increased, as shown on the plots in Fig. 5.3.2.2.1.1(*a*) and (*b*) for the one-floor separation case and in (*c*) and (*d*) for the two-floor separation case. They are the few plots to be depicted in the figure. From many measured data, we can separate the measured data from two different measured areas:

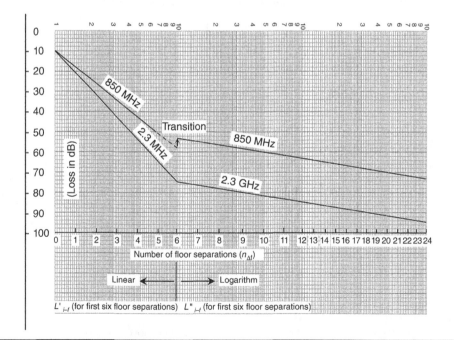

Figure 5.3.2.2.2.1   Interfloor path loss in a conventional office building.

the nondiffraction area and the diffraction area. Two empirical formulas are formed as follows:

$$_1L_n(d) = L(d_0) + {_1}m_n \log d \quad \text{(for diffraction area)} \quad (5.3.2.2.2.10)$$

$$_2L_n(d) = L(d_0) + {_2}m_n \log d \quad \text{(for nondiffraction area)} \quad (5.3.2.2.2.11)$$

where n associated with m is the number of floor separations, m_n is the path slope, and d is the distance from the transmitter to the receiver; d_0 is the minimum distance from the transmitter to the receiver. The numbers 1 and 2 associated with symbols are used to distinguish two different areas:

1. Moving on the one-separated floor ($n = 1$):

   $_1m_1 = 14$ dB/dec

   $_1m_2 = 17$ dB/dec

2. Moving on the two-separated floor ($n = 2$):

   $_2m_1 = 22$ dB/dec

   $_2m_2 = 25$ dB/dec

From the above empirical formulas, we have found that on both one-separated and two-separated floors, the slopes measured in the diffraction area are less attenuated than in the nondiffraction area by 3 dB/dec along the distance from the transmitter to the receiver.

Also, the slope of path loss when moving on the one-separated floor is less attenuated than on the two-separated floor by 6 dB/dec along the distance from the transmitter to the receiver.

The total interfloor path loss can be summed up the floor separation losses from three scenarios; the floor separation loss from Eq. (5.3.2.2.2.1) and Eq. (5.3.2.2.2.5), and the moving-on separated floor loss from Eq. (5.3.2.2.2.10) or Eq. (5.3.2.2.2.11) as

$$L_{\text{interfloor}} = L'_{i\text{-}f} + L''_{i\text{-}f} + {_j}L_n(d) \quad (5.3.2.2.2.12)$$

where $j = 1$ is for measuring at the diffraction area and $j = 2$ is for measuring at the nondiffraction area.

In every case, the user may want to convert the transmitter-to-receiver distance to the distance measured from the minimum distance d_0 on the floor where the receiver is located. These measured data have highlighted some thoughts in predicting the interfloor cases.

5.3.2.3 An Interbuilding Case

5.3.2.3.1 Description of the Case

Measurement data were collected at the north wing from the APs situated in the north wing. Again, due to the limitation of the ERP allowed (1 mw), only limited measurement data were collected.

Based on the specific layout of this building, from Fig. 5.3.2.3.1, the corridor junction has the same floor heights as the north and south wings. Thus, measured data points were divided into two groups: one collected from the propagation with shadow loss and the other from the propagation without shadow loss. The measured data at 2.4 GHz were plotted and are shown in the next section.

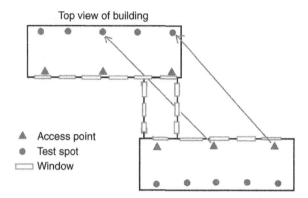

Top view of building

▲ Access point
● Test spot
▭ Window

Figure 5.3.2.3.1   Interfloor layout, top view.

5.3.2.3.2 Interbuilding Measured Data

Figure 5.3.2.3.2.1 shows the path loss between two wings in different situations. One is the path loss received at the same floor, such as from second-floor north wing to the second-floor south wing in the same-floor case; the other is the path loss from the second-floor north wing to the third-floor south wing in the one-floor difference case; and the third is the path loss from the second-floor north wing to the fourth-floor south wing in the two-floor difference case. As expected, generally the path losses of the one-floor separation case were higher than the same-floor case. But the path loss of the two-floor separation case is not necessarily higher than the one-floor separation case. Many deflections, refractions, and penetrations prevailing in different ways and in different cases may be the answer.

In this interbuilding case, we can not take the floor numbers as a direct parameter in calculating the path loss because in a real-world scenario, once a radio signal propagates from one building to another, it is hard to determine how many floors it has penetrated through or whether it has met an obstacle. Otherwise, it would add some complexity to the analysis from the layout of the building. Hence, the method of plotting the path loss as a function of the number of floors would be meaningless without analyzing the exact cause due to the number of floors. We have concluded that there is no simple model for the interbuilding case. However, in Sec. 5.3.4.3, we have modified a formula that is used for the external building wall case, to handle the interbuilding case.

5.3.3 Comparison of Measured Data and Predicted Data

5.3.3.1 Same-Floor Case

We have compared the Lee model results and the measured data in six different locations on three floors, from the second to the fourth, in each wing of the building and on the same floor in the RP building, as shown in Fig. 5.3.3.1.1. At each floor of the building, we do not separate the NLOS condition from the LOS in the measured data. In all cases, the deviation between the Lee model results and the measured data is roughly 5 dB at 60 percent of all the data and 8 dB at 80 percent of all the data. It is shown that the Lee model is a fairly accurate tool to use.

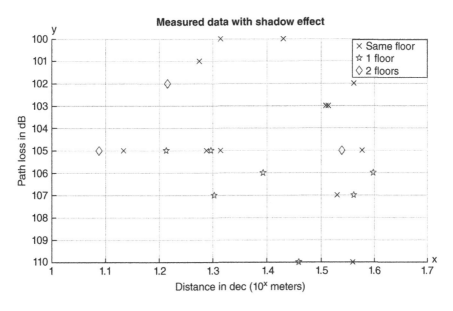

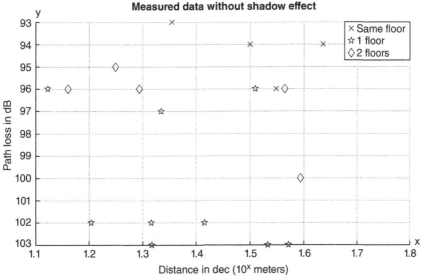

FIGURE 5.3.2.3.2.1 Interbuilding measurement data.

5.3.3.1.1 FDTD Model and Ray Tracing Model versus Lee Model

In this same-floor case, the FDTD model[37] was applied in the RP building. The FDTD model is described in Sec. 5.6.2. We set the whole floor area into 31×8 grids, and each grid was 1×1 m square. The simulation of FDTD model and the data measurement were done for just one room—the second floor of north wing.

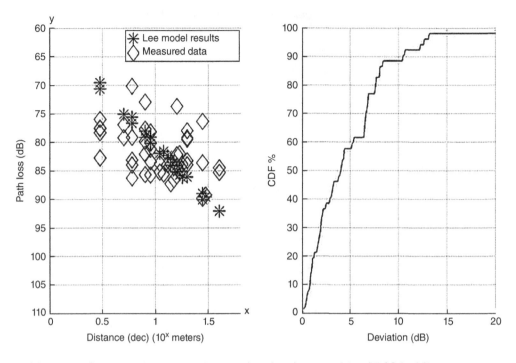

(a) North 2nd floor comparison between Lee model result and measured data (NLOS & LOS)

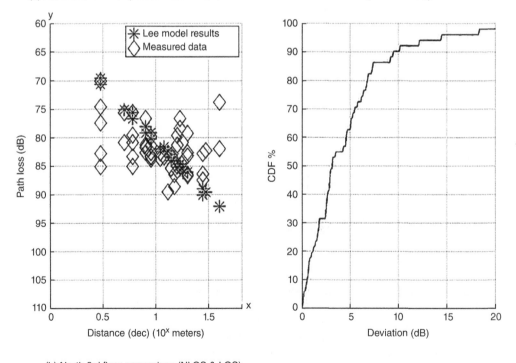

(b) North 3rd floor comparison (NLOS & LOS)

FIGURE **5.3.3.1.1**   Predicted versus measured (each floor—LOS and NLOS).

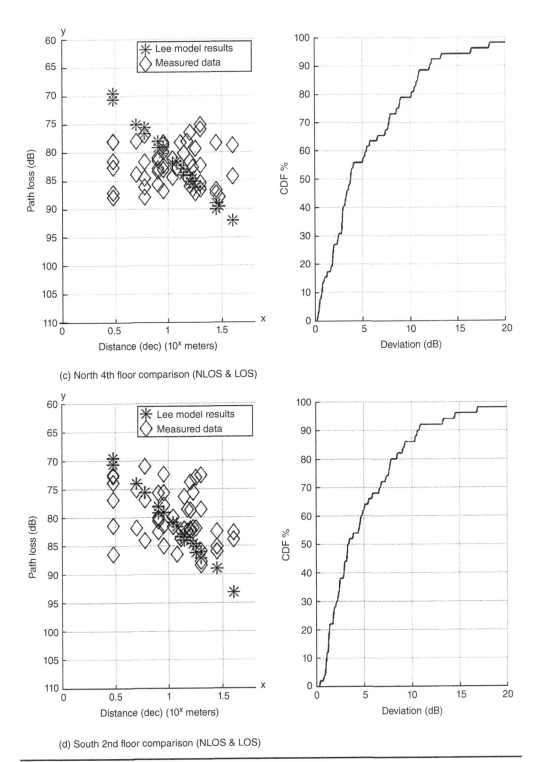

(c) North 4th floor comparison (NLOS & LOS)

(d) South 2nd floor comparison (NLOS & LOS)

Figure 5.3.3.1.1   *(Continued)*

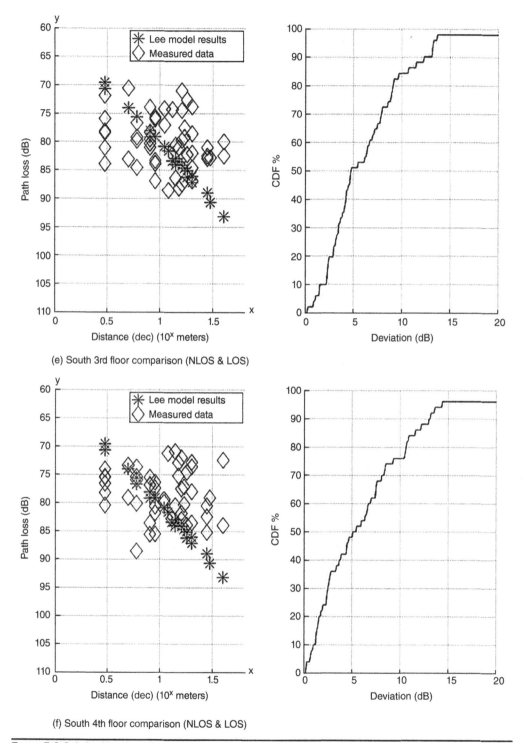

(e) South 3rd floor comparison (NLOS & LOS)

(f) South 4th floor comparison (NLOS & LOS)

FIGURE 5.3.3.1.1 *(Continued)*

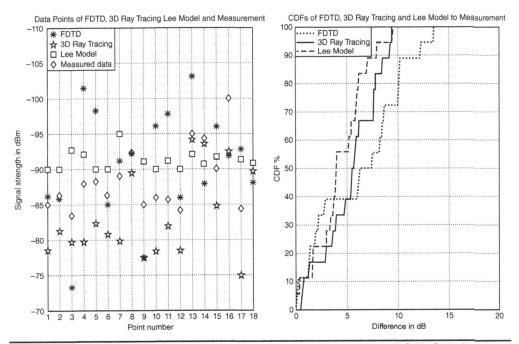

Data Points of FDTD, 3D Ray Tracing Lee Model and Measurement

CDFs of FDTD, 3D Ray Tracing and Lee Model to Measurement

FIGURE 5.3.3.1.2   FDTD, 3D ray tracing, Lee, and measurement. (A color version of this figure is available at www.mhprofessional.com/iwpm.)

From Fig. 5.3.3.1.2, we have calculated that at an 8 dB deviation between the predicted and measured, the CDF of FDTD is around 67 percent while the Lee model in this case is above 80 percent. Thus, the Lee model gives a better prediction than the FDTD model.

To represent the data in a visual display, we created two signal strength coverage plots: one for FDTD coverage and one for the Lee model coverage, as shown in Figs. 5.3.3.1.3 and 5.3.3.1.4, respectively. In these two figures, we have seen that the coverage is bigger, as the red color spreads to more area in the Lee model prediction than the FDTD prediction. It tells that if we use the FDTD model, the predicted signal strength would be underestimated yet the prediction accuracy is not as good as that of the Lee model.

We also simulated a room with the measured conditions by 3D ray tracing.[38,39] The ray-tracing model for in-building is described in Sec. 5.6.1. The results of comparisons between these three models—FDTD, 3D ray tracing, and the Lee model—with measured data are shown in Fig. 5.3.3.1.5. From the CDF plot of deviations between predicted and measured, the Lee model and 3D ray tracing have the same CDF of 70 percent at the deviation of 5 dB. The Lee model has a better prediction than 3D ray tracing when at a deviation of 8 dB, that is, CDF of 95 percent for the Lee model and 79 percent for 3D ray tracing.

FIGURE 5.3.3.1.3   FDTD coverage. (A color version of this figure is available at www.mhprofessional.com/iwpm.)

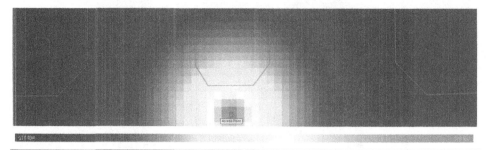

FIGURE **5.3.3.1.4**   Lee model coverage. (A color version of this figure is available at www.mhprofessional.com/iwpm.)

Data points of FDTD, 3D ray tracing, Lee model, and measured data of same floor

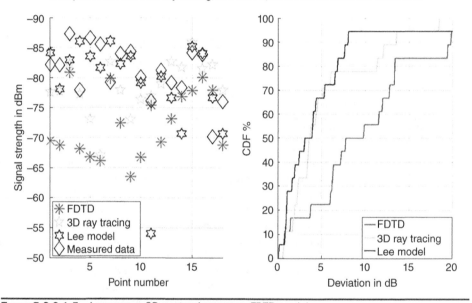

FIGURE **5.3.3.1.5**   Lee versus 3D ray tracing versus FDTD model.

5.3.3.2   Interfloor Case

Figure 5.3.3.2.1 shows four scenarios in the interfloor case. In each of them, the measured data and the predicted result from Eqs. (5.3.2.2.2.9) and (5.3.2.2.2.10) are compared, and the CDF curves are yielded. In all four scenarios; the CDF reached above 80 percent within 8 dB. It is verified that the Lee model is a fairly good prediction tool to use.

In the interfloor case, only one scenario is used to compare with 3D ray tracing, the FDTD model, and Lee model—the transmitter is on the third floor of north wing, and the receiver is on the second floor of north wing.

The results are shown in Fig. 5.3.3.2.2. Among the prediction results from three models, we see that the Lee model has less deviation values than the other two at 80 percent of CDF.

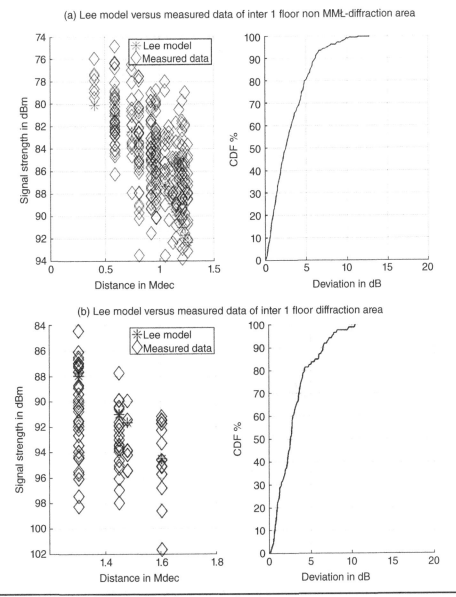

FIGURE 5.3.3.2.1 Interfloor measured versus predicted.

5.3.4 Using Measured Data to Customize the Lee Model

In the previous work, for minimizing the deviation between the measurement and prediction of the Lee model, we pick the best-fit values for the parameters in the Lee model through a regression process. For example, in the same-floor case (i.e., single-floor case), the receiver is not in the Fresnel zone but is under a NLOS path-loss case. The best-fit values are used as follows.

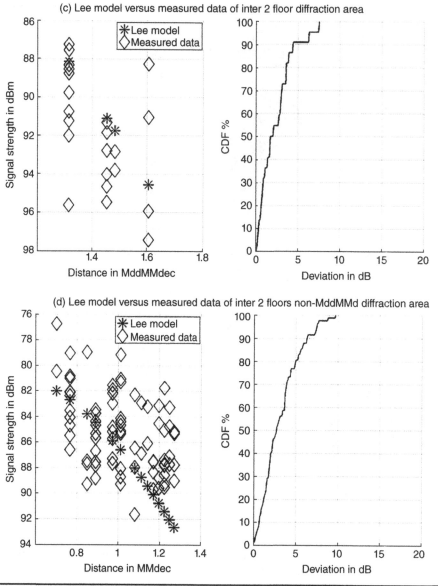

FIGURE 5.3.3.2.1   *(Continued)*

The path loss consists of two components, L_{MMdLO} and L_{room}, as shown in Eqs. (5.2.2.2.1) and (5.2.2.2.3), respectively; m_{room} is the slope of room attenuation shown in Eq. (5.2.2.2.3) of L_{room}. Its value is what we get from the measurement via a best fitting.

The slope m_{room} should be easily derived through a linear regression. First we plotted all the data points as a function of distance and made the best linear fit, finding that the closest slope is 20.378 and 20.261 dB/dec for the north and the south wing, respectively, as shown in Fig. 5.3.2.1.3. Then submitting the slopes back into the

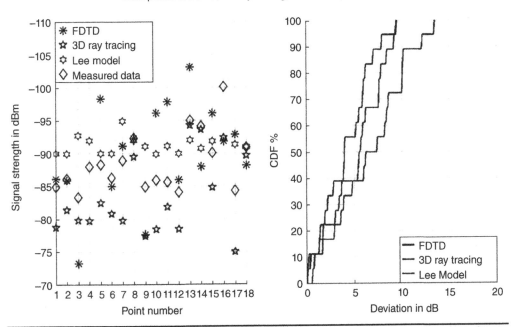

FIGURE 5.3.3.2.2 Predicted Lee versus predicted FTDT versus measured. (A color version of this figure is available at www.mhprofessional.com/iwpm.)

equation of L_{room} in Eq. (5.2.2.2.3), we could calculate the prediction values in a group of different scenarios that are neither in LOS nor in the close-in one. Meanwhile, from the same process, which is based on the best-fitted slope from the linear regression, we are also able to predict values in the LOS group of different scenarios as well.

Now, based on Devasirvatham et al.,[8] we can also find a best-fit parameter from a regression other that a linear regression. This will be discussed in detail for different cases in a following section.

5.3.4.1 On the Same Floor—The Single-Floor Model

Because in the LOS case calculating the path loss is simple, we discuss only the NLOS case here. The total path loss L_{pi} for a single-floor model at a point i is given by

$$L_{pi} = L_{LOS_i} + L_{room_i} = 20 \times \log \frac{4 * \pi * d_{i_1}}{\lambda} + F_{los} + m_{room} * \log\left(1 + \frac{d_{i_2}}{d_{i_1}}\right) \qquad (5.3.4.1.1)$$

where L_{LOS} is shown in Eq. (5.2.2.2.1) and L_{room} is shown in Eq. (5.2.2.2.3), d_{i_1} is the distance from the transmitter to the first-room intersection while the signal is along the radio path to the receiver at point i, and d_{i_2} is the distance from the first room intersection to the receiver at point i:

$$\delta_i = L_{mi} - L_{pi} \qquad (5.3.4.1.2)$$

where δ_i is the deviation at point i, L_{mi} is the path loss from the measurement at point i, and L_{pi} is the path loss from the prediction at point i.

By booking all the deviation points from Eq. (5.3.4.1.2), we could come up a matrix as

$$\delta = y_m - \vartheta A \qquad (5.3.4.1.3)$$

and ϑ is the parameter matrix we try to figure out in order to minimize δ. And in this case

$$\vartheta = [m_{room}] \qquad (5.3.4.1.4)$$

$$A = \left[\log\left(1 + \frac{d_{12}}{d_{11}}\right), \log\left(1 + \frac{d_{22}}{d_{21}}\right), \ldots \log\left(1 + \frac{d_{N2}}{d_{N1}}\right) \right] i = 1, N \qquad (5.3.4.1.5)$$

$$y_m = \left[y_{m1} - 20 \times \log\frac{4 * \pi * d_{11}}{\lambda} - F_{LOS1}, \ y_{m2} - 20 \times \log\frac{4 * \pi * d_{21}}{\lambda} - F_{LOS2} \ldots, \right.$$

$$\left. y_{mN} - 20 \times \log\frac{4 * \pi * d_{N1}}{\lambda} - F_{LOSN} \right]^T i = 1, N \qquad (5.3.4.1.6)$$

Here we are trying to minimize δ by the following technique.

We can use Joint Local Optimization (JLO) to optimize Eq. (5.3.4.1.2). In this approach, the optimized values of the parameters in the prediction model are obtained through the minimization of the cost function given by[40]

$$J(\delta) = \frac{1}{2}\delta^T\delta \qquad (5.3.4.1.7)$$

where

$$\delta = (A^T A)^{-1} A^T y \delta \qquad (5.3.4.1.8)$$

The JLO method solves for all parameters of the model simultaneously and generally provides the best-fit results from the measured data. However, although numerically optimal, the values for the parameters may not be physically interpreted and may vary substantially from cell to cell. Therefore, a final examination on the results is needed.

5.3.4.2   Interfloor Case

As defined in the Lee empirical model (Sec. 5.3.2.2.2), the interfloor case consists of two scenarios: one is measured in the diffraction area and the other in the nondiffraction area on the separated floor where the receiver is located. The total path loss $L_{interfloor}$ is expressed in Eq. (5.3.2.2.2.12) as

$$L_{interfloor} = L'_{i-f} + L''_{i-f} + {}_jL_n(d) \qquad (5.3.2.2.2.12)$$

where $j = 1$ is when the receiver is moving on the separated floor and measuring at the diffraction area and $j = 2$ is measuring at the nondiffraction area. The three components are

$$L'_{i\text{-}f} = m_{f1} \cdot n_{\Delta f} - L_{s\text{-}f} \tag{5.3.2.2.2.1}$$

$$L''_{i\text{-}f} = m_{f2} \cdot \log [n_{\Delta f}] + A \quad \text{in dB} \tag{5.3.2.2.2.5}$$

$$_1Ln (d) = L(d_0) + {}_1m_n \log d \quad \text{(for diffraction area)} \tag{5.3.2.2.2.10}$$

$$_2Ln (d) = L(d_0) + {}_2m_n \log d \quad \text{(for nondiffraction area)} \tag{5.3.2.2.2.11}$$

The detailed description of Eq. (5.3.2.2.2.12) has been shown in Sec. 5.3.2.2.2.

5.3.4.3 The External Building Wall Case—Using a Modified Formula

The focus of Reference 41 is considered the incident angle between the ray path and the external building wall where the receiver was located and found that besides the distance, the incident angle could be an important factor affecting path loss. According to their theory, the path loss in terms of distance and angle could be given by

$$L = 32.4 + 20\log(f_{\text{GHz}}) + 20\log(S) + L_e + L_{Ge}(1 - \sin\theta)^2 \quad \text{(in dB)} \tag{5.3.4.3.1}$$

where S is the physical distance between the transmission antenna and the external wall. This path is assumed as LOS, as shown in Fig. 5.3.4.3.1. θ is the incident angle, L_e is the loss in dB from the externally illuminated wall at perpendicular illumination ($\theta = 90°$), and L_{Ge} is the additional loss in the dB when illuminated at perfectly grazing conditions ($\theta = 0°$). The recommended values of L_e and L_{Ge} are also given that $L_e = 4$ to 10 dB (concrete with normal window size 7 dB, wood 4 dB) and $L_{Ge} =$ about 20 dB.

However, this theory and formula shown in reference[9] would have to be modified before it can fit the situation in which our measured data were collected.

First, in our case, the transmitter antenna did not have a LOS path to the receiver while both the transmitter and the receiver were inside the building, as shown in Fig. 5.3.4.3.2. So we need an extra loss term for the loss of propagation passing through the external wall from the transmitter's building. Second, in our case, there is no window on the walls that the signal would pass through; thus, L_e and L_{Ge} are higher than the given recommended values from reference.[9]

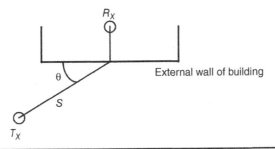

Figure 5.3.4.3.1 Penetration into building from external transmitter.

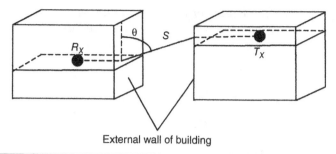

External wall of building

FIGURE 5.3.4.3.2   Penetration into building from external transmitter in another building.

However, the theory and formula shown in reference[40] is used for the scenario that the distance S is from the outside transmitter to the external wall of the building. We may need to modify Eq. (5.3.4.3.1) to cover the internal receiver located away from the interior wall of the building.

5.3.4.3.1   For the Scenario of Either *Tx* Inside and *Rx* Outside the Building or Vice Versa   The modified equation is

$$L = 32.4 + 20 \log (f_{GHz}) + 20 \log (S_1 + S_2) + L_e + L_{Ge}(1 - \sin \theta)^2 \qquad (5.3.4.3.2)$$

where S_1 is the distance from the transmitter to the building wall and S_2 is the distance from the receiver to the building wall. Both S_1 and S_2 are in meters.

5.3.4.3.2   For the Scenario of the Interbuilding Case (*Tx* Is Inside in One Building and *Rx* is Inside in Another Building)

We may have to modify Eq. (5.3.4.3.1) before it can fit the situation for which our measured data were collected. First, in our case, the transmitter antenna did not have a LOS path to the receiver while both the transmitter and receiver were inside the building, as shown in Fig. 5.3.4.3.2. So we need an extra loss term for the loss of propagation passing through the external wall from the transmitter's building. Second, in our case, there is no window on the walls, that the signal would pass through; thus, L_e and L_{Ge} are higher than the given recommended values from reference.[9]

Based on the above two main observations, we modified Eq. (5.3.4.3.1) to get a new formula that should fit this case better:

$$L = 32.4 + 20 \log (f_{GHz}) + 20 \log (S_1 + S_2 + S_3) + 2L_e + L_{Ge}(1 - \sin \theta)^2 \qquad (5.3.4.3.3)$$

where S_1 is the distance from the transmitter to the interior building wall of building A, S_2 is the distance between two building walls, and S_3 is the distance from the receiver to the interior building wall of building B. All three segmental distances, S_1, S_2, and S_3 are in meters.

The reason for doubling L_e is due to the consideration of the transmitter's external wall, and we chose $L_e = 15$ dB and $W_{Ge} = 25$ dB; thus, the calculating equation is given by

$$L = 32.4 + 20 \log (2.4) + 20 \log (S_1 + S_2 + S_3) + 2 * 15 + 25 * (1 - \sin \theta)^2 \qquad (5.3.4.3.4)$$

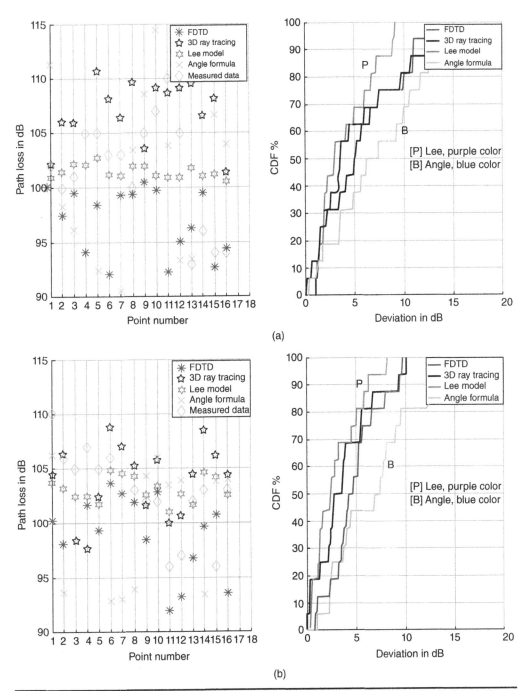

FIGURE 5.3.4.3.3 Comparison of Lee, FDTD, 3D ray tracing, and angle formula models. (*a*) Data points of FDTD, 3D ray tracing, Lee model, angle formula, and measured data of same floor. (*b*) Data points of FDTD, 3D ray tracing, Lee model, angle formula, and measured data of one-floor difference. (*c*) Data points of FDTD, 3D ray tracing, Lee model, angle formula model, and measured data of a two-floor difference. (A color version of this figure is available at www.mhprofessional.com/iwpm.)

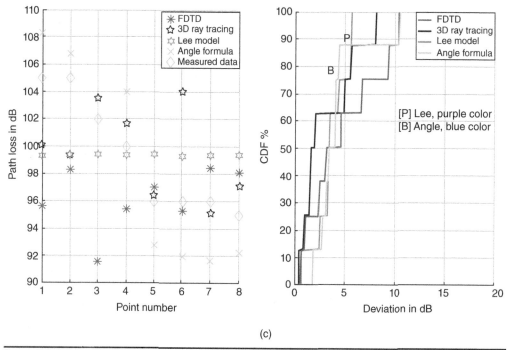

(c)

Figure 5.3.4.3.3   *(Continued)*

We compared the deviations of four models, including the modified angle formula model, using the modified angle formula (Eq. (5.3.4.3.3)) with the measured data. It can be seen that the angle formula model gives decent performance when the number of floors is two or greater—it fits the actual measured data well. Therefore, in (*a*) for the same floor scenario and in (*b*) for the one-floor different scenario, the modified angle formula model does not perform well. In (*c*) at a the two-floor different scenario, the modified angel formula model has the best performance among all the other models. As we described the angle formula model above, the calculation takes into account two elements—distance and angle. So we want to find out which element is the main source that brings such good results.

To find out which element plays an important role, distance or angle, we plot the path loss—both the measured data and the results from the angle formula, with different incident angles in degrees in different number of floors, as shown in Fig. 5.3.4.3.4.

As we can see, the measured path-loss data does not decrease as the incident angle increases calculated from angle formula. Actually, there is no obvious sign of the correlation between the angle and path-loss at each scenario while testing the path loss from the transmitter to the receiver on the different floors, as shown in Fig. 5.3.4.3.4.

The interbuilding propagation experienced that so much path loss is due to the obstacles that have directly impacted along the physical distance, so that the various construction material of concrete wall, and the physical distance play important roles instead of incident angles, in calculating the penetration loss.

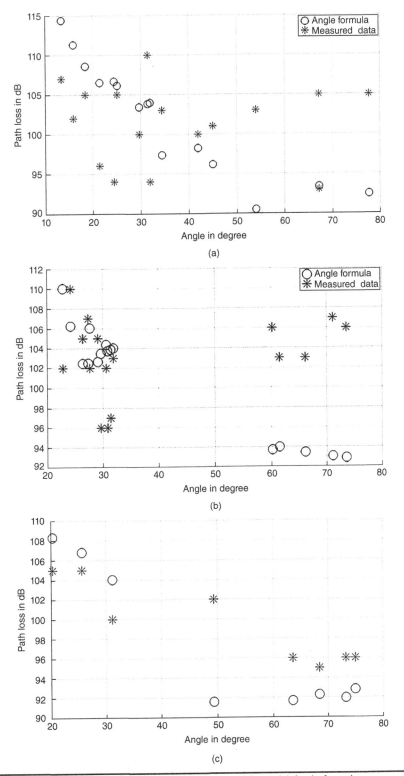

FIGURE **5.3.4.3.4** Angle formula prediction versus measured Data. (*a*) Angle formula versus measured data of same floor. (*b*) Angle formula versus measured data of one-floor difference. (*c*) Angle formula versus measured data of two-floors difference.

5.3.5 The General Formula of the Enhanced Lee In-Building Model

In the Lee in-building model, we introduce a general formula that is similar to the one using for the same-floor scenario as

$$P_r = P_t + G_t - L_{(A)} - L_{(B)} - L_{(c)} - L_{(D)} + G_r \qquad (5.3.5.1)$$

where P_t is the power transmitted, G_t is the transmitter antenna gain, P_r is the receiver antenna gain, and $L_{(A)}, L_{(B)}, L_{(C)},$ and $L_{(D)}$ are the path losses defined for each propagation condition as described below.

5.3.5.1 For the Same Floor

This case is shown in Fig. 5.2.2.1. There are five conditions to be specified as follows:
For considering only the LOS path-loss condition: (see Sec. 5.2.2.1)

$$L_{(A)} = L_{LOS} \qquad \text{(see Eq. (5.2.2.1.1))}$$

$$L_{(B)} = L_{(C)} = L_{(D)} = 0$$

For considering NLOS path-loss conditions: (see Sec. 5.2.2.2)

a. Receiver is in the close-in zone

$$L_{(A)} = L_{LOS} \qquad \text{same as LOS path-loss condition (see Eq. (5.2.2.2.1))}$$
$$L_{(B)} = L_{(C)} = L_{(D)} = 0$$

b. Receiver is not in close-in zone but in a room

$$L_{(A)} = L_{LOS} \qquad \text{(see Eq. (5.2.2.2.1))}$$
$$L_{(B)} = L_{room} \qquad \text{(see Eq. (5.2.2.2.3))}$$
$$L_{(C)} = 0$$
$$L_{(D)} = L_{in\,wall} \qquad \text{(for the enhancement, see Eq. (5.2.8.3.4))}$$

c. Receiver in a special room

$$L_{(A)} = L_{LOS} \qquad \text{(see Eq. (5.2.2.2.1))}$$
$$L_{(B)} = L_{special\,room} \qquad \text{(see Eq. (5.2.2.3.1))}$$
$$L_{(C)} = 0$$

d. Receiver is outside the building

$$L_{(A)} = L_{LOS} \qquad \text{(see Eq. (5.2.2.2.1))}$$
$$L_{(B)} = L_{room} \qquad \text{(see Eq. (5.2.2.2.3))}$$
$$L_{(C)} = L_{outside} \qquad \text{(see Eq. (5.2.2.4.1))}$$
$$L_{(D)} = L_{in\,wall} \qquad \text{(for the enhancement due to the wave passing through interior walls, see Eq. (5.2.8.3.2))}$$

The general formulas of the Keenan–Motley and Lee models are shown in Eqs. (5.2.8.1.1) and (5.2.8.1.2), respectively. In the Keenan–Motley model, there are four attenuation factors. Each factor is used to adjust an in-building path-loss parameter. In the Lee model, there are three different path losses selected due to the different scenarios. All of them come from the measurement data. The implementation of the Lee model is straightforward. The implementation of the Keenan–Motley model may need to account for the sensitivity of adjusting the attenuation factors, as shown in the next section.

5.3.5.2 An Interfloor Case

A general path-loss formula for the interfloor case is

$$L_{interfloor} = L'_{i\text{-}f} + L''_{i\text{-}f} + {}_j L_n (d) \qquad (5.3.2.2.2.12)$$

Equation (5.3.2.2.2.12) is described in Sec. 5.3.2.2.2.

5.3.5.3 Interbuilding Case

In the interbuilding case, we are using our modified formula Eq. (5.3.4.3.2) induced from Borjeson and Backer.[9]

$$L = 32.4 + 20\log(f_{GHz}) + 20\log(S_1 + S_2 + S_3) + 2L_e + L_{Ge}(1 - \sin\theta)^2 \qquad (5.3.4.3.3)$$

Equation (5.3.4.3.3) has been described in Sec. 5.3.4.3.

5.3.5.4 The External Building Wall Case

5.3.5.4.1 Transmitter Is Outside the Building   This case is shown in Fig. 5.3.4.3.1. The transmitter is located outside the building. Use the same formulas shown in Eq. (5.3.5.1), but let $L_{(B)} = 0$.

5.3.5.4.2 Receiver Is Outside the Building   This case is shown in Fig. 5.2.2.4.1. The receiver is located outside the building. Use the same formulas shown in Eq. (5.3.5.1), but let $L_{(D)} = 0$.

5.4   Empirical Path-Loss Models

Path loss is the signal strength attenuated along the radio path when propagating from the transmitter at the base station to the receiver at the mobile. Three forms of modeling are used to analyze these losses: the deterministic model (Maxwell's equation), the statistical model (probability), and the empirical model (measured data). The deterministic model is more accurate in general to find the propagation prediction if the environment can be precisely described. Besides, this type of model is computational complexity. The deterministic model uses Maxwell's equations along with reflection and diffraction laws. The statistical models use probability analysis by finding the probability density function of the path loss deduced from the historical data. The results predicted from this model are fairly accurate, and the computation is simple. The empirical models use empirical equations obtained from the results of several measurement efforts. This model can give accurate results, but the main problem with this type of model is computational complexity.

Empirical path-loss models are usually used to provide a high-level first-order assessment of the design. Once the design is completed, measurement data can be collected to fine-tune the design to provide higher accuracy. The Lee in-building model is one such empirical model. Usually, higher accuracy requires more detailed physical data and algorithms needed. Site-general models (area prediction models in which sites are not specified) are sometimes too generic to provide an efficient system design for a specific building. On the other hand, deterministic (physical) models are sometimes too complex to be implemented to take the full advantage of their capabilities. Even with deterministic models, certain measurement data are needed to ensure the accuracy of the model. There is always a cost associated with accuracy. The model needs to be able to balance how detailed the input data should be, how many measurement data need to be collected, and how accurate and efficient the model needs to be. Usually, site-specific measurements (for point-to-point models) are needed. The data can be used to determine the details of the propagation mechanisms and material parameters for a particular building. These "measurement-based prediction" approaches are described in detail in reference[42] and early sections of this chapter. The Lee picocell model is able to be fine-tuned for accuracy and efficiency from the measurement data. The measurement data should be collected across entire buildings.

There are many different parts of empirical models for indoor propagation prediction; most of which are based on the same propagation path-loss principles. The first approach is introduced from the Keenan model, which uses the free space curve as the baseline and adds additional loss factors relating to the number of floors n_f and walls n_w intersected by the straight-line distance r from the intersection to the receiver.

The second approach is from the Ericsson in-building path-loss model.[43, 44] It is similar to the multiple-breakpoints approach discussed in Chap. 4 on the Lee macro-to-micro integration model. There is a unique slope from one breakpoint to another breakpoint in the Ericsson model. This approach allows the model to be flexible so that field measurement data are gathered more efficiently and move accurately. However, this approach tends to lead to large errors in some areas due to the large variability in propagation mechanisms involved in different building types and in different wave paths within a single building. Although the accuracy and flexibility can be improved by using the multiple-radials approach as adopted by the Lee microcell model and discussed in Chap. 4, the calculation of the Ericsson model becomes complicated and slow for in-building scenarios.

The third approach is from the empirical log-distance path-loss prediction models,[14,45] which should be used to determine a high-level system design. Once the design is complete, the field measurements or an actual deployment can provide a final adjustment to a model's applicability in a given building. A good static design must still include a sufficient margin for deploying a system based on the environmental dynamics. Depending on the bandwidth and data rate, delay spread effects may also need to be considered. As a common practice, the delay spread effects are considered by the radio specification of the receiver and is handled at the equipment requirement.

The Keenan model provides some flexibility and is adopting the measurement data to tune the model with accuracy and speed. The Ericsson multiple-breakpoint model provides an empirical curve of worst-case attenuation versus distance. The attenuation variable in the log-distance model provides some flexibility to the model. The description of log-distance model is not covered in this book because the Lee model does not use it for comparison. The readers can find the details of log-distance model in references.[14,45]

5.4.1 The Motley–Keenan Model (Empirical) and a Comparison with the Lee Model

The Motley-Keenan model considers all walls that are intersecting the direct ray between transmitter and receiver. The user can adjust the attenuations for the walls. The formula for the modified Keenan–Motley model has been shown in Eq. (5.2.8.1.1).

The wall attenuation factor $P \cdot W(K)$ is shown in the formula, where K is the number of floors and P is the number of walls. The formula can be enhanced to count and sum up the loss of each wall and floor if the material is different; thus, the penetration loss is different.

5.4.1.1   Comparison of the Keenan–Motley Model and the Enhanced Lee Model   In this section, the Lee and Keenan–Motley models are made comparison from their path-loss predictions in two situations: LOS and NLOS. Many different values are applied to the Keenan–Motley Model's parameters. In the LOS situation, three different attenuation factors—0.2 dB/m, 0.15 dB/m, and 0.1 dB/m—are applied to the Keenan–Motley model for getting the path losses that are used to compare with Lee model. In the NLOS situation, the wall loss of 10 dB is applied to the attenuation factors of 0.2-dB/m and 0.15-dB/m cases. The path losses from Keenan–Motley model for these two cases are compared with that from the Lee prediction model in two scenarios: the scenario of multiple rooms in the middle of the floor and the scenario of one room at the end of the floor. The general formula of Lee in-building path-loss formula is shown in Eq. (5.3.5.1). The equation of the Keenan–Motley model is expressed in Eq. (5.2.8.1.1) as

$$L\,(\text{dB}) = 32.5 + 20\log(f) + 20\log(d) + K \cdot F(K) + P \cdot W(K) + D(d{-}D_b) \qquad (5.4.1.3)$$

Each symbol of Eq. (5.4.1.3) is described in Eq. (5.2.8.1.1). The first three terms are for LOS only. The entire equation is for NLOS. The results are described below.

5.4.1.2   LOS Situation   Here we show the performances of both models in the LOS situation. The path loss after the breakpoint, using the attenuation factors of 0.2 dB/m and 0.1 dB/m, are compared with the enhanced Lee model with the slope based on the LOS loss formula shown in Fig. 5.4.1.3. Besides, a medium difference is used to compare the predictions between the medium values from the Keenan–Motley model and the values from the Lee model.

As shown in Fig. 5.4.1.1, the medium differences among all the curves are within 8 dB at any spot over 200 m. This means that the differences in predicted values between the two models are small. Therefore, the user can choose to use either one of these. The Lee model appears to be more optimistic than the Keenan–Motley model in the LOS situation. It pretty much follows the 0.1-dB/m attenuation curve. Usually, in the LOS situation, the RF link should suffer less attenuation over a long distance. On some occasions, perhaps the 0.2-dB/m loss may have to calculate the path loss in the scenarios of behind and around building for the Keenan–Motley model, which is not in a LOS situation. The Lee model has predicted these scenarios (around the building), stated in Sec. 5.3.5.3.

The enhanced Lee model shows an 8-dB stronger signal strength at the end of the coverage distance as compared with the result obtained from Keenan–Motley model with a scenario of using a 0.2-dB/m attenuation factor. The Keenan–Motley model sometimes predicts the results conservatively.

5.4.1.3   NLOS Situation   Here we show the performances of both models in the NLOS situation. There are two major scenarios shown in the NLOS situation. One scenario is that to many rooms being located in a straight line along the radial path. The other scenario is that of one room being located at the end of the building.

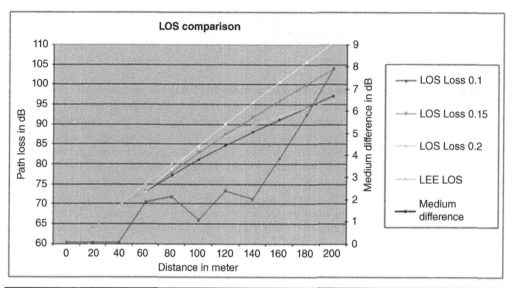

FIGURE 5.4.1.1 Comparison of the enhanced Lee and Keenan–Motley models—LOS. (A color version of this figure is available at www.mhprofessional.com/iwpm.)

Figure 5.4.1.2 shows the one of the NLOS scenarios. Assume there are three rooms in a straight line adjacent to each other (20 m long per room in this example). The maximum differences in predicted values between two models are within 6 dB. The peak of the differences is 6 dB at a distance of 100 m, which is in the middle of the building. In the figure, the Lee model has considered more path loss in the middle

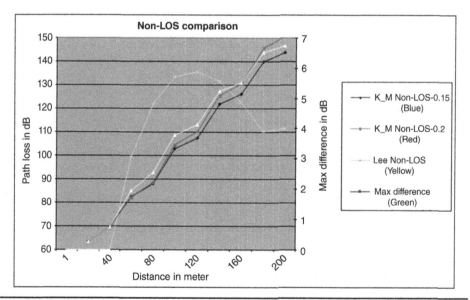

FIGURE 5.4.1.2 Comparison of the enhanced Lee and Keenan–Motley models—NLOS. (A color version of this figure is available at www.mhprofessional.com/iwpm.)

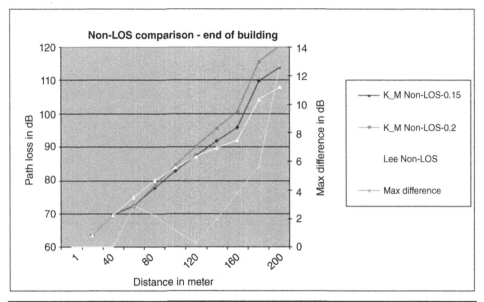

FIGURE 5.4.1.3 Comparison of the enhanced Lee and Keenan–Motley models—LOS. (A color version of this figure is available at www.mhprofessional.com/iwpm.)

of the building than the Keenan–Motley model does. It is very important that the area in the middle of the building be a critical area in designing an in-building network system. The enhanced part of the Lee model also provides flexibility and easier integration.

Figure 5.4.1.3 compares the path loss of the end-of-building scenario. The difference in predictions between the two models grows apart as the distance increases. Again, this phenomenon comes back to remind us of the previous argument on the path loss that the change is too drastic with the linear attenuation factors in the Keenan-Motley model in the LOS situation.

As we can see from the above analysis, the enhancement of the Lee model provides an effective means to deal with two scenarios. One is the LOS scenario, which better addresses the LOS loss outside the close-in zone. The other is the introduction of additional parameters that deal with the RF characteristics inside rooms.

5.4.2 Ericsson Multiple-Breakpoint Model (Empirical)

The Ericsson in-building path-loss model is an empirical model and was established from measurements in one multiple-floor office building by Ericsson Radio system.[43,44] This model is used around 900 MHz. It takes four regions based on the different distances from the base station, as shown in Fig. 5.4.2.1 and Table 5.4.2.1.

Measurements have been made for a picocellular office system in a six-story modern building with a floor plan shown in Fig. 5.4.2.1. Nine fixed stations (FS) for 900 MHz and 14 (FS) per 1800 MHz on each floor are placed, as the *x* and *o* indicate respectively.

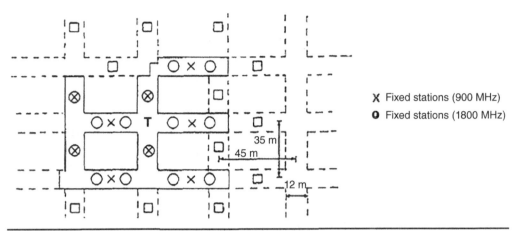

X Fixed stations (900 MHz)

O Fixed stations (1800 MHz)

FIGURE **5.4.2.1** Building plan of the measurements.

Distance (m)	Lower Limit of Path Loss [dB]	Upper Limit of Path Loss [dB]
$1 < r < 10$	$30 + 20 \log r$	$30 + 40 \log r$
$10 \le r < 20$	$20 + 30 \log r$	$40 + 30 \log r$
$20 \le r < 40$	$-19 + 60 \log r$	$1 + 60 \log r$
$40 \le r$	$-115 + 120 \log r$	$-95 + 120 \log r$

TABLE **5.4.2.1** Ericsson Indoor Propagation Model

The two curves are shown in Fig. 5.4.2.2, one for clear case and the other for shadow case. The gap between the two curves indicates a 20-dB shadow variation. The transmitter located in the middle of a corridor and the receiver is in rooms along other corridors on the same story and on other stories. The model covers the conditions from a straight line along a corridor and a 45° angle across the house on the same floor. In the model, an additional isolation between the floors was at least 15 dB on average except for rooms that were located parallel across the open space, which was closed to the next building. In this open space case, an added isolation between the floors was set to 10 dB. In the vertical direction propagation, all channels can be reused after every three floors.

The in-building model has a rapidly increasing path-loss slope, starting from a power exponent of two over a range from 1 to 10 m, and reaches a power exponent of 6.6 from 20 to 40 m. The shadowing distribution in the in-building model is limited by the lower boundaries. A typical prediction from the model is shown in Fig. 5.4.2.2 for 900 MHz. The model can be used at 1800 MHz by adding an extra path loss of 8.5 dB at all distances.

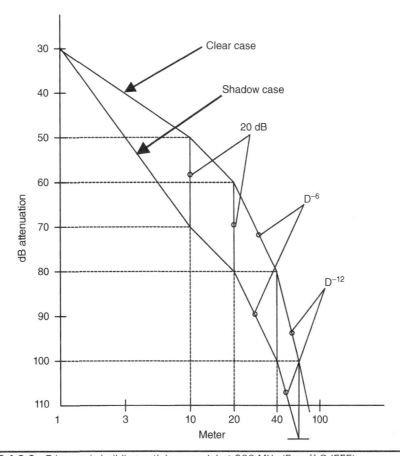

FIGURE 5.4.2.2 Ericsson in-building path-loss model at 900 MHz (From[44] © IEEE).

5.5 ITU Model

5.5.1 COST 231 Multiwall Model (Empirical)

The COST 231 multiwall model is also an empirical model that is used for the propagation within the buildings.[46] This model predicts a linear component of loss in dB, proportional to the number of walls penetrated, plus a term that depends on the number of floors penetrated, as shown below;

$$L_T = L_F + L_c + \sum_{i=1}^{W} L_{wi} n_{wi} + L_f n_f^{((n_f+2)/(n_f+1)-b)} \tag{5.5.1.1}$$

where L_F is the free space loss for the straight-line (direct) path between the transmitter and receiver, L_c is an empirically derived constant, n_{wi} is the number of walls crossed by the direct path of type i, W is the number of wall types, L_{wi} is the penetration loss for a wall of type i, n_f is the number of floors crossed by the path, b is an empirical constant, and L_f

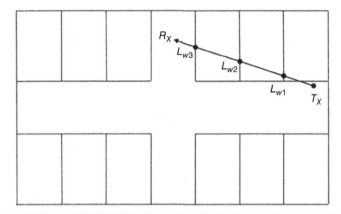

FIGURE 5.5.1.1 Building input for COST 231 model.

is the loss per floor. Some recommended values are $L_w = 1.9$ dB (900 MHz), 3.4 dB (1800 MHz) for light walls, 6.9 dB (1800 MHz) for heavy walls, $L_f = 14.8$ dB (900 MHz), 18.3 dB (1800 MHz), and $b = 0.46$. L_c is empirical derived constant and can be adjusted for each different environment.

The COST 231 model is the most sophisticated empirical model. All walls intersecting the direct ray between transmitter and receiver are considered and for each wall, individual material properties are taken into account, as shown in Fig. 5.5.1.1.

The predictions of COST 231 models are often too pessimistic. Therefore, an extension was added to modify the model: with an increasing number of penetrated walls, the individual attenuations (due to the material properties) of the walls are decreased. With this modification, the COST 231 model achieves good results with very fast computation times.

5.5.2 ITU-R 1238 (Empirical)

The ITU-R 1238 is an empirical model that is used for indoor path-loss prediction[47] assuming that the base station and portable are located inside the same building. The indoor base station to mobile/portable radio path loss can be estimated with either site-general or site-specific models. The models described in this section are considered to be site general, as they require little path or site information. The model accounts for the loss through multiple floors to allow frequency reuse between floors. The model also can estimate the loss for transmission through walls and over and through obstacles within a single floor of a building.

The basic model has the following linear form:

$$L_{total} = 20 \log 10\, f + N \log 10\, d + L_f(n) - 28 \text{ dB} \qquad (5.5.2.1)$$

where N = distance power loss coefficient,
f = frequency (MHz),
d = separation distance (m) between the base station and portable,
L_f = floor penetration loss factor (dB), and
n = number of floors between base and portable.

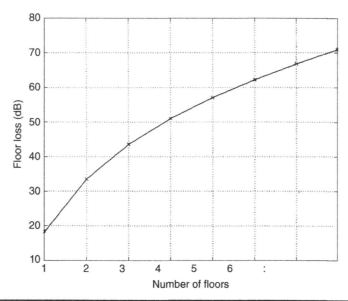

Figure 5.5.2.1 Loss versus number of floors.

Typical parameters, based on various measurement results, are given in Tables 5.5.2.1 and 5.5.2.2. Additional general guidelines are given at the end of the section.

The path-loss exponent is 20 dB/dec for the free space loss and is usually applied to open area. However, the path-loss exponent can be lower than 20 dB/dec based on the measurement results because the corridors might actually cause a tunnel effect. It is around 18 dB/dec. The path-loss exponent for wall penetration and around corners is between 30 and 40 dB/dec, depending on the building material and structure.

Frequency	Environment		
	Residential	**Office**	**Commercial**
900 MHz	—	3.3	2.0
1.2-1.3 GHz	—	3.2	2.2
1.8-2.0 GHz	2.8	3.0	2.2
4.0 GHz	—	2.8	2.2
60 GHz*	—	2.2	1.7

---

*60 GHz values assume propagation within a single room or space, and do not include any allowance for transmission through walls. Gaseous absorption around 60 GHz is also significant for distances greater than about 100 m which may influence frequency re-use distances. (See Recommendation ITU -R P.676.)

Source: Table 5.5.2.1 from,[47] courtesy of ITU.

Table 5.5.2.1 Distance Power Loss Coefficients, *N*, for Indoor Transmission Loss Calculation

	Environment		
Frequency	**Residential**	**Office**	**Commercial**
900 MHz	—	9 (1 floor)	—
		19 (2 floors)	
		24 (3 floors)	
1.8-2.0 GHz	$4n$	$15 + 4(n - 1)$	$6 + 3(n - 1)$

*Note that the penetration loss may be overestimated for large numbers of floors.
Source: Table 5.5.2.2 from,[47] courtesy of ITU.

TABLE 5.5.2.2   Floor Penetration Loss Factors, L_f (dB) with n Being the Number of Floors Penetrated for Indoor Transmission Loss Calculation

5.6 Physical Models—Application of Geometrical Theory of Diffraction (GTD)

Physical models are used when the characteristics of the environment vary more complicated ways. Physical models can be classified into two kinds of models. One is a ray-tracing model for in-building, and the other is FDTD (finite-difference time domain). These two kinds of models are introduced in this section.

5.6.1   Ray-Tracing Model for In-Building (Picocell)

In the past, ray-tracing methods were proposed for propagation prediction in microcellular environments[48–52] and for modeling propagation in rough terrain.[53] Later, ray tracing for indoor propagation has also been proposed.[6,54–59] In this section, we are concentrating for the in-building applications. A 3D ray-tracing-based model is applied on the same floor using both brute force ray tracing based on geometrical optics and the uniform geometrical theory of diffraction.[60]

Ray tracing is a physically tractable method of predicting the delay spread and path loss of an in-building radio signal. The delay spread and path loss can be found by counting the time and loss of arrival of all the possible reflected and diffracted rays. In order to implement site-specific propagation models, accurate site-specific building information is required.

5.6.1.1   Ray-Tracing Technique

Ray tracing is an exhaustive search of a ray tree accounting for the decomposition of the ray at each planar intersection. There are many steps for searching a ray tree:

1. First, the program determines if a LOS path exists and, if so, computes the received field.

2. Next, the program traces a source ray in a previously determined direction and detects if an object intersection occurs.

3. If no intersection is found, the process stops, and a new source ray is initiated.

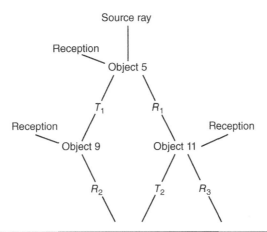

FIGURE **5.6.1.1**   Ray tree that shows how one source ray can be decomposed into many transmitted, reflected, and scattered rays from intersections with planar boundaries.

4. Once the program determines that an intersection has occurred, it checks to see if a specularly reflected or transmitted ray has an unobstructed path to the receiving location.

5. After checking for reception, the program divides the source ray into a transmitted ray and a reflected ray that are initiated at the intersection point on the boundary.

6. The recursion process starts; these rays are treated in a similar fashion to source rays. This recursion continues until a maximum number of ray tree levels is exceeded, the ray intensity falls below a specified threshold, or no further intersections occur.

Figure 5.6.1.1 shows a portion of a ray tree for one source ray. This ray tree shows how one source ray can be decomposed into many transmitted, reflected, and scattered rays from intersections with planar boundaries

5.6.1.2   Path Loss for Various Kinds of Ray Paths

The propagation of a signal from the transmitter to the receiver occurs through various kinds of ray paths, such as direct, reflected, transmitted, and diffracted paths. The path loss of each different kind of paths can be stated as follows:

1. Direct (LOS) rays exhibit a $1/d^2$ power loss with distance according to Friis free space transmission.

2. Specularly reflected and/or transmitted rays follow a $1/d^2$ power loss with distance, where $d = r_1 + r_2$ where r_1 and r_2 are labeled path segments for a reflected ray, as shown in Fig. 5.6.1.2.

3. Diffracted rays are exhibited in the shadow regions. The GTD supplements the geometrical optics (ray tracing) by introducing a diffracted field that accounts for the nonzero fields in shadow regions and modifies the field in the geometrical optics (GO) region so that the total field is continuous. The path loss from a diffracting corner will be shown in Eq. (5.6.1.1).

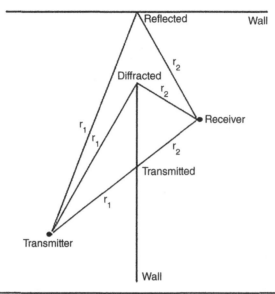

FIGURE 5.6.1.2   Transmitted, reflected, and diffracted ray paths.

5.6.1.3   Consideration in Shadow Regions

This is important since in a mobile and portable radio environment, the receiver is often shadowed from the transmitter. It is important to be able to predict the changes in the propagation as a receiver moves from an unobstructed to an obstructed location.

A single diffraction from diffracting corners in a hallway is considered where the corner is modeled as a dielectric wedge. The received field from the dielectric wedge is determined from the uniform geometrical theory of diffraction.[60,61] The path loss is a power loss with distance r_1 and r_2 as

$$L_i = \sqrt{\frac{r_1}{r_2(r_1 + r_2)}} \tag{5.6.1.1}$$

where r_1 and r_2 are as displayed in Fig. 5.6.1.2 for the diffraction case. The diffraction model accounts for all scattering in nonspecular directions.

5.6.1.4   Comments on Ray Tracing

A ray-tracing algorithm provides a relatively simple solution for radio propagation based on GO (Geometrical Optics) and usually supplemented by uniform theory of diffraction (UTD). However, GO provides good results only for electrically large objects, and UTD is rigorous only for perfectly conducting wedges. For dealing with complex lossy structures with finite dimensions, ray tracing fails to predict the scattered fields correctly, but FDTD, introduced in the next section, can handle this case.

Using a 3D ray-tracing tool for the wave propagation is also based on the GO and UTD. This model includes modified Fresnel reflection coefficients for the reflection from surrounds. The diffraction is calculated from the UTD. The implementation of this wave propagation phenomenon in the ray-tracing tool appeared in references.[62,63]

The 3D ray-tracing tool is used for evaluation in urban wave propagation, and its major advantage is the wideband analysis of the channel. The frequency dependence, such as the time delay spread, and the time variance, such as the Doppler spread of the channel, can be determined.

5.6.2   FDTD

5.6.2.1   What Is FDTD?

FDTD was developed based on the Maxwell's equations to handle a complicated communication environment in which transmitting and receiving antennas are often installed close to structures with complex material properties. Such problems can be solved not by asymptotic solutions but by the numerical solution of Maxwell's equations, called the FDTD method. The FDTD method simultaneously provides a complete solution for all the points on the map that can give signal-coverage information throughout a given area very accurately. In a simple outdoor environment, a 2D FDTD is generally applied.[29,37]

5.6.2.2   General Procedure of the FDTD Algorithm

5.6.2.2.1   Starting from Maxwell's Equations   FDTD was developed based on the Maxwell's equations to cover all the propagation phenomena, such as reflections, diffractions, refractions, and transmission.

Maxwell's equations in the time domain are

$$\nabla \times \bar{E} = -\mu \frac{\partial \bar{H}}{\partial t} \tag{5.6.2.2.1}$$

$$\nabla \times \bar{H} = \partial \bar{E} + \varepsilon \frac{\partial \bar{E}}{\partial t} \tag{5.6.2.2.2}$$

Modify two equations for FDTD:
   Eq. (5.6.2.2.1) becomes

$$\frac{\partial \bar{H}}{\partial t} = \frac{1}{\mu} \nabla \times \bar{E} - \frac{1}{\mu} \sigma \bar{H} \tag{5.6.2.2.3}$$

Eq. (5.6.2.2.2) becomes

$$\frac{\partial \bar{E}}{\partial t} = \frac{1}{\varepsilon} \nabla \times \bar{H} - \frac{1}{\varepsilon} \sigma \bar{E} \tag{5.6.2.2.4}$$

The vector differential equations can be converted into three coupled scalar equations:
   For E-field:

$$\frac{\partial E_x}{\partial t} = \frac{1}{\varepsilon} \left(\frac{\partial H_z}{\partial y} - \frac{\partial H_y}{\partial z} - \sigma E_x \right) \tag{5.6.2.2.5}$$

$$\frac{\partial E_y}{\partial t} = \frac{1}{\varepsilon} \left(\frac{\partial H_x}{\partial z} - \frac{\partial H_z}{\partial x} - \sigma E_y \right) \tag{5.6.2.2.6}$$

$$\frac{\partial E_z}{\partial t} = \frac{1}{\varepsilon} \left(\frac{\partial H_y}{\partial x} - \frac{\partial H_x}{\partial y} - \sigma E_z \right) \tag{5.6.2.2.7}$$

For *H*-field:

$$\frac{\partial H_x}{\partial t} = \frac{1}{\mu}\left(\frac{\partial E_y}{\partial z} - \frac{\partial E_z}{\partial y}\right)$$

(5.6.2.2.8)

$$\frac{\partial H_y}{\partial t} = \frac{1}{\mu}\left(\frac{\partial E_z}{\partial x} - \frac{\partial E_x}{\partial z}\right)$$

(5.6.2.2.9)

$$\frac{\partial H_z}{\partial t} = \frac{1}{\mu}\left(\frac{\partial E_x}{\partial y} - \frac{\partial E_y}{\partial x}\right)$$

(5.6.2.2.10)

Maxwell's equations from Eq. (5.6.2.2.5) to Eq. (5.6.2.2.10), these six equations are used for FDTD algorithm. In the next, section, we review the Discretize Space and Discretize Time named by Yee.[64]

Figure 5.6.2.2.1 gives us a clean picture of the "Yee" cell, named after Kane Yee, who developed this method in 1966. In the figure, the E, H components are located at different locations and are a half cell apart, as shown in Fig. 5.6.2.2.2.

In Fig. 5.6.2.2.2, the E and H components will be at different times (E_x, E_y, and E_z are separated at times $n \pm t$, and H_x, H_y, H_z are separated at times $(n + 1/2) \pm t$. This is "leapfrogging in time" and is useful for defining the central differences in time (d/dt).

Then the derivative terms can be replaced by difference terms according to the central difference formulas in space and time:

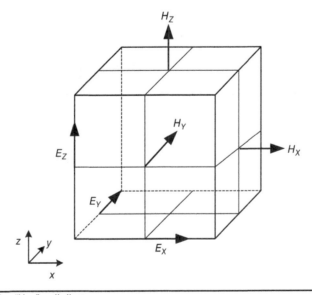

Figure 5.6.2.2.1 "Yee" cell diagram.

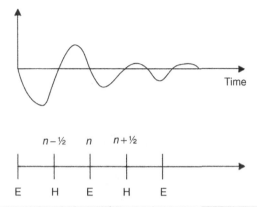

FIGURE 5.6.2.2.2   The Yee cell demonstrated.

For Eq. (5.6.2.2.5), we transform the E-field equation to the different form:

$$\frac{E_x^{n+1}(i,j,k)-E_x^{n}(i,j,k)}{\Delta t}=\frac{1}{\varepsilon}\left[\frac{H_z^{n+\frac{1}{2}}(i,j,k)-H_z^{n+\frac{1}{2}}(i,j-1,k)}{\Delta y}\right.$$

$$\left.-\frac{H_y^{n+\frac{1}{2}}(i,j,k)-H_y^{n+\frac{1}{2}}(i,j,k-1)}{\Delta z}-\frac{\sigma(E_x^{n+1}(i,j,k)-E_x^{n}(i,j,k))}{2}\right] \qquad (5.6.2.2.11)$$

Then Eq. (5.6.2.2.11) can be rewritten as

$$E_x^{n+1}(i,j,k)=\frac{E_x^{n}(i,j,k)\left[1-\dfrac{\sigma\Delta t}{2\varepsilon}\right]+\dfrac{\Delta t}{\varepsilon}\left[\dfrac{H_z^{n+\frac{1}{2}}(i,j,k)-H_z^{n+\frac{1}{2}}(i,j-1,k)}{\Delta y}-\dfrac{H_y^{n+\frac{1}{2}}(i,j,k)-H_y^{n+\frac{1}{2}}(i,j,k-1)}{\Delta z}\right]}{\left[1+\dfrac{\sigma\Delta t}{2\varepsilon}\right]}$$

$$(5.6.2.2.12)$$

For Eq. (5.6.2.2.8), we transform the H-field equation to the different form:

$$\frac{H_x^{n+\frac{1}{2}}(i,j,k)-H_x^{n-\frac{1}{2}}(i,j,k)}{\Delta t}=\frac{1}{\mu}\left[\frac{E_y^{n}(i,j,k+1)-E_y^{n}(i,j,k)}{\Delta z}-\frac{E_z^{n}(i,j+1,k)-E_z^{n}(i,j,k)}{\Delta y}\right] \quad (5.6.2.2.13)$$

And Eq. (5.6.2.2.13) can be rewritten as

$$H_x^{n+\frac{1}{2}}(i,j,k)=H_x^{n-\frac{1}{2}}(i,j,k)+\frac{\Delta t}{\mu}\left[\frac{E_y^{n}(i,j,k+1)-E_y^{n}(i,j,k)}{\Delta z}-\frac{E_z^{n}(i,j+1,k)-E_z^{n}(i,j,k)}{\Delta y}\right] \quad (5.6.2.2.14)$$

Assume that we know the E and H values at the very beginning of the first cell (in 3 D) or grid (in 2 D) at the time ($n = 0$). Depending on the number of cells or grids, we

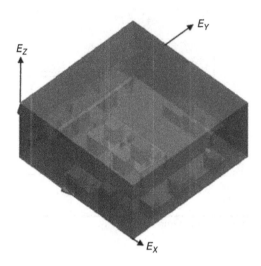

Figure 5.6.2.2.3 Data model for FDTD simulation. (A color version of this figure is available at www.mhprofessional.com/iwpm.)

can do the iterations within one time step to gain all the E and H values of all the cells or grids. Since the iteration is processing one by one in time steps, we can reach any time's state of the whole system after repeating the same procedure. This is the key part of the FDTD procedure. The initial E and H are equal to zero when the procedure starts from the very beginning. Therefore, we may not need to start at the very beginning. If we have knowledge of the past and current state of the system at any time point, we can start from that time point to predict the future state of the system. The iterations make up the procedure of the FDTD.

Figure 5.6.2.2.3 shows the data model for the FDTD calculation obtained from a picocell. Based on the input data, the FDTD can calculate the expected values for each point. Again, this depends on how much detail and on the granularity to derive the final value. Figure 5.6.2.2.4 shows electromagnetic field simulated using the FDTD technique. The electric field is strong at the corner of the picocell, as shown in Fig. 5.6.2.2.4.

5.6.2.3   3D Ray-Tracing and FDTD Models versus the Lee Model
In this case, the 3D ray-tracing and FDTD models were implemented in the same scenarios for comparison with the Lee Model. The room is assumed to have a uniform dielectric environment, and the simulation was done for just one room case—the second floor of north wing at Republic Polytechnic (RP) in Singapore (see Fig. 5.3.2.1.2).

The deviation is the difference between the predicted and the measured values, as shown in Fig. 5.6.2.3.1. As shown, the ray tracing (with and without furniture) and Lee models gave a decent prediction in this scenario. However, the FDTD model did not show its performance to be as good as that of the other two models. The main possible reason is that the 1×1 m-square grid is too large to be accurate for the FDTD model in this case. Furthermore, with the consideration of furniture, the gird should be even smaller. This is also the reason that applying the FDTD model in a furnished room here gives the worst performance. However, as we mentioned earlier, if we put more computational resources into the FDTD model, the results should be better.

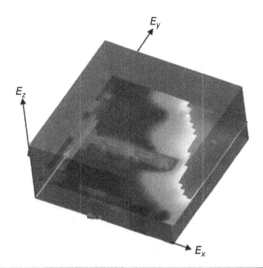

FIGURE 5.6.2.2.4 Electromagnetic field simulated using FDTD technique. (A color version of this figure is available at www.mhprofessional.com/iwpm.)

Data points of FDTD, ray tracing, Lee model, and measurement of north 2nd floor

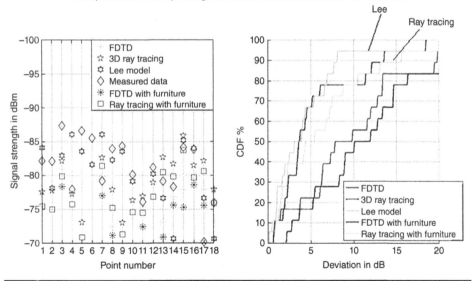

FIGURE 5.6.2.3.1 3D ray tracing and FDTD models versus the Lee model. (A color version of this figure is available at www.mhprofessional.com/iwpm.)

5.7 Summary and Conclusions

Different models for in-building (picocell) predictions are discussed in this chapter. In dense urban areas, the interference between all different sizes at a cell site is difficult to predict and manage. Propagation in in-building scenarios is especially complicated and unexpected. Every building has its own structured characteristics with different materials and layouts and different surroundings.

Comparing the predicted results with the measured data, the deviation between the two at any measured spot is plotted. From all the deviation values, a map of deviations is formed. By applying a linear regression, we can find the best-fit slope of each element to replace the existing one in the formulas. Through this procedure of fine-tuning the model, we do not worry about the specific details of the floor or building. The Lee model deals with single-floor, interfloor, single-building, interbuilding, and intersites. The model, integrated with measured data, provides better accuracy and efficiency. It is simple and cost effective to use. The general formula of the Lee in-building model is discussed in Sec. 5.3.5.

Chapter 6 will describe how to integrate the three models of the Lee model—macro-, micro-, and picocell—into one Lee integrated model by combining and transiting the different individual models in different areas. The Lee model tries to find a balance among many different parameters, including accuracy, run time, data granularity, complexity, scalability, and a means of fine-tuning the model. The most common practice is measurement integration.

As demands increase for more integration and more accuracy, propagation-predicted models are needed to provide ubiquitous coverage, throughput, and capacity. The Lee models have evolved and developed to meet the demands on capabilities as well as accuracy, speed, required data input, and flexibilities. The Lee model started from a theoretical and statistically based model, then the measured data are integrated in the model to further improve the accuracy of the model. In addition, different cell size models need different input parameters to improve the accuracy of the models. In the macrocell model, we need terrain and clutter data; in the microcell model, we need building, terrain, and attribute data; and in the Picocell model, we need to have wall, building material, window, and room dividers. Also in the next chapter, we will discuss some relatively new activities in the propagation prediction field with new and old technologies. The chapter also provides a more detailed and insightful view of the Lee comprehensive models.

References

1. Lee, W. C. Y. *Wireless and Cellular Telecommunications*. 3rd ed. New York: McGraw-Hill, 2005: 389–391, 639–640, 655–656, 674–677.
2. Dobkin, D. "Indoor Propagation Issues for Wireless LANs." *RF Design Magazine* (September 2002): 40–46.
3. Lee, D. J. Y., and Lee, W. C. Y. "Propagation Prediction in and through Buildings." *IEEE Transactions on Vehicular Technology* 49 (2000): 1529–33.
4. Honcharonko, W., Bertoni, H. L., Dialing, I., Qian, J., and Yee, H. D. "Mechanisms Governing UHF Propagation on Sinfle Floor in Mordern Office Buildings." *IEEE Transactions on Vehicular Technology* (May 1992): 77–82.
5. Rappaport, T. S. "Indoor Radio Communication for Factories of the Future.' *IEEE Communications Magazine* (May 1989): 15–24.
6. Molkdar, J. D. "Review on Radio Propagation into and within Buildings." *IEE Proceedings-H* (February 1991): 61–73.
7. Lafortune, J. F., and Lecours, M. "Measurement and Modeling of Propagation Losses in a Building at 900 MHz." *IEEE Transactions on Vehicular Technology* 39 (1990): 101–8.
8. Devasirvatham, D. M. J., Banerjee, C., Krain, M., and Rappaport, D. "Multi-Frequency Radiowave Propagation Measurements in the Portable Radio Environment." in *Proceedings of the IEEE ICC '90*, April 1990, 1334–40.

9. Austin, Andrew C. M., Neve, Michael J., Rowe, Gerard B., and Pirkl, Ryan J. "Modeling the Effects of Nearby Buildings on Inter-Floor Radio-Wave Propagation." *IEEE Transactions on Antennas and Propagation* 57, no. 7 (July 2009): 21–55.

10. Horikishi, J., et al. "1.2 GHz Band Wave Propagation Measurements in Concrete Buildings for Indoor radio communications." *IEEE transactions on vehicular technology* 35, no. 4 (1986): 146–52.

11. Cheung, K.W., Sau, J. H. M., and Murch, R. D. "A New Empirical Model for Indoor Propagation Prediction." *IEEE Transactions on Vehicular Technology* 47 no. 3 (1998): 996–1001.

12. Keenan, M., and Motley, A. J. "Radio Coverage in Buildings." *British Telecommunications journal* 8, no. 1 (January 1990): 19–24.

13. Devasirvatham, D. M. J. "A Comparison of Time Delay Spread and Signal Level Measurements within Two Dissimilar Office Buildings." *IEEE Transaction on Antennas and Propagation* 35, no. 3 (1998): 319–24.

14. Seidel, Scott Y., and Rappaport, Theodore S, "914 MHz Path Loss Prediction Models for Indoor Wireless Communications in Multifloored Buildings." *IEEE Transactions on Antennas and Propagation* 40, no. 2 (February 1992): 207–17.

15. Barry, P. J., and Williamson, A. G. "Modeling of UHF Radiowave Signals within Externally Illuminated Multi-Storey Buildings." Journal of Institute of Electronic and Radio Engineering (IERE) 57, no. 6 (1987): S231–40.

16. Turkmani, A. M. D., Parsons, J.D., and Lewis, D. G. "Measurement of Building Penetration Loss on Radio Signals at 441, 900 and 1400 MHz." Journal of Institute of Electronic and Radio Engineering (IERE) 58, no. 6 (1988): S169–74.

17. Lee, W. C. Y. " Finding the Statistical Properties of the Mean Values of a Fading Signal Directly from Its dB Values." *IEEE Proceedings* 58, no. 2 (February 1970): 287–8.

18. Lee, W. C. Y. "Estimate of Local Average Power of a Mobile Radio Signal." *IEEE Transactions on Vehicular Technology* 34, no. 1 (February 1985): 22–7.

19 Lee, W. C. Y., and Lee, D. J. Y. "Computer Implemented In-Building Prediction Modeling for Cellular Telephone Systems," US Patent 6.388,031, January 8, 2002.

20. Lee, W. C. Y. *Mobile Communications Design Fundamentals.* 2nd ed. New York: John Wiley & Sons, 1993: 59–61.

21. Lee, W. C. Y., *Mobile Communications Engineering.* 2nd ed. New York: McGraw-Hill, 1998: 105–108.

22. Ibid: 106.

23. Zhang, J. T., and Huang, Y. "Indoor Channel Characteristics Comparisons for the Same Building with Different Dielectric Parameters." IEEE International Conference on Communications, ICC April 28–May 2, 2002, vol. 2, 916–20.

24. Suzuki, Hajime. "Accurate and Efficient Prediction of Coverage Map in an Office Environment Using Frustum Ray Tracing and In-Situ Penetration Loss Measurement." IEEE Vehicular Technology Conference Proceedings, VTC 2003-Spring, April 22–25, 2003, vol. 1, 236–40.

25. Jemai, Jaouhar, Piesiewicz, Radoslaw; and Thomas, Kürner. "Calibration of an Indoor Radio Propagation Prediction Model at 2.4 GHz by Measurements of the IEEE 802.11b Preamble." IEEE Vehicular Technology Conference Proceedings, VTC 2005-Spring, May 30, June 1, 2005, 111–5.

26. Jemai, Jaouhar, Kurner, Thomas, Varone, Alexandre, and Wagen, Jean-Frederic. "Determination of the Permittivity of Building Materials through WLAN Measurements at 2.4GHz." IEEE 16th International Symposium on Personal, Indoor and Mobile Radio Communications, 2005, 589–93.

27. Janssen, G. J. M. "Short Range Propagation Measurement at 2.4, 4.5, 11.5 GHz in Indoor Environment." TNO Physics and Electronics Laboratory, Report FEL-92-B154, 2509 JG, The Hague, The Netherlands, May 1992.

28. Lee, W. C. Y. *Mobile Communications Design Fundamentals*. 2nd ed. New York: John Wiley & Sons, 1993.

29. Nagy, L. "FDTD Field Strength Prediction for Mobile Microcells." ICECom 2005, Dubrovnik, October 2005, 1–3.

30. Seidel, S. Y., and Rappaport, T. S. "Site-Specific Propagation Prediction for Wireless In-building Personal Communication System Design." IEEE *Transactions on Vehicular Technology* 43, no. 4 (November 1994): 879–91.

31. Bertoni, H. L., and Walfisch, J. "Theoretical Model of UHF Propagation in Urban Environments." *IEEE Transactions on Antennas and Propagation* 36 (December 1988): 1788–96.

32. Parsons, D. *The Mobile Radio Propagation Channel*. New York: John Wiley & Sons, 1992.

33. Saunders, Simon R., and Aragon-Zavala, Alejandro. *Antenna and Propagation for Wireless Communication Systems*. 2nd ed. New York: John Wiley & Sons, 2007.

34. Bertoni, Henry L. *Radio Propagation for Modern Wireless Systems*. Englewood Cliffs, NJ: Prentice Hall, 2000.

35. deToledo, A. F., and Turkmani, M. D. "Propagation into and within Buildings at 900, 1800, and 2300 MHz." IEEE 42nd Vehicular Technology Conference Proceedings, vol. 2, 633–6.

36. deToledo, A. F., Turkmani, A. M. D., and Parsons, J. D. "Estimating Coverage of Radio Transmission into and within Buildings at 900, 1800 and 2300 MHz." *IEEE Personal Communication*, 5 no. 2 (1998): 40–7.

37. Eswarappa, Channabasappa, and Hoefer, Wolfgang J. R., "Bridging the Gap between TLM and FDTD. "*IEEE Microwave and Guided Wave Letters* 6, no. 1 (January 1996): 4–6.

38. Fugen, Thomas, Maurer, Jurgen, Kayser, Thorsten, and Wiesbeck, Werner. "Capability of 3-D Ray Tracing for Defining Parameter Sets for the Specification of Future Mobile Communications Systems." *IEEE Transactions on Antennas and Propagation* 54 no. 11 (November 2006): 3125–37.

39. Catedra, M. F., Perez, J., deAdana, F.S., and Guutierrez, O. "Efficient Ray-Tracing Techniques for Three-Dimentional Analyses of Propagation in Mobile Communication: Application to Picocell and Microcell Scenarios." IEEE Antennas and Propagation Magazine 14 (April 1998): 15–28.

40. Kostanic, I., Guerra, I., Faour, N., Zec, J., and Susani, M. "Optimization and Application of W. C. Y. Lee Microcell Model in 850 MHz Frequency Band." Proceedings of Wireless Networking Symposium, Austin, TX, October 22–24, 2003.

41. Börjeson, H., and de Backer, B. "Angular Dependency of Line-of-Sight Building Transmission Loss at 1.8 GHz." In *Proceedings of the IEEE Ninth International Symposium* PIMRC (PIMRC'98), Boston, 1998, 466–70.

42. Lee, W. C. Y., and Lee, D. J. Y. "Pathloss Predictions from Microcell to Macrocell." IEEE VTC, Tokyo, Spring, 2000, 1988–92.

43. Akerberg, D. "European Advances in Cordless Telephony." National Communications Forum '88, (NCF 88), Chicago 1988.

44. Akerberg, D. "Properties of a TDMA Pico Cellular Office Communication System." IEEE GLOBECOM '88, Hollywood, FL, December 1, 1988, 1343–9.

45. Turin, G. L., Clapp, F. D., Johnston, T. L., Fine, S. B., and Lavry, D. "A Statistical Model of Urban Multipath Propagation." *IEEE Transactions on Vehicular Technology* 21, no. 1 (February 1972): 1–9.

46. COST231 Final Report. "Digital Mobile Radio towards Future Generation Systems." European Cooperation in the Field of Scientific and Technical Research, http://www.lx.it.pt/cost.231, 1999.

47. ITU-R Recommendations. "Propagation Data and Prediction Methods for the Planning of Indoor Radiocommunication Systems and Radio Local Area Networks in the Frequency Range 900 MHz to 100 GHz. "ITU-R P.1238-2, Geneva, 2001.

48. Ikegami, F., Takeuchi, T., and Yoshida, S. "Theoretical Prediction of Mean Field Strength for Urban Mobile Radio." *IEEE Transactions on Antennas and Propagation* 39 (March 1991): 299–302.

49. Rossi, J. P., Bic, J. C., Levy, A. J., Gabillet, Y., and Rosen, M. "A Ray Launching Method for Radio-Mobile Propagation in Urban Area." In *IEEE* Antennas and Propagation Symposium, London, Ontario, June 1991, 1540–3.

50. Takeuchi, T., Sako, M., and Yoshida, S. "Multipath Delay Prediction on a Workstation for Urban Mobile Radio Environment." In IEEE Globecom '91, Phoenix, AZ, December 1991, 1308–12.

51. Rossi, J. P., and Levy, A. J. "A Ray Model for Decimetric Radiowave Propagation in an Urban Area." *Radio Science* 27, no. 6 (November-December 1992): 971–9.

52. Schaubach, K. R., Davis, N. J., and Rappaport, T. S. "A Ray Tracing Method for Predicting Path Loss and Delay Spread in Microcellular Environments." IEEE 42nd Vehicular Technology Conference Proceedings, Denver, May 1992, 932–5.

53. Bisceglia, B., Franceschetti, G., Mazzarella, G., Pinto, I. M., and Savarese, C. "Symbolic Code Approach to GTD Ray Tracing." IEEE Transactions on Antennas and Propagation 36, no. 10 (October 1988): 1492–5.

54. Driessen, P. F. "Development of a Propagation Model in the 20-60 GHz Band for Wireless Indoor Communications." IEEE Pacific Rim Conference, Victoria, BC, May 1991, 59–62.

55. McKown, J. W., and Hamilton, R. L. "Ray Tracing as a Design Tool for Radio Networks." *IEEE Nemork Magazine* 5, no. 6 (November 1991): 27–30.

56. P. F. Driessen *et al.* "Ray Model of Indoor Propagation." In Proceedings of the 2nd Annual. *Virginia Tech Symposium on Wireless Personal Communications* Blacksburg, VA, June 17–19, 1992, 17.1–17.12.

57. Kiang, J. "Geometrical Ray Tracing Approach for Indoor Wave Propagation in a Corridor." *ICUPC Int. Conf. Univ. Personal Commun.*, Dallas, TX, September 29–October 2, 1992, 04/05/1–04.05/6.

58. Lawton, M. C., and McGeehan, J. P. "A Deterministic Ray Launching Algorithm for the Prediction *of* Radio Channel Characteristics in Small Cell Environments." *IEEE Transaction on Vehicular Technology* 43, no. 4 (November 1994): 955–69.

59. Kouyoumjian, R. G., and Pathak, P. H. "A Uniform Geometrical Theory of Diffraction for an Edge in a Perfectly Conducting Surface." *Proceedings of the IEEE* 62, no. 11 (1974): 1148–61.

60. Bumside, W. D. "High Frequency Scattering by a Thin Lossless Dielectric Slab." *IEEE Transactions on Antennas and Propagation* 31, no. 1 (January 1983): 10, 110.

61. Zhang, Y. P., Hwang, Y., and Kouyoumjian, R. G. "Prediction of Radio Wave Propagation Characteristics in Tunnel Microcellular Environments, Part 2: Analysis

and Measurements." IEEE Transactions on Antennas and Propagation 46 (1998): 1337–45.

62. Maurer, J., Knorzer, S., and Wiesbeck, W. "Ray Tracing in Rich Scattering Environments for Mobile-to Mobile Links." International Conference on Electromagnetics in Advanced Applications, September 2005, 1073–6.

63. Laki, I., Farkas, L., and Nagy, L. " Rough Surface Scattering Evaluation in Urban Wave Propagation." European Conference on Wireless Technology, 2002.

64. Yee, Kane, "Numerical Solution of Initial Boundary Value Problems Involving Maxwell's Equations in Isotropic Media." *IEEE Transactions on Antennas and Propagation* 14, no. 3 (May 1966): 302–7.

CHAPTER 6

The Lee Comprehensive Model—Integration of the Three Lee Models

6.1   Introduction

The fast growth of wireless systems has changed our lifestyles in both our work and our play. The Lee propagation models have been through generations of revolutions as well as evolutions. At the very beginning, terrain was the dominant factor, and then the human-made environment became important. Based on the power of computation and the degree of accuracy and granularity of the input data used in the propagation models, such as terrain and building maps, three different models were developed for different environments and focuses.

In the previous chapters, the Lee models have been compared with other popular models. Based on the predicted accuracy and granularity, there are area-to-area and point-to-point propagation models. Based on the size of radius of coverage area, we have pico- (in-building), micro-, and macrocell sites. The propagation model can also be categorized as theoretical, empirical, or statistical to deterministic site-specific models. A model is characterized by the input data, the computational power needs, and the demand of accuracy and speed for solving theories and techniques.

The Lee model has developed based on the theoretical, statistical, and empirical approaches and validated the Lee macrocell, microcell, and in-building (picocell) prediction models with a large amount of field data, as shown in Chaps. 3 to 5, respectively. The Lee models have been used by the industry and recognized in academic publications.[1-6] Many continuous integrations and enhancements of the Lee model have been shown.[7-15] In this chapter, we present the Lee comprehensive model. It explains how to integrate the three flavors of Lee models to make a complete system design, from picocell (in-building) to microcell to macrocell. Each cell-specific model can be fine-tuned with field-measured data. The Lee comprehensive model is used to design a general large system with all three models used simultaneously in the simulated deployment environment. The predicted results in this deployment modeling are compared with other models. Also, the predicted results are compared with newly developed models, such the ray-tracing and FDTD techniques,[16-21] to validate its applicability, as shown in Sec. 5.6.2.3 for the in-building environment. Results show that the Lee comprehensive model is as good as any other model.

Also in this chapter, we introduce four different transmission media that are different from our terrestrial mobile transmission medium. They are satellite communication signal, underwater communication signal, aeronautical communication signal, and bullet train communication signal. The readers can understand and compare the differences among the different media.

6.2   Integrating the Three Lee Models

The Lee model is based on the different types of human-made environment and the terrain variations. Under the terrain variations, the slope and intercept are considered as the baseline. The effect of human-made constructions can be translated to a slope of the path-loss curve. The Lee macrocell model has been implemented in AT&T markets since 1984. However, with the fast growth of femto-, pico-, and micro-cells, new models were needed to accurately predict the propagation in smaller and highly impacted human-made environment. Lee has developed the micro- and pico-cell models, which can be used in the femtocell prediction. The goal of this section is to identify a way to integral all three models together to solve all system deployment issues.

The generalized equation of received signal strength is expressed in Eq. (3.1.2.1) and restated as

$$P_\gamma = P_{r_0} - \gamma \cdot \log\left(\frac{r}{r_0}\right) + A_f + G_{\text{effh}}(h_e, h_2) + L + \alpha$$

The Lee single-breakpoint model is derived from Eq. (3.1.2.1) and is composed of four components, as shown in Sec. 3.1.2:

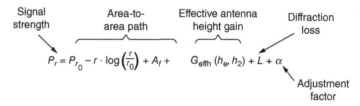

The causes of these components are described as follows;

1. The area-to-area path-loss slope γ is affected by the human-made environment, such as suburban, urban, open area, metropolitan, in-building, and other human-made construction areas.

2. The effective antenna height gain varies due to the fluctuation of terrain contour. The effective antenna height gain varies as the mobile moves along the mobile path. The larger the fluctuation of the terrain contour, the more rapid the change of the effective antenna height gain.

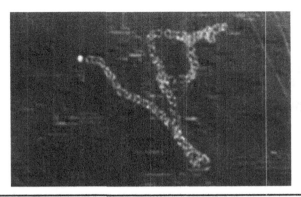

FIGURE **6.2.1.1.1** Macrocell drive test routes.

3. Diffraction loss occurs when the mobile is in the shadow region.

4. An adjustment factor is used for the input data other than the default values.

The received signal strength can be predicted at each point on the signal path (radial). Also, the Lee model can use a signal-smoothing process to find an optimum prediction value.

The human-made environmental factors used by the micro- and picocell models can be easily integrated into the macrocell model. A seamless and integrated prediction model can be applied to cover femto-, pico-, micro- and macrocells.

6.2.1 Validation of the Macrocell Model

6.2.1.1 Collection of Measured Data

The measurement data for a single cell was collected in the Ivrea, Italy, area (see Fig. 6.2.1.1.1). There are a total 10 drive routes. This cell site had an elevation of 246 m above sea level with a transmitter height of 50 m and ERP of 45.3 W. The measured data were collected by driving away from and toward the cell sites. Figure 6.2.1.1.1 shows the actual physical locations of few of these 10 routes, which covered most of the characteristics of this cell site.

Because the morphology was drastically different at different spotted area in the same cell-site coverage, it is important to be able to apply different slopes and intercepts for different areas covered by the same cell site. This situation has been handled by applying the method shown in Chaps. 3 to 5.[7-15]

6.2.1.2 Measured versus Predicted

A cell site was located in a hilly area where there were some known unreliable terrain data problems. Also, morphology data, tunnel, elevated highway, and bridge information was not always available. The measured data were collected without screening these restrictions. Figures 6.2.1.2.1, 6.2.1.2.2, and 6.2.1.2.3 show some samples of measured versus predicted data for the Lee macrocell model. The deviations between the prediction results and the measured data are 6 to 9 dB at

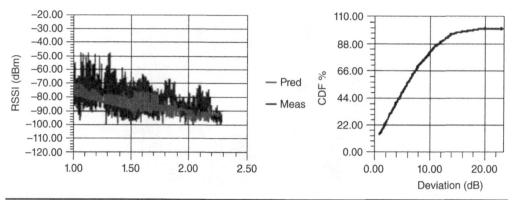

FIGURE 6.2.1.2.1   Drive route #1—measured versus predicted.

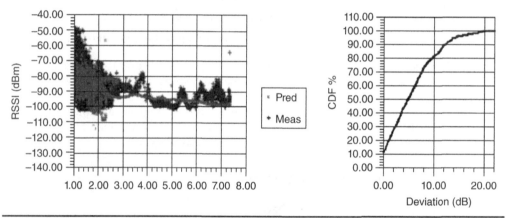

FIGURE 6.2.1.2.2   Drive route #2—measured versus predicted.

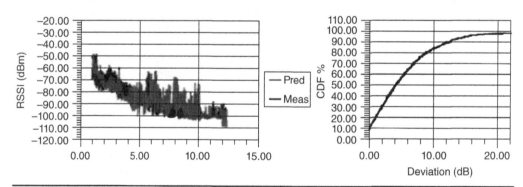

FIGURE 6.2.1.2.3   Drive route #3—measured versus predicted.

60 percent of the measured data. This is fairly good performance in a macrocell prediction model.

6.2.2   Validation of the Microcell Model

In Sec. 4.2.7, measurement data were collected by different groups in various countries/ cities for selected applications with unique frequencies, varying transmitter heights, and differing cell site parameters within a range of mobile communication environments. We must also consider different buildings that have unique shapes with individual heights and widths and built with various materials and different spacing. Measurement data were carried out in San Francisco, Japan, and Spain.

Comparisons with collected measurement data demonstrate that the model performed exceptionally well in varying mobile environments with different cell site parameters, as shown in Sec. 4.2.7.

6.2.3   Validation of the In-Building Model (Picocell Model)

In Sec. 5.2.5, measured data were collected at the Qualcomm building. This set of tests examined some special cases in the building, such as the impact of elevators on the signal attenuation. LOS, NLOS, single room, and multiple rooms were considered.

Figure 5.2.6.1 shows the measured versus predicted charts. More than 85 percent of the predicted values were within a delta value of 4 dB.

6.3   System Design Aspects Using Different Prediction Models

6.3.1   Preparing to Design a System

There are many technologies that can help to improve and enhance the CDMA network's coverage and capacity. However, the key consideration is the trade-off among cost, benefit, and risk. In this section, three different new technologies are analyzed and applied to a specific study area. The deployment of a microcell provides much enhanced capacity and coverage. The implementation of repeaters can improve limited coverage but lose certain capacities. The masthead low-noise amplifier (LNA) improves both the coverage (and with higher sensitivity) and the capacity. First, we analyze the benefits from these technologies and then consider the trade-off on the cost of improvement on a per-decibel basis. A system design is conducted based on the terrain and ETAK data. Then the design criteria and link budgets are analyzed and balanced. The Lee propagation model was applied to come up with radio coverage plots from each cell site within the study area. Then the CDMA coverage plots (E_c/I_O and E_b/I_O) for forward and reverse links are derived. Based on these plots, the analysis on effects of applying these new technologies can be further studied and compared. As shown from these studies, each technology has its own pros and cons.

The goal here is to study the modeled design and applying the new technology to see if there are any of the following:

- Performance improvement
- Reduction of capital without affecting the service
- Associated cost and risk

6.3.2   Design Parameters and Input Data

6.3.2.1   Design Parameters

The following design parameters are used as a default set in the prediction model to design a system in a special area:

Transmitter height:	30 to 40 m
Antenna gain:	17 dBi
Output power:	13.5 W
Cable loss:	2 to 3 dB
Vertical beamwidth:	4° to 5°

The antenna pattern used in the model is shown later in Fig. 6.3.3.2.1. The coverage plot was generated based on a 144-dB link budget. This translates to a receiver level about −85 dBm forward link coverage. This value is reasonable for providing ground car coverage. USGS terrain (a 3 × 3-sec map) data were used for this study.

6.3.2.2   Terrain Data

The 3D terrain of a special area for designing a system is shown in Fig. 6.3.2.2.1. As we can see, the study area is pretty flat. With a 30- to 60-m antenna height, the signal coverage should be very good. From the measured data, the 1-mile intercept value is 3 dB below that of the typical suburban areas. The standard 1-mile intercept in the suburban area from the Lee model is −61.7 dBm under the default condition. The 1-mile intercept in this special area after normalization with the Lee model's standard intercept would be −64.7 dBm, different from the default value shown in Fig. 3.1.3.1.1. The terrain undulation of the study area is between 10 and 200 ft, as shown in Fig. 6.3.2.2.1.

6.3.2.3   Cell Sites and Vector Data

Designing a system in a study area is shown in Fig. 6.3.2.3.1. The locations of cell sites are specified by longitude and latitude. The vector data in the study area include shoreline, political boundary, major and minor streets, and so on. These are very important data when deploying a cellular CDMA system. The vector data, such as major and minor streets, in a studied area are indicated on the map, as shown in Fig. 6.3.2.3.1.

FIGURE 6.3.2.2.1   Top view of the study terrain with 5× magnification.

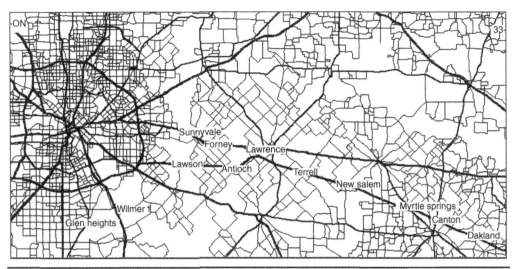

Figure 6.3.2.3.1   Cell-site, major and minor street data.

6.3.3   System Coverage in General

6.3.3.1   Forward Link Coverage
Forward link coverage plot is the groundwork for analyzing performance of any cellular design. Since the propagation characteristics are critical in mobile communication, it is important to ensure that the coverage requirement is met in the first place.

Figure 6.3.3.1.1 shows the coverage map for the study area, which is in suburban areas along different highways. Figure 6.3.3.1.2 shows the coverage map overlay on a 3D terrain.

6.3.3.2   Antenna Patterns
The characteristics of an antenna are also an important factor when dealing with the coverage and interference. Figure 6.3.3.2.1 shows the antenna pattern with 45° mechanical downtilt. We have mentioned the advantage of using a mechanical downtilt of an antenna in Sec. 3.1.7.1.

6.3.4   CDMA Coverage
From the CDMA plots, how to generate the usage of channels and other parameters is discussed in this section. There are several different CDMA plots that are needed to evaluate the performance of the CDMA system from both forward and reverse links.

6.3.4.1   Forward Link E_c/I_o Plot
The forward link of the CDMA E_c/I_o plot is equivalent to the normal AMPS/TDMA/GSM forward link E_b/I_o coverage plot. However, the E_c/I_o is a negative dB value, but E_b/I_o is a positive dB value. The E_c/I_o value defines the coverage of the plot channel of the CDMA system.

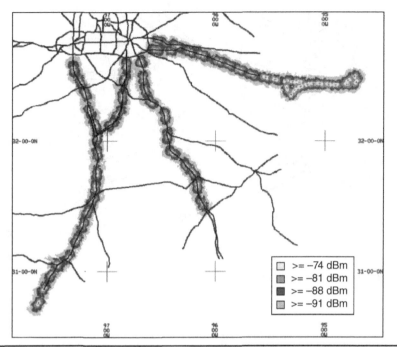

Figure 6.3.3.1.1   Forward link coverage. (A color version of this figure is available at www.mhprofessional.com/iwpm.)

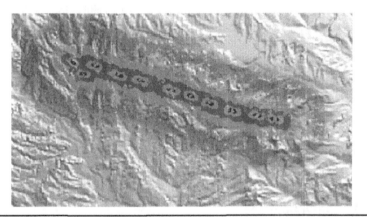

Figure 6.3.3.1.2   3D coverage on 3D terrain—top view.

As shown in Fig. 6.3.4.1.1, most areas are signal covered with E_c/I_o above −11 dB, which is the threshold level for providing good coverage and decent handoff support for the system.

6.3.4.2   Handoff (HO) Plots

The E_c/I_o plot is also used to calculate the HO regions and figure the HO threshold by the system engineers. Any bin on the map that is in the HO region is further calculated

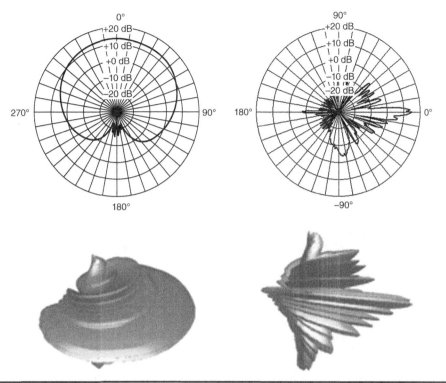

FIGURE 6.3.3.2.1   Antenna pattern: horizontal; vertical; 3D, 45° 0 tilt; 3D, front view, 5° tilt.

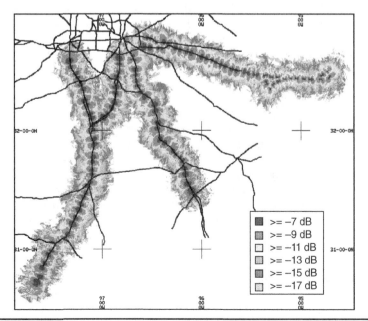

FIGURE 6.3.4.1.1   CDMA E_c/I_o plot. (A color version of this figure is available at www.mhprofessional.com/iwpm.)

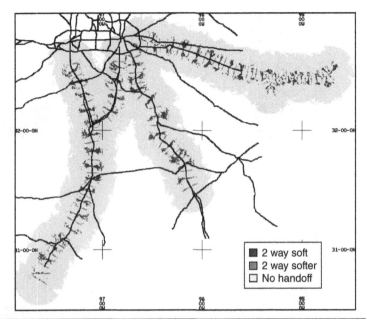

FIGURE 6.3.4.2.1   CDMA HO plot. (A color version of this figure is available at www.mhprofessional.com/iwpm.)

to define which bin is in which HO region, for example, softer HO between sectors in a serving cell or soft HO between cells. The time window of HO for this plot is generated with T_{ADD} being −12 dB, as shown in Fig. 6.3.4.2.1.

The HO region will be bigger if the T_{ADD} value changes to a lower E_c/I_o value. For example, the HO region for T_{ADD} of −13 dB, which is bigger than a T_{ADD} value of −12 dB. The E_c/I_o plot is used to evaluate the performance of the CDMA system in the HO region effectively.

6.3.4.3   Reverse Link E_b/N_o Plot

The reverse link E_b/N_o plot presents the system performance of the CDMA system on the reverse channel. It is generated by first calculating the CDMA system noise level based on either the output of system simulation or the user's input. Once the CDMA noise level is acquired, by placing an imaginary mobile at every bin with a maximum reverse link ERP, the E_b/N_o value at that bin can be calculated. When the mobile is in the HO region, the CDMA reverse link model calculates the signal gain in this region. The reverse link plot here is generated by using a requirement of the E_b/N_o level equal to 9 dB. In the CDMA system, 9-dB noise figure was added to the noise floor as the design guideline. The HO threshold was set to be −11dB for T_{ADD} (call add) and −12 dB for T_DROP (call drop).

As shown in Fig. 6.3.4.3.1, most areas meet the reverse link coverage criteria with E_b/N_o above 10 dB.

6.3.4.4   An Overall Plot

An overall plot basically combines the forward link E_c/N_o and reverse link E_b/N_o displaying on one plot so that engineers can easily identify the area that might have problems on either forward or reverse links, as shown in Fig. 6.3.4.4.1.

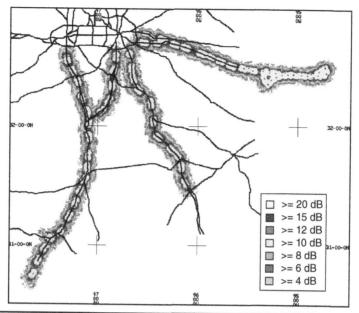

FIGURE 6.3.4.3.1 Reverse link E_b/N_o plot. (A color version of this figure is available at www.mhprofessional.com/iwpm.)

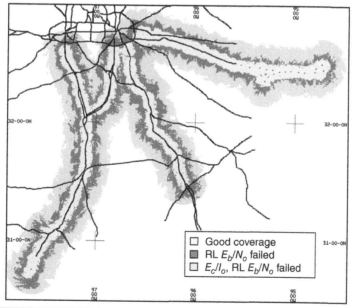

FIGURE 6.3.4.4.1 Combined plot. (A color version of this figure is available at www.mhprofessional.com/iwpm.)

6.3.5   System Design in Special Areas with New Technologies

6.3.5.1   Special Area Design Using Repeaters

Repeaters are used to cover those areas that are normally difficult to cover or become the extended coverage area.[22,23] Such as tunnels, valleys, terrain-limited, or morphology-limited areas. It can be divided into three different types—on-frequency repeaters, off-frequency repeaters, and fiber-optical repeaters. The on-frequency repeater is used basically as a signal enhancer. However, this kind of repeater usually has a limited output power (maximum of 2 W) and provides only a limited coverage area. Although using an on-frequency repeater can boost the normal RF signal strength in a CDMA coverage area, it does not seem to be an effective solution for resolving the highway coverage issue.

Assume the following required power levers for repeaters and base stations: In the forward link, we need a 33-dBm output power from a repeater when BS the base station is at its maximum ERP 50 dBm in order to match the system, shown in Fig. 6.3.5.1.1. At the forward link, the signal strength and the noise level can be calculated as: The transmitting signal power at the output of the repeater is

$$p_t = 50 \text{ dBm} - PL + G_2 + G_3 = 33 \text{ dBm}$$

The required gains G_2 and G_3 in the repeater are

$$G_2 + G_3 = PL - 17 \text{ dB} \tag{6.3.5.1.1}$$

If $PL = 100$ dB, then $G_2 + G_3 = 83$ dB and the total reverse link noise level at point A is given by:

$$N_t = \frac{NF_r \cdot N_{th} \cdot G_2 \cdot G_3}{PL} \cdot G_1 + NF_{BS} \cdot N_{th}$$
$$= N_{th} \left(\frac{NF_r \cdot G_1 G_2 G_3}{PL} + NF_{BS} \right) \tag{6.3.5.1.2}$$

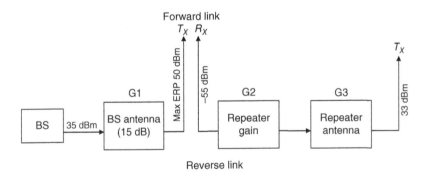

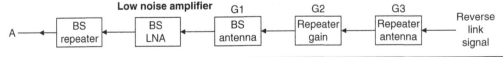

FIGURE 6.3.5.1.1   Reverse link.

where NF_r is the repeater noise figure of the receiver, NF_{bs} is the base station noise figure, and N_{th} is the thermal noise level. The description of noise figure appeared in Sec. 6.5.3. Using the parameters in Table 6.3.5.1.1, the dB value of total noise from Eq. (6.3.5.1.2) at point A is

$$N_t = 10 \log\left[N_{th}\left(\frac{63 \times 31.6}{50.11} + 6.3 \right) \right] = 10 \log(N_{th} \cdot 10.27)$$

$$= N_{th} + 10.1\,\mathrm{dB} \quad (\mathrm{dB})$$

(6.3.5.1.3)

This result of Eq. (6.3.5.1.3) clearly illustrates that the installation of a repeater will increase the noise level of the base station by 2.1 dB, as the noise figure without the repeater is 8 dB. The signal-to-noise ratio of the base stations and consequently the reverse link coverage will be decreased. Note that the forward link is not affected by the result in a link imbalance from the existing system. In a typical suburban environment (35 dB/dec), a 2.1-dB increase in base station noise level can be translated into a 13 percent shrinkage in coverage radius at the reverse link as follows:

$$\left(\frac{R_1}{R_2} \right)^{3.5} = 2.1\,\mathrm{dB} => \frac{R_2}{R_1} = 0.87$$

(6.3.5.1.4)

where R_1 and R_2 are reverse link coverage radii before and after the installation of a repeater. If the original reverse link coverage is a radius of 5 miles, then it will be reduced to $5 \times 0.87 = 4.36$ miles. Therefore, when we install repeaters into an existing wireless system, they should not be placed at the cell boundary. A 13 percent margin is necessary to keep the reverse link coverage in a suburban area. When designing a new system with repeaters, the increase of base station noise figures is always be taken into account when determining the link budget.

The cascade of repeaters will introduce a significant increase in noise level. A base station with n cascaded repeaters will represent a noise level of

$$N_t = N_{th}\left(\frac{NF_r \cdot G_1^n}{50.11} + NF_{BS} \right)$$

(6.3.5.1.5)

Repeater RF input	−55 dBm
Repeater power output	−33 dBm
Repeater noise figure	8 dB
Repeater gain	G_2
Repeater donor antenna gain	G_3
Base-station coverage antenna gain	15 dB
Base-station receiver noise figure	8 dB
BS maximum erp	50 dBm

TABLE 6.3.5.1.1 Parameters of Repeater and Base Station

6.3.5.2 Studying Design Plots in a Special Area before Using Repeaters

In this section, several CDMA design plots are shown in Figs. 6.3.5.2.1 to 6.3.5.2.5. before the repeater is implemented that is used as the benchmark for comparison

Figure 6.3.5.2.2 shows the normal coverage plot for the special-area design. The two way soft HO and softer HO are indicated in the plot.

Figure 6.3.5.2.3 shows the E_c/I_o plot on the forward link for the special-area design. The terrain undulation effect is shown in the coverage plot.

Figure 6.3.5.2.4 shows the E_b/N_0 plot on the reverse link for the special-area design. The signal strength shown in the plot does not cover the whole highway.

Figure 6.3.5.2.5 shows the combined CDMA coverage plot, both FL (forward link) and RL (reverse link), for the special-area design. In the red area, the RL signal has failed. In the green area, both FL and RL have failed.

6.3.5.3 Studying Coverage Plots in a Special Area Using Repeaters

The plots shown in Figs 6.3.5.2.1 to 6.3.5.2.5 without repeaters and in Figs. 6.3.5.3.1 to 6.3.5.3.5 with repeaters were deployed in the special design area.

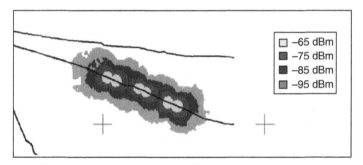

FIGURE 6.3.5.2.1 Special area coverage without repeaters. (A color version of this figure is available at www.mhprofessional.com/iwpm.)

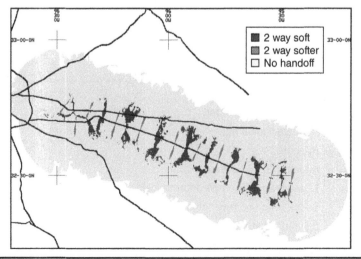

FIGURE 6.3.5.2.2 Special area CDMA HO plot without repeaters. (A color version of this figure is available at www.mhprofessional.com/iwpm.)

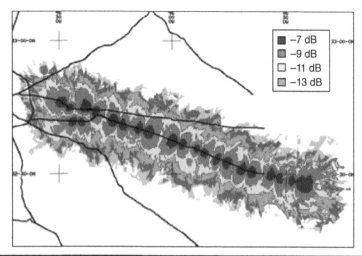

FIGURE 6.3.5.2.3 Special area CDMA E_c/I_o Plot without repeaters. (A color version of this figure is available at www.mhprofessional.com/iwpm.)

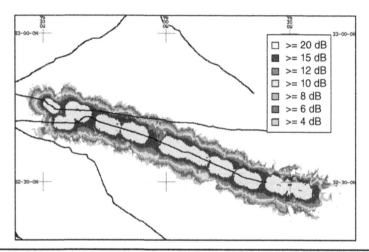

FIGURE 6.3.5.2.4 Special-area CDMA reverse link E_b/N_0 without repeaters. (A color version of this figure is available at www.mhprofessional.com/iwpm.)

Figure 6.3.5.3.1 shows the normal coverage plot for the special-area design with two repeaters to replace a Base Transceiver Station (BTS). The two small green areas show the coverage where the two repeaters were installed.

Figure 6.3.5.3.2 shows the CDMA HO plot for the special-area design with two repeaters to replace a BTS. The extended coverage area compared in Fig. 6.3.5.2.1 is due to the repeaters.

Figure 6.3.5.3.3 shows the E_c/I_o plot of FL for the special-area design with two repeaters to replace a BTS. The smaller blue areas are the spots where two repeaters were installed.

Figure 6.3.5.3.4 shows the reverse link E_b/N_o plot for the special-area design with one repeater to replace a BTS. The repeater is installed in the center yellow area connected to the two big yellow areas.

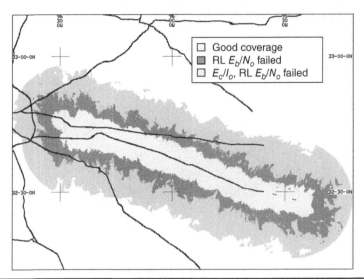

FIGURE 6.3.5.2.5 Special-area combined plot without repeaters. (A color version of this figure is available at www.mhprofessional.com/iwpm.)

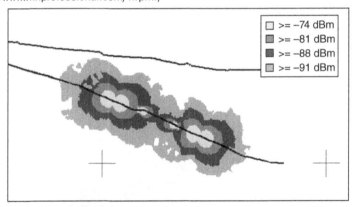

FIGURE 6.3.5.3.1 Special-area coverage with the repeaters. (A color version of this figure is available at www.mhprofessional.com/iwpm.)

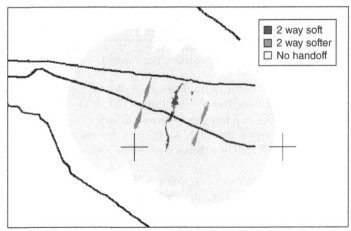

FIGURE 6.3.5.3.2 Special-area CDMA HO plot with repeaters. (A color version of this figure is available at www.mhprofessional.com/iwpm.)

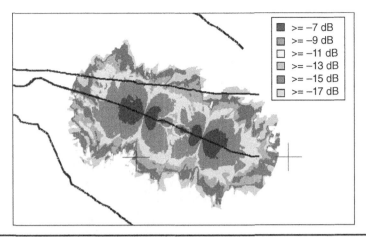

FIGURE 6.3.5.3.3   Special-area CDMA E_c/I_o plot with repeaters. (A color version of this figure is available at www.mhprofessional.com/iwpm.)

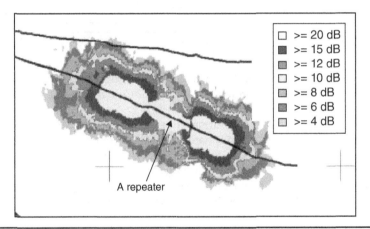

FIGURE 6.3.5.3.4   Special-area CDMA reverse link E_b/N_o with repeaters. (A color version of this figure is available at www.mhprofessional.com/iwpm.)

Figure 6.3.5.3.5 shows the combined CDMA coverage plot of both FL and RL for the special-area design with repeaters trying to enhance a BTS.

As we can see from our analysis from the plots, the coverage increases, and the capacity does not decrease by using repeaters. However, this is more an application limitation. It is usually used for tunnel or a specific area coverage.

6.3.5.4   Special-Area Design Using Microcell Systems[24-26]

The same area is used to evaluate the difference of applying microcell technology compared with the use of repeaters. Microcell technology was invented by Lee[25] at Pactel. This microcell system has been deployed in the Los Angeles and San Diego areas beginning in 1992. The microcell system uses either microwave or fiber optics to communicate between the cell site and microcell site. A brief description of microcell systems is given in Sec. 3.1.7.2.2. The biggest difference between using repeaters and the microcell

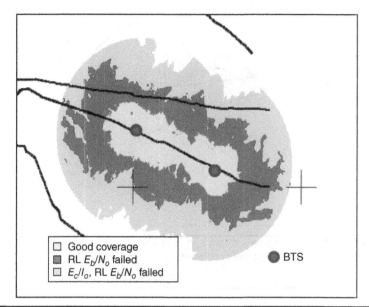

FIGURE 6.3.5.3.5 Special-area combined plot with repeaters. (A color version of this figure is available at www.mhprofessional.com/iwpm.)

system is the output power. The microcell system supports much higher power than the repeater (usually the maximum output power for a repeater is 2 W and for microcells can be up to a much higher power of 10 W).

Figure 6.3.5.4.1 shows the normal coverage plot for the special-area design with the Lee microcell system trying to enhance a BTS. The center weak signal strength area was enhanced by two repeaters, as shown in Fig. 6.3.5.3.1. Now, as shown in the figure, the signal strength is much stronger in this area by installing the microcell system.

In Fig. 6.3.5.4.2, the CDMA HO plot shows the normal coverage plot for the special-area design with the Lee microcell system trying to enhance a BTS. Comparing Fig. 6.3.5.4.2

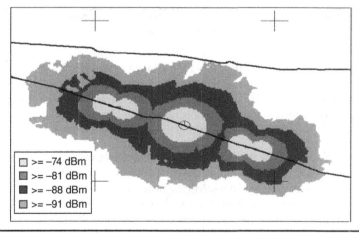

FIGURE 6.3.5.4.1 Special-area coverage with the Lee microcell system. (A color version of this figure is available at www.mhprofessional.com/iwpm.)

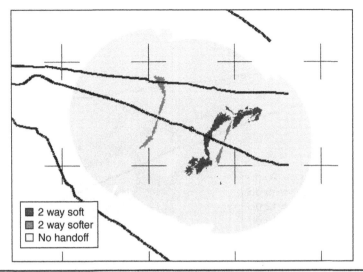

Figure 6.3.5.4.2   Special-area CDMA HO plot with microcell. (A color version of this figure is available at www.mhprofessional.com/iwpm.)

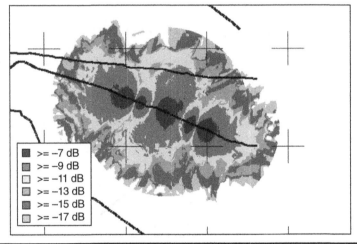

Figure 6.3.5.4.3   Special-area CDMA E_c/I_o plot with microcell. (A color version of this figure is available at www.mhprofessional.com/iwpm.)

using the microcell system and Fig. 6.3.5.3.2 using repeaters, we notice that in the former figure, the two-way soft HO area is more extended so that the call drop rate is reduced. Also, the capacity of the microcell system always increases due to the interference reduction.

Figure 6.3.5.4.3 shows the normal E_c/I_o plot of FL for the special-area design with the Lee microcell system trying to enhance a BTS. Comparing Fig. 6.3.5.4.3 with Fig. 6.3.5.3.3, we notice that the signal in the center area was weak with the repeater and becomes strong when the Lee microcell system is installed.

Figure 6.3.5.4.4 shows the reverse link E_b/N_o plot for the special-area design with the microcell system trying to enhance a BTS. Comparing Fig. 6.3.5.3.4 with Fig. 6.3.5.4.4, we

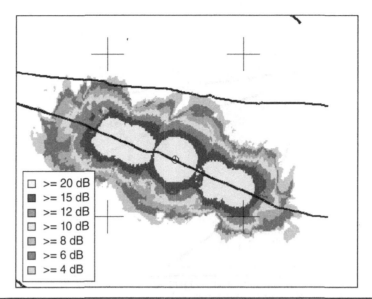

FIGURE 6.3.5.4.4 Special-area CDMA reverse link E_b/N_o with microcell. (A color version of this figure is available at www.mhprofessional.com/iwpm.)

notice that in the former, the signal at the center area between two strong yellow areas, which were served by two base stations, was not covering in the defined area well by repeaters. In the latter, the center area is well covered by the microcell system.

Figure 6.3.5.4.5 shows the combined coverage plot of both FL and RL for the special-area design with the Lee microcell system to replace a BTS. Comparing the plot using repeaters shown in Fig. 6.3.5.3.5 with the plot using the Lee microcell system shown in Fig. 6.3.5.4.5, the latter plot with the microcell system shows more coverage, although using the Lee microcell system, the fiber optic, or the off-frequency repeater can provide larger coverage but at a certain cost.

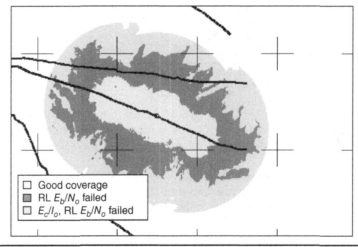

FIGURE 6.3.5.4.5 Special-area combined plot with microcell. (A color version of this figure is available at www.mhprofessional.com/iwpm.)

6.3.5.5 Special-Area Design Using Masthead LNA[23,27]

The masthead LNA can provide a better noise figure value in a long cable (big cable loss) extension environment. In this environment, the antenna is usually high, and it would improve the performance of a rural cell site.

Before designing a system by implanting masthead LNA, a noise figure (NF) model should be used. This NF model can calculate the noise figure and gain in different set-ups of a cable length with the LNA and the filter. We can show a quick analysis of the performance of a specific setup before the actual hardware is deployed. It can also provide solutions based on a specific noise figure requirement.

As shown Fig. 6.3.5.5.1, incoming signals go through the antenna, cable 1, masthead front-end unit, cable 2, and finally into the base station. Each element shown in the figure has gain (or loss) and the noise figure. The overall gain and noise figure from the antenna to the measurement point (just before getting into the base station) can be calculated in a spreadsheet, shown in Table 6.3.5.5.1 at 900 MHz.

The formula for obtaining the noise figure is shown below. The values of the parameters shown in the formula are listed in the manufacturer's spreadsheet:

$$NF = NF_{cable1} + (NF_{masthead} - 1) * Loss_{cable1} + \frac{(NF_{cable2} - 1) * Loss_{cable1}}{G_{masthead}} \quad \text{in linear scale}$$

(6.3.5.5.1)

where in most cases

$$NF_{cable}(\text{dB}) = Loss_{cable}(\text{dB})$$

(6.3.5.5.2)

Based on our field trial result at Qualcomm Lab. of UC Davis, using a 200-ft cable, a 3-dB noise figure improvement can be achieved. Now taking the noise figure as a base, we can start to calculate E_b/N_o in respective reverse links. Figures 6.3.5.5.2 and 6.3.5.5.3

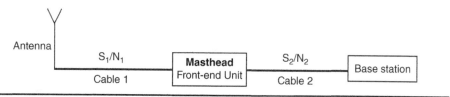

Figure 6.3.5.5.1 Noise figure model diagram.

Cable 1 loss (dB)	0.4
Cable 2 loss (dB)	3.0
Masthead gain (dB)	16
Masthead NF (dB)	2.2
Noise figure at cell (dB)	5.59
Noise figure at roof (dB)	2.65
Improvement (dB)	2.94

Table 6.3.5.5.1 Noise Figure Model Spreadsheet

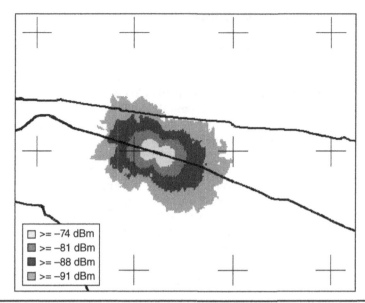

FIGURE 6.3.5.5.2 Special-area coverage without conventional masthead. (A color version of this figure is available at www.mhprofessional.com/iwpm.)

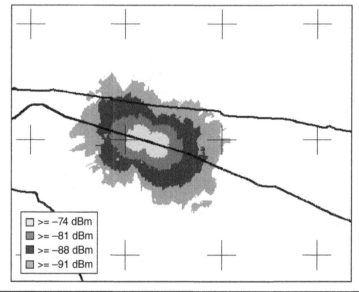

FIGURE 6.3.5.5.3 Special-area coverage with conventional masthead. (A color version of this figure is available at www.mhprofessional.com/iwpm.)

show the reverse link coverage plot before and after we apply the masthead technology. As we can see by comparing these two plots, the reverse link E_b/N_o coverage plot extends with masthead LNA.

We have made three different setups for our measurement. One was at a cell site, another was on the rooftop, and the third one was on the rooftop with diversity.

The following observations were made after analyzing the measured data collected during the field trials.

As shown in Fig. 6.3.5.5.4 and 6.3.5.5.5, the reverse link E_b/N_o from all test cases, the medium value (50 percent CDF) was maintained at a level about 8 dB, and the overall frame error rate (FER) on average is at 10 percent except the case of applying diversity. When the diversity is applied, the FER drops to 7 percent. No significant variations in these two key parameters were noticed across the test cases. The quality of service was maintained at the same level during the test. Thus, an adjustment of mobile transmission gain will be an excellent parameter to compare with the sensitivity of E_b/N_o at the base station in different configurations.

As shown in Figure 6.3.5.5.6, the medium (50 percent CDF of mobile gain adjust) mobile transmission power can be reduced by around 2 dB if the LNA/filter is installed at the rooftop. This is because the high gain of filter/LNA reduces the effect of cable loss in the overall noise figure.

Based on the analysis and simulation discussed in the earlier section, it is clear that the new technology that provides the most dB or dollar will be the masthead LNA. The coverage can extend by 15 percent.

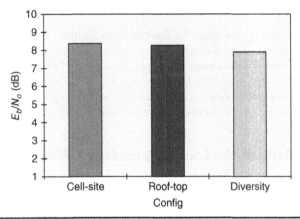

FIGURE 6.3.5.5.4   Medium (50 percent CDF) reverse link E_b/N_o comparison.

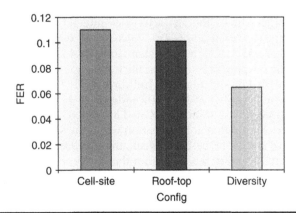

FIGURE 6.3.5.5.5   Reverse link FER for all rates comparison.

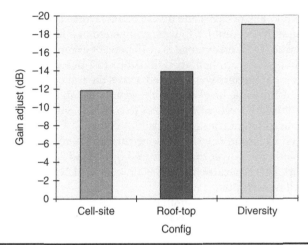

Figure 6.3.5.5.6   Mobile transmission gain adjust (dB) at 50 percent CDF.

6.3.5.6   Selecting Antennas for Special-Area Design

An ideal antenna is an isotropic radiator that radiates or receives equally in all directions, with a spherical pattern. However, in reality, all antennas have their own characteristics. Therefore, choosing the right one and introducing it in a system in a special area can enhance the system performance. The antenna characteristics are the parameters requested by the prediction model as the inputs.

6.4   User's Menu of the Lee Comprehensive Model

6.4.1   The Overall System Design Chart from the Lee Comprehensive Model

This section discusses the integrated Lee pico-, micro-, and macrocell prediction models, which can be used to do a complete system design based on the path-loss slopes acquired from measured data.

A virtual environment is set up to demonstrate the use of the comprehensive Lee model in a complete system design.

Figure 6.4.1.1 shows the characteristics interactive in a typical wireless system design with the combination of pico-, micro-, and macrocell. Most existing prediction models focus on one specific area/system. In Fig. 6.4.1.1, the distance scale is not drawn linearly for the purpose of illustrating the three regions and covers, three sizes of cells: macrocell, microcell, and pico- (in-building) cell. The macrocell and the microcell base stations are located at the right side of each region boundary shown in the figure. The picocell base station is located at the left side of its region boundary. From the figure, we can see that at the location when the signal strength received from the macrocell base station BS becomes weak, the microcell base station starts to take over. When the signal strength received from the microcell base station becomes weak, the picocell base station from the left side of its region starts to take over. Also there is a high-rise building at point A shown in the figure. In this case, a picocell base station is needed, and the strong signal strength coverage will extend to the right side of point A.

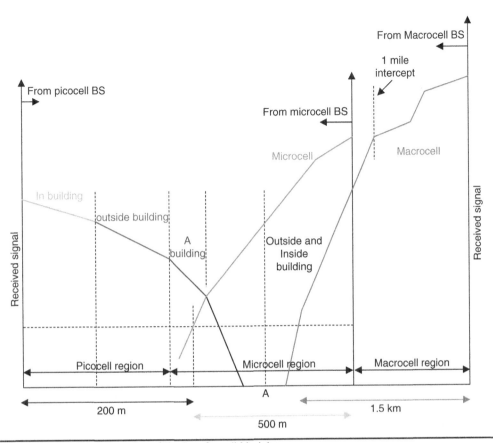

FIGURE 6.4.1.1   The Interaction of the Lee Overall Model.

The Lee comprehensive model has developed, validated, and integrated the pico-, micro-, and macrocell models for a complete and accurate system design. The performance of each cell size–specific model has been shown in Chaps. 3, 4, and 5, respectively.

In the next three subsections, we give examples of calculating the path-loss components in each of three Lee cell-specified models. For demonstrating how the Lee comprehensive model works, a virtual environment is created and shown in Figs. 6.4.1.2 and 6.4.1.3. There are three cell sites deployed in this virtual environment. The macrocell is located at the far left, the microcell in the middle, and the picocell at the far-right building. The distance from macrocell to microcell sites is about 2.7 km (1.677 miles), and the distance from microcell to picocell sites is 300 m. There are several locations identified to perform the detail prediction analysis using the integrated Lee model. Some of these locations, belonging to macro-, micro-, and picocells, are calculated and discussed in later sections.

The picocell BTS is located at the top floor of the most right-side building transmitting at 0 dBm at 2.4 GHz. The building is 15 m in height and 90 m in width and about 3 m tall per floor. One microcell sites deployed at the roof of the building on the left-hand side of the terrain. It is transmitting at 1.9 GHz with 5 dBm ERP and is 33 m in height (30 m from the building and 3 m above the building for the antenna). The macrocell site is located at a tower that is 50 m tall with 10 dBm ERP at 1.9 GHz. Each red dot associated

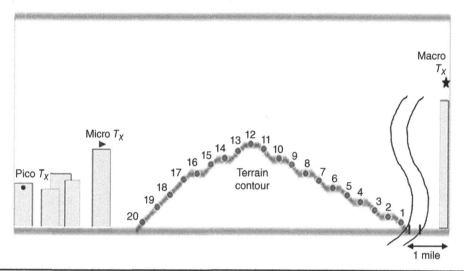

Figure 6.4.1.2 The interaction of the Lee overall model—test environment. (A color version of this figure is available at www.mhprofessional.com/iwpm.)

with a number in Fig. 6.4.1.2 represents a calculation point, and each is about 50 m apart. The calculation of points 12 and 14 are described in detail later.

The test sample points are chosen, some of them in picocell, some of them in microcell, and some of them in macrocell. These details are shown and discussed in the following sections. The star symbol shown on the far right is the macrocell site, the triangle symbol shown in the middle is the microcell site, and the circle symbol shown in the building on the far left is the picocell site. Note the scale does not reflect the real distance and height of the terrain, building, and cell sites.

The test environment is broken down into two separate environments for easier analysis: the macro and the pico combined with the micro. The pico- and microenvironment is shown in Fig. 6.4.1.3(*a*) and (*b*).

The red circles show the prediction locations from the picocell site, and the green circles show the prediction from the microcells site.

As shown in Fig. 6.4.1.3(*a*), the red circle shows the prediction locations from the picocell and the green circles show the prediction location from the microcell sites. This is from the side view.

We will start discussing pico, micro, and then macro and finally integrate the coverage plot from all three different models.

6.4.2 In-Building Cell—Point-by-Point Analysis for the Lee In-Building Model

Figure 6.4.2.1 (*a*) shows the top view of a picocell site, and Fig. 6.4.2.1 (*b*) shows a microcell site. The selected 15 sample points (red dots in picocell and green dots in microcell) are shown in the figure. Coverage of each region, either picocell or microcell, is calculated, and then two coverage regions are combined by selecting the best values of signal at each sample.

Three points—5, 7, and 13—are selected from Fig. 6.4.2.1(*a*) and calculated for demonstrating the prediction values from the Lee picocell model.

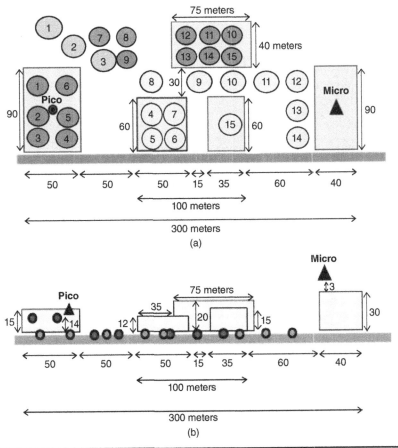

FIGURE 6.4.1.3 The pico- and micro-test environment. (a) top view. (b) vertical view. (A color version of this figure is available at www.mhprofessional.com/iwpm.)

6.4.2.1 Finding RSSI at Point 5

Test point 5 is four floors above the transmitter. Its vertical distance is 12 m (four floor separations), and its horizontal distance is 0 m from the transmitter, meaning that they are at the same location if we look from the top-view perspective. The transmitting ERP is 0 dBm (1 mw) at 2.4 GHz. The received signal at the same floor is a 10-dB loss, or 10 dBm. The receiver moved on the fifth floor, and the distance between it and the transmitter is 14 m in a penetration area.

Following Eq. (5.3.2.2.2.1), where $n_{\Delta f} = 4$, $L_{s-f} = 10$ and m_{f1} is obtained from Eq. (5.3.2.2.2.2) for $f_{GHz} = 2.4$,

$$m_{f1} = 1.93 f_{GHz} + 6.36 = 11$$

then

$$L'_{i-f} = m_{f1} \cdot n_{\Delta f} + L_{s-f} = 11 \times 4 + 10 = 54 \text{ dB}$$

$$= L(d_0)$$

The received signal on the fifth floor (four floor separations) at d_0 is –54 dbm. We also can estimate the received signal while the receiver is moved to a new location,

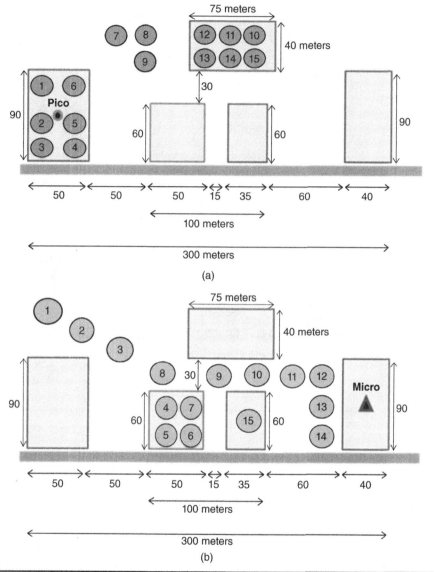

FIGURE 6.4.2.1   Top view of test points of pico- and microcells. (*a*) Test points of picocells (in-building cells). (*b*) Test point of microcells. (A color version of this figure is available at www.mhprofessional.com/iwpm.)

which is 14 m between the transmitter and receiver ($d_{T\text{-}R}$ = 14). From Eq. (5.3.2.2.2.11),

$$_2L_n(d) = L(d_0) + {}_2m_n \log(d/d_0) \qquad \text{(for penetration area)}$$

where the path-loss slope $_2m_2$ is 25 dB/dec when the receiver moves on the fifth floor ($n = 2$):

$$_2L_n(d) = L(d_0) + {}_2m_2 \log(d/d_0) = 54 + 25 \times \log(14/12)$$
$$= 54 + 1.67 = 55.67 \text{ dB}$$

The received signal at a new location ($d_{T\text{-}R}$ = 14) is 0 dBm − 55.67 dB = −55.67 dBm.

6.4.2.2 Finding RSSI at Point 7

Point #7 is located outside the building. Along the radio path from the transmitter to the receiver point 7, we use the inter-building formulas shown in Eq. (5.3.4..3.2):

$$L = 32.4 + 20\log(f_{GHz}) + 20\log(S_1 + S_2) + L_e + L_{Ge}(1 - \sin\theta)^2 \qquad (5.3.4.3.2)$$

where θ is the incident angle between the external building wall and the wave path. When $\theta = 90°$, the wave is perpendicular to the wall, and when $\theta = 0°$, the wave is at perfectly grazing conditions.

S_1 is the first segment of the distance (path) from the transmitter inside the building to its building wall, that is, $S_1 = \sqrt[2]{25^2 + 45^2} = 51.5$ m. The path-loss slope m is assumed 20 dB/dec because it is in a free space condition.

The second segment of the distance is the outside-building distance S_2 to the receiver, which is about $\sqrt[2]{30^2 + 25^2} = 39$ m. Since the radio wave passes through just one wall, a path loss causes by one wall (at $\theta = 90°$) is about 10 dB, that is, $L_e = 10$ dB. L_{Ge} is an additional loss when the incident angle $\theta = 0°$. Usually, $L_{Ge} = 20$ dB. In this case, as shown from Fig. 6.4.2.1(a), $\theta = 135°$. Thus, from Eq. (5.3.4.3.2),

$$L = 32.4 + 20\log(f_{GHz}) + 20\log(S_1 + S_2) + L_e + L_{Ge}(1 - \sin\theta)^2$$

$$= 32.4 + 20 \times \log(2.4) + 20\log(51.5 + 39) + 10 + 20(1 - \sin 135°)^2$$

$$= 32.4 + 7.6 + 39.1 + 10 + 1.175 = 90.27 \text{ dB}$$

The RSSI at point 7 is

$$\text{Transmitter power (0 dBm)} - L = -90.27 \text{ dBm}$$

6.4.2.3 Finding RSSI at Point 13:

Point #13 is located in the another building. Along the radio path from transmitter to the receiver at point 13, the first segment S_1 of the distance (path) is the distance from the transmitter inside the building to its building wall, that is, $\sqrt[2]{25^2 + 9^2} = 26.57$ m.

The second segment S_2 of the distance is the outside distance, which is about $\sqrt[2]{(45 - 9)^2 + 100^2} = 106.28$ m. The radio wave in this case passes through 2 walls, each wall causes about 10 dB loss, $2L_e = 20$ dB

The third segment S_3 of the distance is from the outside wall of the receiver building to the receiver inside the building, which is about 10 m. The free-space path-loss slope of 20 dB/dec is assumed. Also, L_{Ge} has a default number 20 dB. The incident angle $\theta = \tan^{-1}\left(\dfrac{45}{125}\right) = 19.8°$ can be found from Fig. 6.4.2.1(a).

The total loss L calculated from Eq. (5.3.4.3.3) is

$$L = 32.4 + 20\log(f_{GHz}) + 20\log(S_1 + S_2 + S_3) + 2L_e + L_{Ge}(1 - \sin\theta)^2$$

$$= 32.4 + 20\log(2.4) + 20\log(26.57 + 106.28 + 10) + 20 + 20(1 - \sin 19.8°)^2$$

$$= 32.4 + 7.6 + 20\log(142.85) + 20 + 20 \times (0.661)^2$$

$$= 32.4 + 7.6 + 43 + 20 + 8.73 = 111.73 \text{ dB}$$

The RSSI at Point 13 is

$$\text{Transmitter power (0 dBm)} - L = 0\ \text{dBm} - 111.73\ \text{dB} = -111.73\ \text{dBm}$$

Follow the same calculations to find the, predicted RSSIs of all the 15 sample points—interbuilding, four points; interfloor, six points; and throughout from building, five points—from the Lee in-building model. Since from Sec. 5.6.2.3 we have seen the performance of the Lee in-building model is matched better with the measured data than that of the ray-tracing model and FDTD model, here we just give the readers a picture by calculating the RSSI values of 15 sample points from both the 3D ray-tracing and the FDTD models using the same virtual environment.

Because there are no measured data available from this virtual environment, Fig. 6.4.2.2 compares the RSSI from both the 3D ray-tracing and FDTD models with that from the Lee in-building model. The RSSI values of the 3D ray-tracing model are closed to that of the Lee model, but the RSSI values of the FDTD model are farther apart from that of the Lee model. These findings are agreeable with the results shown in Fig. 5.6.2.3.1, which is backed up by the measured data.

6.4.3   Microcell—Point-by-Point Analysis for the Lee Microcell Model

As specified earlier, 15 points were identified to do the point-by-point analysis of the Lee microcell model. Two sample points, 10 and 15 are selected from Fig. 6.4.2.1(b) and

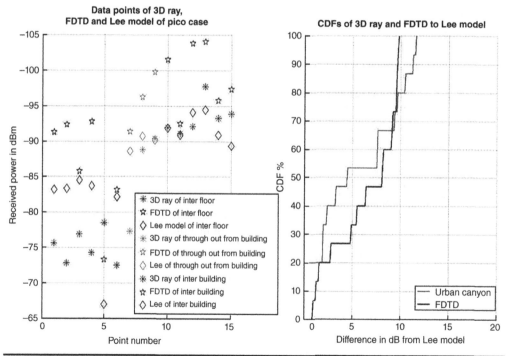

Figure 6.4.2.2   Picocell—Lee versus FDTD and ray tracing.

calculated with the Lee model. The antenna height at the transmitter is 33 m (100 ft), and the mobile antenna height is 1.5 m. The RSSI can be found from Eq. (4.2.1.2.1):

$$P_r = P_t - L_{LOS} - L_B + G_A + G_a \qquad (6.4.3.1)$$

where P_r = received signal strength, P_t = transmitted power, L_{LOS} = loss under line of sight, L_B = loss due to building, G_a = antenna height gain and G_A = additional gain.

If L_B is obtained from Eq. (4.2.1.2.3), then we have to use Eq. (4.2.1.2.4) for G_a when the actual antenna height is different from 20 ft and $G_A = 0$ in these case below.

Finding RSSI at Points 10 and 15—For finding RSSIs of these two points in the microcell virtual environment, Eq. (4.2.1.2.1) is used.

6.4.3.1 RSSI at Point 10

Point 10 is in a LOS condition. At this point, the transmitter antenna height h_a is 33 m (100 ft), the mobile antenna height h_2 is 1.5 m, and the distance from the transmitter to the point is $\sqrt[2]{80^2 + (75-45)^2} = 85.44$ on the ground distance. The radial distance is $\sqrt[2]{85.44^2 + (33-1.5)^2} = 91$ m. Using the near-in distance formula from Eq. (4.2.1.1.4),

$$df = \frac{4h_a h_2}{\lambda} \qquad (6.4.3.2)$$

The near-in distance at a frequency of 2.4 GHz ($\lambda = 0.125$ m) is

$$d_f = 1584 \text{ m} > 82 \text{ m}$$

Sample point 13 is within the near-in distance. Using the free-space path-loss formula;

$$L_{LOS} = 20\log\frac{4\pi d}{\lambda} = 20\log\frac{4\pi * 91}{0.125} = 79.22 \text{ dB} \qquad (6.4.3.3)$$

So, the RSSI at point 13 is P_r = 20 dBm 79.22 dB = –59.22 dBm.

6.4.3.2 RSSI at Point 15

Point 15 is 10 m inside of another building. The ground distance from the transmitter to point 15 is $\sqrt[2]{(10+60+20)^2 + (45-30)^2} = 91.24$ m. The radial distance is $d = \sqrt[2]{91.24^2 + (33-1.5)^2} = 96.53$ m. Although point 15 is within the near-in distance, it is in the NLOS condition due to the wall of the building. There is one wall between transmitter and point 15. The additional loss due to one wall is about 10 dB (can be obtained from the measurement data). Then the total loss is calculated from Eq. (6.4.3.1):

$$P_r = P_t - L_{LOS} - L_B + G_a$$
$$= 20 \text{ dbm} - 20\log\frac{4\pi \times 96.53}{\lambda} - 10 + 0$$
$$= 20 - 79.73 - 10 = -69.73 \text{ dBm}$$

The received signal at point 15 is −69.73 dBm.

Figure 6.4.3.1 compares FDTD and 3D ray tracing models with Lee micro model as the baseline for the selected 15 points. The signal strengths of Lee microcell model from all the 15 points are stronger than those of other two models.

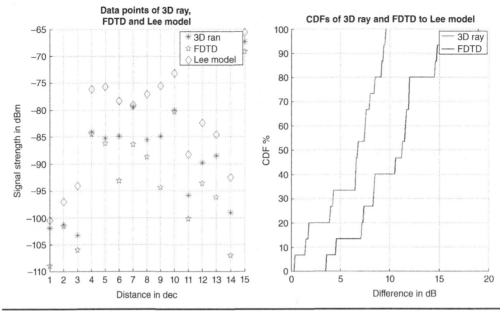

Figure 6.4.3.1   Microcell—Lee, FDTD, and ray tracing.

As shown in Figure 6.4.3.2, there are potential interferences between the pico- and microcells if they are on the same frequency.

6.4.4   Macrocell—Point-by-Point Analysis for the Lee Macrocell Model

The Lee macrocell prediction model is described in Chap. 3. The point-to-point path-loss formulas are shown in Sec. 3.1.2:

$$P_r = P_{r_0} - \gamma \cdot \log\left(\frac{r}{r_0}\right) - A_f + G_{\mathit{efft}}(h_e) - L + \alpha \tag{3.2.2.1}$$

where P_{r_0} = the received power at the intercept point r_o in dBm

$$L = L_D - 20\log\left(\frac{h'_e}{h'_1}\right) \qquad \text{(shadow condition)} \tag{3.1.2.4.2}$$

Figure 6.4.3.2   Best servers for pico and micro.

The effective antenna height h_e' is measured the height at the base station from the intercepted point of a line that is drawing from the tip of the hill along the slope of the hillside to the base station:

$$A_f = \text{the frequency offset adjustment in dB} = 20 \log\left(\frac{f}{850}\right) \qquad (3.1.2.2)$$

$$\alpha = (g_b' - 6) + (g_m - 0) + 20 \log\left(\frac{h_1'}{31.5}\right) + 10 \log\left(\frac{h_2'}{1.5}\right)$$

$$= \Delta g_b + \Delta g_m + \Delta g_{h1} + \Delta g_{h2} \qquad (3.1.2.6)$$

$$G_{effh}(h_e) = 20 \log\left(\frac{h_e}{h_1'}\right) \qquad \text{(no-shadow condition)} \qquad (3.1.2.3)$$

where h_1' and h_2' are in meters.

Figure 6.4.4.1 shows the radio distance versus elevation profile. The distance between each point is 50 m (164 ft). The transmitter height is 45 m. Point 12 is the peak of the terrain, and it is 30 m in elevation. Note it is not scale proportional so that we can fit all points into one figure. Two points, 12 and 14, are selected from Fig. 6.4.4.1.

Assume this is suburban area and we use the signal receipt $P_{r_0} = -61.7$ dBm at the 1-mile (or 1.6-km) intercept and the slope of 38.4 dB/dec based on a given set of standard conditions:

$$\text{Path loss} = -61.7 - 34.8 \log (r/\text{intercept distance}) \qquad (6.4.4.1)$$

where d is the distance from the transmitter to the receiver in miles and r_0 is the intercept. The actual antenna height $h_1' = 45$ m, $h_2' = 1.5$ m, $g_b' = 6$ dBd, and $g_m' = 0$ dBd.

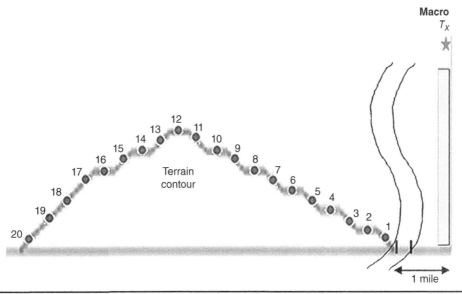

Figure 6.4.4.1 Vertical view of the macrocell test environment.

6.4.4.1   Finding RSSI For Points 12 and 14

6.4.4.1.1   RSSI at Point 12   For point 12, the mobile distance is 1 mile plus 0.373 mile, and it is in LOS. The effective antenna height is 90 m, while the height of actual antenna h_1' is 45 m. Thus:

$$G_{effh}(h_e) = 20 \log (h_e/h_1') = 20 \log (90/45) = 6 \text{ dB}$$

$$\alpha = \Delta g_b + \Delta g_m + \Delta g_{h1} + \Delta g_{h2}$$

$$= (6-6) + (0-0) + 20 \log \frac{45}{31.5} + 0 = 3.1 \text{ dB}$$

$$A_f = 20 \log \left(\frac{2400}{850} \right) = 9 \text{ dB}$$

$$L = 0 \qquad \text{(no-shadow condition)}$$

Then the received signal strength is found from Eq. (3.2.1.1):

$$P_r = P_{r_0} - \gamma \cdot \log \left(\frac{r}{r_0} \right) - A_f + G_{effh}(h_e) - L + \alpha$$

$$= -61.7 \text{ dBm} - 34.8 \log \left(\frac{1.373}{1} \right) - 9 + 6 - 0 + 3.1 = -66.39 \text{ dBm}$$

6.4.4.1.2   RSSI at Point 14   For point 14, the terrain height is 25 m, and mobile distance is 1.437 miles. In this case, the signal is blocked by the peak of hill and is under a shadow condition. The base station antenna height as well as the ERP of point 14 is the same as that of Point 12.

From Fig. 6.4.4.2, we have found that $r_1 = 2200$ m, $r_2 = 100$ m, and $h_p = 1$ m. We use the diffraction parameter $v = (-h_p) \sqrt{\left(\frac{2}{\lambda} \right) \left(\frac{1}{r_1} + \frac{1}{r_2} \right)}$, and the diffraction factor v is 0.4. From Fig. 1.9.2.2.1.2, the diffraction loss curve shows the loss is about 10 dB at $v = 0.4$.

The maximum effective antenna height gain defined in Eq. (3.1.2.4.2) is

$$\text{Max. } G_{effh}(h_e') = 20 \log \left(\frac{h_e'}{h_1'} \right)$$

where the maximum effective antenna height h_e' is the same as h_e of point 12 since point 12 is on the top of the hill. Therefore, $h_e' = 90$ m, and Max. $G_{effh}(h_e') = 6$ dB. From Eq. (3.1.2.4.2), the diffraction loss is

$$L = L_D - 20 \log \left(\frac{h_e'}{h_1'} \right) = 10 - 6 = 4 \text{ dB}$$

$$G_{effh}(h_e) = 0$$

$$\alpha = (6-6) + (0-0) + 20 \log \frac{45}{31.5} + 0 = 3.1 \text{ dB}$$

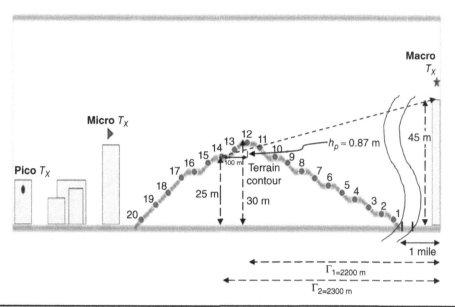

FIGURE 6.4.4.2 Calculation illustrated for test point 14.

$$A_f = 20 \log\left(\frac{2400}{850}\right) = 9 \text{ dB}$$

The received signal P_r at point 14 is found as

$$P_r = P_{r_0} - \gamma \cdot \log\left(\frac{r}{r_0}\right) - A_f + G_{effh}(h_e) + L + \alpha$$

$$= -61.7 \text{ dBm} - 34.8 \log\left(\frac{1.437}{1}\right) - 9 + 0 - 4 + 3.1$$

$$= -61.7 \text{ dBm} - 5.48 - 9 + 0 - 4 + 3.1 = -77 \text{ dBm}$$

Figure 6.4.4.2 shows how the test point 14 is calculated for the diffraction loss.

Figure 6.4.4.3 shows the complete "test environment" of this study with the terrain data. The size of the test environment was limited to a manageable scope so that effort can be focused on benchmarking Lee's three different models.

Figure 6.4.4.3 shows the coverage areas from each cell site. This plot demonstrated the importance of integrated models and the value of the propagation model based on the radio wave traveling along the human-made environment and natural terrain.

6.5   How to Use Prediction Tools

In previous sections, we have given examples of design aspects in each particular scenario. In this section, we give a systematic procedure to guide the user in using the prediction tools that have been developed from the prediction models. Once we choose

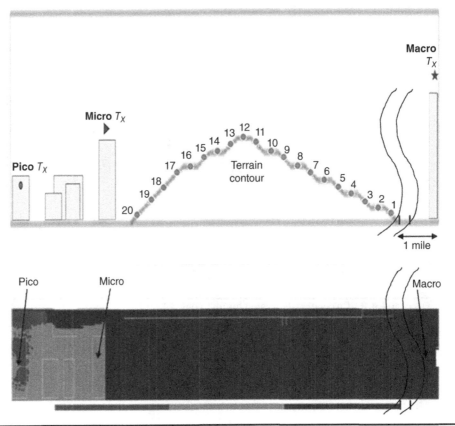

FIGURE 6.4.4.3   Best server plot for the three cell sites.

a cell size–specific prediction model to use, then we have to know how to use it. First, we have to find the link budget for a particular system. Before doing so, there are many factors to he found over a radio communication link, such as signal power, noise, and interference.

6.5.1   Radio Communication Link—The Channel

The radio propagation medium is the electromagnetic field. The propagation path that connects the transmitter and receiver through the medium is a radio communication link, and we call it the *channel*. The radio communication link propagates through the atmosphere and near the ground results in four waves: direct, reflection, refraction, and diffraction waves. In the direct wave, there are two conditions: LOS NLOS. Chapter 1 has described their nature. After the signal is received at the mobile over a radio communication link, many different factors appear.

6.5.2   Types of Noise, Losses, and Gain

To transmit a signal power from the base station, there are noises, interference, and losses. When the signal is received at the mobile, the signal power after its path loss and any other kind of losses will still be strong enough, and the noise floor must be kept as low as possible. Therefore, we have to count all the noise sources, including the interference sources, as one kind of noise. Usually, the interference is much higher than the thermal noise.

6.5.2.1 Thermal Noise—The thermal noise power is

$$N = kTB \text{ watts} \qquad (6.5.2.1)$$

where k = Boltzmann's constant = –174 dBm/Hz at 17°C, T = temperature in Kelvin, and B = bandwidth, in hertz.

6.5.2.2 Transmitter and Receiver Noises

The local oscillator phase noise, the AM-FM conversion noise in nonlinear devices, and the IM noise, which the intermodulation (IM) products, appear within the bandwidth region.

6.5.2.3 Feeder Line Loss

The waveguide or cables between the receiving antenna and the receiver front contribute both signal attenuation and thermal noise.

6.5.2.4 Atmosphere Loss

A primary atmosphere loss is from rainfall. The more intense the rainfall and the higher the frequency, the more signal energy will be absorbed. Operating below 10 GHz, the loss can be negligible.

6.5.2.5 Interference

Interference can be considered as one kind of noise. Based on the interfered power linking into the signal channel, the interference plus noise level would be used to compare with the received signal level. Usually, the interference level is much higher than the thermal noise in the cellular system due to the co-channel reuse scheme. The required signal-to-interference ratio can provide an error probability from transmitting a signal that meets our system requirement. The signal-to-interference ratio is a parameter to calculate the link budget. System Interferences—System interference is formed by adjacent channel interference and co-channel interference. These types of interference can be predicted by the prediction model as long as the input data are complete and accurate. Those data will be found in Sec. 6.5.5.

6.5.2.6 Antenna Efficiency and Gain

Antenna efficiency is the ratio of its effective aperture to its physical aperture. The reduction of antenna efficiency is due to aperture tapering, aperture blockage, scattering, re-radiation, spillover, edge diffraction, and dissipative loss. Antenna gain is the ratio of the maximum radiation intensity from the subject antenna to the radiation intensity from isotropic source with the same power input. The relationship between antenna effective aperture A_e and antenna gain G is

$$G = 4\pi A_e / \lambda^2$$

All of these items would be used to calculate the link budget.

6.5.3 Received Signal Power and Noise Power

The received signal power after a distance d is

$$P_r = \frac{P_t G_t G_m L_o}{A \, d^\gamma} \qquad (6.5.3.1)$$

where A is a constant and γ is the path-loss exponent. L_0 is the loss of the signal strength from any cause while propagating along the radio path. When in a free space propagation condition, $A = \left(\dfrac{4\pi}{\lambda}\right)^2$, and $\gamma = 2$ and $L_0 = 0$.

Equation (6.5.3.1) is a general equation, that can be represented in all the cell size–specific path-loss formulas.

The noise power is the sum of all the noise sources:

$$N = \text{Thermal noise} + \text{amplifier noise} + \text{human-made noise} \qquad (6.5.3.2)$$

The noise figure F is the ratio of two signal-to-noise ratios (SNR), that is $(SNR)_{in}$ and $(SNR)_{out}$. It can be expressed as $(SNR)_{in}$ at the input of a network to $(SNR)_{out}$ at the output of the network, as shown in Fig. 6.5.3.1.[28]

$$F = \frac{(SNR)_{in}}{(SNR)_{out}} = \frac{(S_i/N_i)}{GS_i/G(N_i+N_a)} = \frac{N_i+N_a}{N_i} = 1 + \frac{N_a}{N_i} \qquad (6.5.3.3)$$

where S_i = signal power at the input of the network, such as at the amplifier input port; N_i = noise power at the output of the network, such as at the amplifier output port; N_a = network noise referred to the input of the network, such as the amplifier noise; and G = network gain, such as the amplifier gain.

Figure 6.5.3.1 shows that S_i/N_i is 40 dB and that S/N_o becomes 35 dB after passing through an amplifier with a gain of 20 dB and an amplifier noise of 3 dB above the thermal noise, which is $N_i = -100$ dBm. The noise figure as 5 dB can be found from Eq. (6.5.3.3).

In a cascaded system shown in Figure 6.5.3.2, the composite noise figure can be obtained as

$$F_{comp} = F_1 + \frac{F_2-1}{G_1} + \frac{F_3-1}{G_1 G_2} + \cdots + \frac{F_n-1}{G_1 \dots G_{n-1}} \qquad (6.5.3.4)$$

where F_n is the noise figure of the nth network.

Usually, we will use the received signal power and noise figure to calculate the link budget.

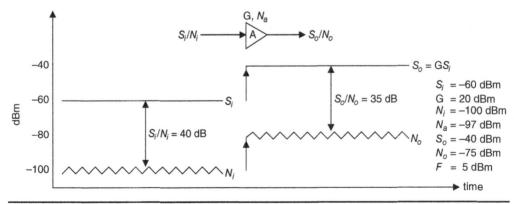

FIGURE 6.5.3.1 S_o/N_o after gain and amplifier noise.

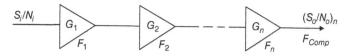

FIGURE **6.5.3.2** A cascaded system.

6.5.4   Required Information for Calculating Link Budget

The Lee comprehensive prediction model needs to have the items listed as follows:

- Size of cells, macrocell, microcell, or in-building (picocell)
- How many cells and the cell's locations
- Frequency 400 MHz to 2.5 GHz
- Bandwidth 1.25 MHz to 30 MHz
- Range (feet or miles, meters or kilometers)
- A few runs of measurement data taken at the system-deployed area
  Human-made environment, free space, open, rural area, suburban area, urban area, metropolitan area, microcell, in-building (picocell)
- Terrain condition, degree of undulation, LOS, NLOS, shadow condition
- Antenna gain at the transmitter side and the receiver side
- Noise floor, $N = kTB$: Use the room temperature $T = 290\ K = 17\ °C$, then
  $kT = -204\ dBW = -174\ dBm$
  If bandwidth $B = 1.25\ MHz$, the noise floor $N = -174\ dBm + 81\ dB = -93\ dBm$
- Feed Line Loss: The length and the diameter of the cable
- Macrocell
  1. Human-made environment: free space, open, rural area, suburban area, urban area, and metropolitan area
  2. Digitized terrain maps
  3. Specifications of the equipment and the deployed system
  4. Special situations: over the water, tunnel, and so on
- Microcell
  1. Street layout
  2. Building layout
  3. Air view map
  4. Small-scale terrain map
- In-building
  1. Building layout
  2. Floor layout
  3. Building and wall material
  4. Glass loss

6.5.5   Link Budget Analysis

6.5.5.1   Required $(E_b/N_o)_{req}$ and Planned $(E_b/N_o)_{pl}$

We refer to E_b/N_o as the value of bit energy per noise power spectral density, That is bit per sec/noise per hertz required to achieve a specified error probability. There are a required E_b/N_o and a planned E_b/N_o. The required one is set by the system spec. The planned one is the one that ensures there is a margin in dB to be added so that under any circumstances, the planned one will never go under the required one. Figure 6.5.5.2.1 illustrates the two levels and the margin shown in dB.

The difference in dB between $(E_b/N_o)_{pl}$ and $(E_b/N_o)_{req}$ yields the link margin:

$$M\ (\text{dB}) = (E_b/N_o)_{pl} - (E_b/N_o)_{req}\ (\text{dB}) \tag{6.5.5.1.1}$$

where M is expressed as a link margin or a safety factor, which is a ratio between $(E/N)_{pl}$ and $(E/N)_{req}$. When it expressed in dB, M is the difference between the two, as shown in Eq. (6.5.5.1.1).

6.5.5.2   The Formula for Link Margin M

Figure 6.5.5.2.1 shows the margin M between two E_b/N_o. The link margin can be derived from

$$\frac{C}{N} = \left(\frac{E_b}{N_o}\right)_{pl} \times \frac{R_b}{B} = \left(\frac{E_b}{N_o}\right)_{req} \cdot M \times \frac{R_b}{B} \tag{6.5.5.2.1}$$

From Eq. (6.5.3.1), we can come up the following equation:

$$\frac{C}{N} = \frac{P_t G_t G_m}{A\ d^\gamma \cdot N_o B \cdot L_o} \tag{6.5.5.2.2}$$

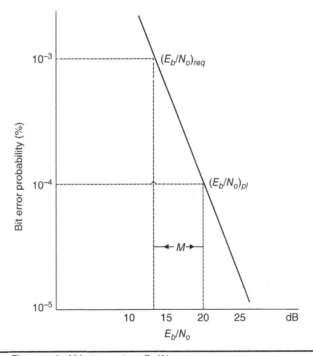

Figure 6.5.5.2.1   The margin M between two E_b/N_o.

Since Eqs. (6.5.5.2.1) and (6.5.5.2.2) are equal, we can solve M as

$$M = \frac{P_t G_t G_m}{A \, d^\gamma \cdot N_o \cdot L_o} \times \frac{1}{\left(\dfrac{E_b}{N_o}\right)_{req} \cdot R_b} \qquad (6.5.5.2.3)$$

We denote
ERP $= P_t G_t$
$D = A \, d^\gamma$
Then expressing the link margin M in dB is

$$M \text{ (in dB)} = \text{ERP} + G_m - D - \left(\frac{E_b}{N_o}\right)_{req} - R_b - N_o - L_o \text{ (in dB)} \qquad (6.5.5.2.4)$$

From Eq. (6.5.5.2.4), all the parameters are known, and we can obtain a derived value of M by adjusting ERP, that is, either the transmitting power or the transmitting antenna gain.

6.6   How to Plan and Design a Good Wireless System

Designing a good wireless system is just like playing a game. There are many challenging factors. Two important factors are the two Cs: coverage and capacity. These are related. You may sacrifice capacity for coverage or vice versa. They all involve cost. Therefore, in designing a macrocell system, coverage is more important than capacity. Then in terms of both coverage and capacity, the microcell system should be deployed. In the microcell system, reducing interference is a key task. The in-building (picocell) system will be used for the need of system capacity only. These are the general concepts of designing a system.[29]

6.6.1   Understanding the System Requirement

The system requirement would include the budget, the size of deployed area, the quality of the system needed, the number of carrier channels, the frequency reuse plan, foreseen system growth, the competition factor, the characteristic of the equipment, the capability of the designing engineers, and the zoning issues from the local government.

6.6.2   Choosing the Right Prediction Model

For exploring a general system deployed area without going into detail, an area-to-area model is suitable. The area-to-area prediction model is used when both locations of base station and mobile are not specified. The other two kinds of prediction models are point-to-area and point-to-point. The point-to-area prediction model is that the location of base station is specified but not the mobile. The point-to-area model is not very practical to be used and seldom developed. The third one is the point-to-point prediction model. Both locations of the base station and the mobile are specified. The point-to-point prediction model is very useful in choosing the right locations for the base stations when a system is deployed.

6.6.2.1   For Macrocell Systems

The simple but fairly accurate prediction models would be chosen for the macrocell system. The prediction model has to have a system design capability. Check for the strength of an interested model, such as digital terrain maps, a few runs of measured data, and so on, and be sure it meets our requirement. Also understand what the outputs are from the prediction models.

Once the prediction model is chosen, we first have to find the path-loss slope and intercept in the area or in different angles from a site-specific base station from the prediction model. Determine the gains and losses over the radio path before reaching the mobile receiver. Then the noise floor is calculated, including the thermal noise, equipment noise, human-made noise, and interference, and the C/N or C/I can be found. Taking the link margin into account, we can determine the $(C/N)_{pl}$. To use the planned $(C/N)_{pl}$ as a parameter, we can plot the boundary around each base station. At this point, we adjust and move around the locations of the base stations to see the coverage of the system. Those planned base station sites will be rented or negotiated if there is a zoning issue. Also, we can install the sites owned by the antenna tower companies to be the first choices. Whenever base stations, are moved, the effectiveness of interference in the system changes. The HO areas are critical and need to be well planned to avoid call drops. Also, the HO regions should be as small as possible so that capacity will not be hurt.

6.6.2.2   For Microcell Systems

In the microcell system, we have to realize that the building structures in microcell systems, the unique landscape of the small area beging to be a key component in the model. The building layout along the streets and in each street block can be used to find the building density in each square block around four streets. The prediction model can find the path loss of the radio wave due to the building blocks. The small-scale terrain tour map may be used to calculate the effective antenna gain or shadow loss. Sometime if we do not have enough information, we may use the macrocell model to predict the path loss. The deviation between the predicted results and the measured data may be still within 8 dB in 60 percent of CDF, which is acceptable.

6.6.2.3   For In-Building (Picocell)

The Lee in-building model is a simple and cost-effective prediction model. The layout of the inside as well as the outside of the building is required. The wall material is must be known. The attenuation factor of the walls comes from the wall material. A few runs of measured data for the same floor, inter-floor, inter-building, and outside building should be collected. All the path-loss slopes are come from the measured data. If no measured data are available, then the default values can be used. Because the measured data are used for enhancing the model, the deviation between the predicted results and measured data should be less than 6 to 8 dB in 60 percent of CDF. The ray-tracing model and the FDTD model are physically based prediction models. They can be used when no measured data are available. However the input data for running these models are more complicated. Some models need 3D display. The costs and time needed are high and making the models is not very attractive. These models are good to use to find the physical explanations of many ray path phenomena. There are some military applications of using these physically based prediction models.

6.6.2.4   For the Lee Comprehensive Model

The Lee comprehensive model can provide an overall signal coverage map of a wireless system in a large operating area. Because this comprehensive model integrates three cell size–specific Lee models, this coverage map can include the macrocells, microcells, and in-building (picocell) cells. This comprehensive model is very useful for planning a mature system while all the different cell size–specific models have been used in the system. From this overall-view map provided by this model, the interference, area of interface, Handoff (HO), and further growth among them can be solved.

6.7 Propagation Prediction on Different Transmission Media

In this section, must call to the reader's attention that there are four different transmission media—satellite communication, underwater communication, bullet-train communication, and aeronautical communication—in which that their transmitted signals are predicted are different from those of predicting terrestrial mobile propagation. We may highlight just the differences, so the interested reader will early understand the differences and search for the right sources.

6.7.1 Prediction of Satellite Communication Signals

In satellite communications, space propagation phenomena affect the earth space link, mainly propagation through two sky layers—troposphere and ionosphere. Most of the propagation effect on the operational frequency in the ionosphere occurs below 1 GHz, and most of the propagation effect on the troposphere occurs above 1 GHz.

The ionosphere involves interactions between the layers of charged particles around the earth, the earth's magnetic field, and the radio waves. Ionospheric effects are Faraday rotation, group delay, dispersion, and ionospheric scintillation.

The tropospheric ozone layer involves interactions between the waves and the lower layer of the earth's atmosphere, including the effects of the gases composing the air and hydrometeors such as rain. Tropospheric effects are path loss, rain attenuation, gaseous absorption, tropospheric refraction, tropospheric scintillation, depolarization, and sky noise.

Also, the satellite communication signal is affected by local environment in the vicinity of the earth station, such as terrain, trees, and buildings. These effects may be a significant impairment and need to provide some site shielding of fixed earth stations from terrestrial interferers. However, in mobile satellite systems, the direct path may frequently be wholly or partially obscured from the surroundings. Local effects have been treated in previous chapters.

6.7.1.1 Path-Loss Models, Channel Models, and Overall Path-Loss Model

Here we are introducing the path-loss models (long-tern fading) and the channel models (short-tern fading) of satellite communications.[30]

6.7.1.1.1 Path-Loss Models for $f > 1$ GHz—Based on Atmosphere Conditions Usually, the operational frequency of the satellite communication is around 10 GHz, and then the tropospheric effects are a big concern. The path-loss models are applied to the frequency above 3 GHz.

Two different modeling approaches have been considered. The first approach considers all attenuation effects as being correlated; That is, the total path loss is given by

$$L_{tot} = A_{0_2} + A_{H_2O} + A_C + A_R + A_{ML} + A_S \tag{6.7.1.1}$$

where A_{0_2}, A_{H_2O}, A_C, A_R, A_{ML} and A_S are the attenuation effects due to oxygen, water vapor, cloud, rain, melting layer, and scintillation, respectively.

The second approach considers all attenuation effects being partially uncorrelated; therefore, RMS summing is used for the total path loss:

$$L_{tot} = \left[A_{0_2}^2 + A_{H_2O}^2 + A_C^2 + A_R^2 + A_{ML}^2 + A_S^2 \right]^{\frac{1}{2}} \tag{6.7.1.2}$$

A combination method reflects the independence of various attenuation factors and better considers some of the propagation effects uncorrelated.[31,32] The corresponding ITU-R recommendation adopts the following formula for total path loss:

$$L_{tot} = A_{0_2} + A_{H_2O} + \sqrt{(A_C + A_R)^2 + A_S^2} \tag{6.7.1.3}$$

6.7.1.1.2 Channel Models—Long-Term and Short-term Fading Due to the Effects of Surroundings   The channel models for the satellite communication channel, the LMS channel (land-mobile-satellite channel), are statistical models, that can be characterized into two categories: single state and multistate models.[33] The single-state models are described by single statistical distributions and are valid for fixed satellite scenarios where the channel statistics remain constant over the areas of interest. The multistate or mixture models are used to demonstrate non-stationary conditions where channel statistics vary significantly over large areas for particular time intervals in nonuniform environments.[34]

1. Single-State Model   Loo Model:[35] The Loo model is one of the most primitive statistical LMS channel models with applications for rural environments specifically with shadowing due to roadside trees. In this model, the shadowing attenuation affecting the LOS signal due to foliage is characterized by log-normal pdf, and the diffuse multipath components are described by Rayleigh pdf. The model illustrates the statistics of the channel in terms of probability density and cumulative distribution functions under the assumption that foliage not only attenuates but also scatters radio waves. The resulting complex signal enveloper is the sum of correlated lognormal and Rayleigh processes. The pdf of the received signal envelope $p(r)$ is given by Loo and Butterworth:[36]

$$p(r) = \begin{cases} \dfrac{1}{r\sqrt{2\pi\sigma_0}} \exp\left[-\dfrac{(\ln r - \mu)^2}{2\sigma_0} \right] & \text{for } r \gg \sqrt{\sigma_0} \\[4mm] \dfrac{r}{b_0} \exp\left(-\dfrac{r^2}{2b_0} \right) & \text{for } r \ll \sqrt{\sigma_0} \end{cases} \tag{6.7.1.4}$$

where μ and σ_0 are the mean and standard deviation, respectively. The parameter b_0 denotes the average scattered power due to multipath effects. If attenuation due to shadowing (lognormal distribution) is kept constant then the pdf in Eq. (6.7.1.4) simply yields a Rician distribution.

There are other models. The Corraza–Vatalaro model[37] has modeled effects of shadowing on both the LOS and diffuse components; the extended Suzuki model[38] is a statistical channel model for terrestrial communications characterized by Rayleigh and lognormal process; the Xie–Fang model[39] is based on propagation scattering theory and deals with the statistical modeling of propagation characteristics in low-earth-orbit and middle-earth-orbit satellites communication systems; and the Abdi model[33] is convenient for performance predictions of narrowband and wideband satellite communication systems.

2. Multistate Models   There are two-state (good state and bad state) models, three-state (clear state, shadowing states, and blocked state) models, five-state model.

Five-State Model: This channel model is based on the Markov modeling approach in which two-state and three-state models are extended to five-state models under different time share of shadowing.[40]

We may illustrate the three-state model as follows:

Three-State Model: This statistical channel model,[41] based on three states, namely, clear or LOS state, the shadowing state, and the blocked state, provides the analysis of improvement in nongeostationary LMS communication systems. The clear state is characterized by Rice distribution (Sec. 1.8.4). The shadowing state is described by Loo's pdf in Eq. (6.7.1.4), and the blocked state is illustrated by Rayleigh fading. The pdf of the received signal envelope is a weighted linear combination of these distributions:

$$p\,(r) = M\,p_{\text{Rice}}\,(\text{r}) + L\,p_{\text{Loo}}\,(\text{r}) + N\,p_{\text{Rayleigh}}\,(r) \qquad (6.7.1.5)$$

where M, L, and N are the time share of shadowing of the Rice, Loo, and Rayleigh distributions, respectively. The three states are as follows:

State A: LOS path plus scattered and reflected paths

State B: Shadow path plus scattered and reflected paths

State C: Block path plus scattered and reflected paths

The time-share of three states can be operated by the Markov process. The distribution parameters for the model were found by means of the data obtained from measurements using the "INMARSAT" satellite and other available data sets. The model was validated by comparing the theoretical cumulative distributions with those obtained from measurement data.

Another model also can be formed by a five-state model. This channel model is based on Markov modeling approach in which two-state and three-state models are extended to five-state models under different time share of shadowing.[40]

The model is basically a composition of the Gilbert–Elliot channel model and the three-state Markov channel model, in which shadowing effects are split into five states:

State 1: good state—clear LOS without shadowing, and LOS state with low shadowing

State 2: good state—clear LOS with low shadowing

State 3: not good not bad state—characterizes moderate shadowing

State 4: bad state—describes heavy shadowed area

State 5: bad state—describes completely shadowed or blocked areas

The transitions of the states can take place from low and high shadowing conditions to moderate shadowing conditions but cannot occur directly between low and high shadowing environments.

6.7.1.1.3 The Overall Path-Loss Formula   The overall path loss can be obtained from three approaches:

1. The overall path loss is the sum of two sources. One is from the path loss caused by the atmosphere condition, and the other is the path loss to be predicted from the causes of terrestrial surroundings:

$$L_{overall} = L_{tot} + L_{terrestrial} \qquad \text{in dB} \qquad (6.7.1.6)$$

where $L_{terrestrial}$ can be predicted based on models described in previous chapters of this book.

2. We also can calculate the mean square of the received signal r by taking the measured data $r(t)$ to process to get the measured local mean $m'(t)$ as follows:

$$m'(t) = \frac{1}{2L} \int_{r-L}^{r+L} r(t) \, dt \qquad (6.7.1.7)$$

where the length L is around 40λ (see Sec. 1.6.3.1). The path loss $L_{overall}$ is obtained by as subtracting the local mean square from the transmitted ERP, P_s, as

$$L_{overall} = P_s - [m'(t)]^2 \qquad (6.7.1.8)$$

3. We can calculate the mean power P_r from the measured pdf, $p'(r)$, as

$$P_r = \int_{-\infty}^{\infty} r^2 \, p'(r) \, dr \qquad (6.7.1.9)$$

Then the overall path loss is

$$L_{overall} = P_s - P_r \qquad (6.7.1.10)$$

We have noted that the channel models described in Sec. 6.7.1.1.2 are not considered the atmosphere effects as described in Sec. 6.7.1.1.1. Therefore, the signal strength obtained from the predicted pdf of the models has to add the additional atmosphere effect to be the overall path loss.

6.7.2 Prediction of Underwater Communication Signals[42-44]

The electromagnetic waves propagate over only extremely short distances under water. The underwater signals that are used to carry digital information through an underwater channel are not radio signals but acoustic waves, which can propagate over long distances. However, an underwater acoustic channel presents three distinguishing characteristics: frequency-dependent propagation loss, severe multipath, and low speed of sound propagation.

Another unique characteristic is geometric spreading. The spreading of the sound energy results in the expansion of the wave front. There are two spreadings:

1. One is cylindrical spreading. It is at horizontal radiation only, which characterizes shallow-water communication.

2. Another is spherical spreading. It is from an omnidirectional point source, which characterizes deep-water communication.

These characteristics make underwater wireless communication extremely difficult. To solve these difficulties, two tools can be used. One is the underwater propagation path-loss model created by Urick, and another is the propagation speed of the acoustic signal underwater.

6.7.2.1   Urick Propagation Path-Loss Model

One underwater propagation model, called the Urick propagation path-loss model, is introduced:

$$L(d, f) = \beta \cdot \log(d) + \alpha(f) \cdot d + A \qquad (6.7.2.1)$$

where β is the geometric spreading, $\beta = 10$ (shallow water); and $\beta = 20$ (deep water); $\alpha(f)$ is medium absorption obtained from the experiments; and A is transmission anomaly in dB a degradation of acoustic intensity due to multipath refraction, diffraction, and scattering of the sound; $A = 5$ to 10 dB [deep water]; $A > 10$ dB [shallow water].

6.7.2.2   Propagation Speed of Acoustic Signal Underwater

The propagation speed of the acoustic signal (S) underwater is a function of three parameters the temperature T, the salinity (Sa), and the depth of the water (D):

$$S = f(T, Sa, D) \qquad (6.7.2.2)$$

Because of the three effects, the signals bend toward the region of slower sound speed. There are four main types of thermal structure: isothermal gradient (uniformed), negative gradient (sound speed increases as water becomes shallower), positive gradient (sound speed increases as water becomes deeper), and negative gradient over position (change speed gradient at a certain depth). The four types of thermal structure are shown in Fig. 6.7.2.2.1 due to the effects of temperature, salinity, and the depth of the water, as shown in Eq. (6.7.2.1).

6.7.2.3   Shallow-Water Communication—Channel Model

In shallow water, multipath occurs due to signal reflection from the surface and bottom, as illustrated in Fig. 6.7.2.3.1. Shallow-water communication is greatly affected by the multipaths due to multiple reflected rays from either the sea bottom or the surface, ray bending as a result of sound speed variations with respect to depth of the water and cylindrical spreading to expand the wave front horizontally. The shadow water

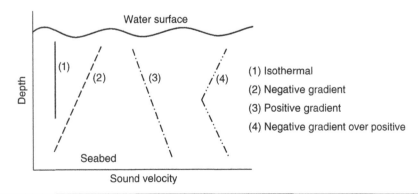

FIGURE 6.7.2.2.1   Four main types of thermal structure for the sound speed under deep water.

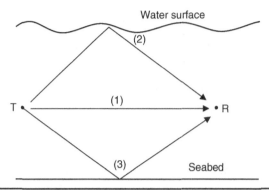

Figure 6.7.2.3.1 Three paths conduct shallow-water communication.

communication is performed through three paths, as shown in Fig. 6.7.2.3.1; (1) direct path; (2) reflection from the surface, and (3) reflection from the bottom as—well as from other objects underwater.

The shallow-water communication channel is modeled as a multiray Rayleigh fading channel. Each of the three paths is modeled as a Rayleigh fading channel. The composition of three rays at the receiver forms a shallow water channel model.[42]

6.7.2.4   Deep-Water Communication

In deep water, the bending of acoustic waves occurs, that is, the tendency of acoustic waves to travel along the axis of lowest sound speed. Figure 6.7.2.4.1 shows an ensemble of channel responses obtained in deep water. The multipath spread, measured along the delay axis, is on the order of 10 ms in this example. The channel response varies in time. The deep-water propagation channel is not as affected by multipaths but due to the spherical spreading of the acoustic signal. The four main types of thermal structure of the sound wave are the unique characteristics for the deep-water communications, as shown in Fig. 6.7.2.2.1. The path of the acoustic signals underwater is also unique, as shown in Fig. 6.7.2.4.1. The Urick propagation path-loss formula expressed in Eq. (6.7.2.1) is used to evaluate the communication performance. The sound signal in deep water also exhibits randomness in the propagation waves and is also modeled using the Rayleigh fading model.[43]

6.7.3   Prediction of Aeronautical Communication Signal

The aeronautical mobile radio channel appears in many applications. Air ground radio communication between aircraft and ground radio sites is one of the most important applications. In this section, a stochastic model is proposed.[45–50]

In the aeronautical VHF band, ionospheric effects usually can be neglected. Aeronautical communication is a different medium than terrestrial mobile medium. There are two kinds of interferences. One is interferential propagation due to the direct and reflected signal, and the other is diffraction and scattering. Sometimes, these two interferences are mixed.

The transmission is carried by troposcatters; and those small, random irregularities or fluctuations in the refractive index of the atmosphere dominate. The refraction index changes with seasons and also varies according to altitude and decreases

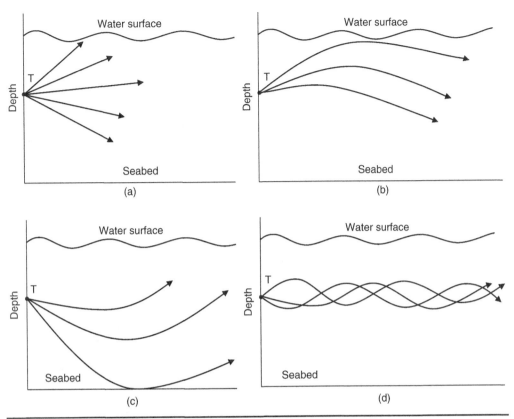

FIGURE 6.7.2.4.1 Sound propagation paths under deep water.

as the altitude increases. Thus, when the aircraft fly at high altitude, the equivalent earth radius at the altitude becomes smaller than that assumed on the earth.

6.7.3.1 Air-to-Ground Path-Loss Formula (LOS)

The received signal $r(t)$ received the signal from the transmitted signal $s(t)$ through the air-to-ground channel $h(\tau, t)$ after time τ can be expressed as,

$$r(t) = \int_{-\infty}^{\infty} s(t-\tau)h(\tau,t)\,d\tau \tag{6.7.3.1.1}$$

where the time variance $h(\tau, t)$ consists of two parts, the large-scale part and small-scale part, as

$$h(\tau, t) = h_{large\ scale}\,(\tau, t) + h_{small\ scale}\,(\tau, t) \tag{6.7.3.1.2}$$

The propagation channel characteristics may equivalently be defined in the frequency domain. The time-varying transfer function of the channel $H\,(f, t)$ is simply the Fourier transform of $h\,(\tau, t)$, that is,

$$H\,(f, t) = \int_{-\infty}^{\infty} h(\tau, t)\,e^{-\,j2\pi f\,\tau}\,d\tau \tag{6.7.3.1.3}$$

When both transmitter and receiver antennas are the same polarization, we have the simple expression for the large-scale transfer function:

$$H_{LS}(f,t) = \frac{\lambda}{2\pi} g_T g_R \left\{ \phi_T^{Dir} \phi_R^{Dir} \frac{e^{-jkr_D}}{r_D} + \rho \, R_{vert \, smooth} \, \phi_T^{ref} \, \phi_R^{ref} \frac{e^{-jk \, r_R}}{r_R} \right\} \qquad (6.7.3.1.4)$$

Where g_T and g_R are the maximum field gains of transmitting and receiving antennas; ϕ_T^{Dir} and ϕ_R^{Dir} are complex valued functions describing amplitude/phase variation of the field radiation diagram of these transmitted and received antennas over a direct path. Similarly, ϕ_T^{ref} and ϕ_R^{ref} are complex valued functions related to the transmitted and received antennas over a reflected path, and $R_{vert \, smooth}$ is the smooth earth reflection coefficient in vertical polarization depending on grazing angle and dielectric (ε, σ) parameters of the ground.[49]

ρ is the reduction factor due to ground roughness as

$$\rho = \exp\left\{ -\frac{1}{2}\left(4\,\pi \frac{\Delta h \sin \alpha}{\lambda}\right)^2 \right\} \qquad (6.7.3.1.5)$$

The received signal power is

$$P_r = P_t \, L_T \, L_R \, |H_{LS}|^2 \qquad (6.7.3.1.6)$$

The global transmitter losses L_T are calculated by taking the transmitter antenna vertical electric dipole (with reflector) maximum gain assessed at $g_T = 3$ dB, the mismatch loss 0.1 dB, the cable loss 1.3 dB and the splitter loss 2 dB; hence, the global transmitter losses $L_T = -3.4$ dB.

The global receiver losses L_R and calculated by taking the receiver vertical electric half wavelength dipole antenna maximum gain was theoretical $g_R = 2.14$ dB, the mismatch loss 0.5 dB, and the cable loss 1.5 dB; hence, the global receiver losses were $L_R = -2$ dB.

The overall path loss in dB can be expressed as

$$L_{overall} = L_T + L_R + |H_{LS}|^2 \text{ in dB} \qquad (6.7.3.1.7)$$

6.7.3.2 Prediction versus Measured Data

6.7.3.2.1 LOS Scenario Two experiments were carried out. One is at 118 MHz (VHF) In the plot, the SWR, the splitter, and cable loss were set to unity. A Beechcraft 90 aircraft was at a flight level of 6000 m and was flying along the runway axis from 20 nm north to 20 nm south. One nautical mile = 1852 m = 6080 ft. The nature of the ground was found to be smooth ($\Delta h = 0$ in Eq. [6.7.3.1.5]) and dry ($\varepsilon_r = 2$, $\sigma = 0.05$ S/m). Figure 6.7.3.1 shows two models: the free space large-scale model and the direct-reflected large-scale model were used to compare with the measured data. The direct-reflected large-scale model is in very good agreement with the large-scale measured fading data. But the free space model cannot predict the large-scale fading.

The next experiment was at a lower flight level but kept the same conditions. Figure 6.7.3.2 shows the power measurements and predictions for a receiver altitude of 1500 ft (460 m). We see that the direct-reflected model prediction is less

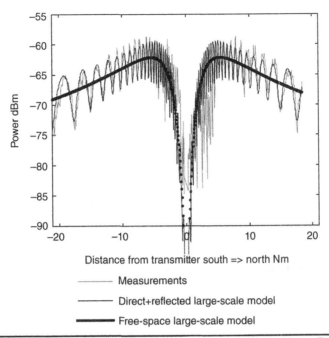

FIGURE 6.7.3.1 Power measurements and predictions for an in-flight receiver at FL 6000 m.

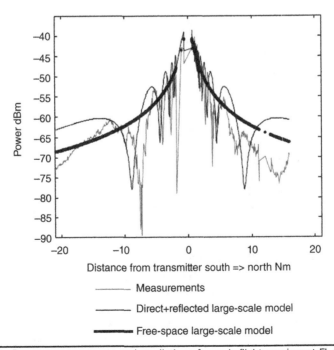

FIGURE 6.7.3.2 Power measurements and predictions for an in-flight receiver at FL 1500 ft.

precise than at a high flight level as shown in Fig. 6.7.3.1, but still is in rather good agreement with the measured results. Also, a greater azimuthal asymmetry between the left side and the right side of the measured curve was observed.

6.7.3.2.2   NLOS Scenario   In the case of the less frequent NLOS scenario, the spherical diffraction waves occur when the signal is in the NLOS and propagate mainly in the diffraction region. In this region, signal strength decreases rapidly as distance increases. A deterministic approach has been done[51] to compute large-scale fading with the uniform theory of diffraction, and the experiment/theoretical model comparison with the measured data showed errors between 5 and 15 dB. Air-to-ground radio communication usually avoids this condition.

6.7.3.2.3   A LOS Channel—Satellite Aircraft Channel   The satellite aircraft channel is always in a LOS condition. The atmosphere condition causes additional losses from free space path loss, a described in Sec. 6.7.1.1.1. Most of the published literature,[45,52] has focused on the satellite aircraft channel.

6.7.4   Prediction of Bullet Train Communication Signal

The bullet train communication medium is an unchanged medium. The train is always running on a fixed route. Therefore, the path loss of this medium can be calculated straightforward without difficulty. The path losses of their signals follow the free space path loss.

However, because of the fast speed of the train, the received signal will be considered differently than the signal received from land mobiles. The bullet train systems in different countries are different. China and Germany[53,54] use remote radio heads, as in the cloud-based high-speed train communication system architecture, as shown in Fig. 6.7.4.1, or radio over fiber, which consists of radio access units along the rail for wideband communications system.

In Japan,[55] most bullet train communications use coaxial cables. Outer conductors of the coaxial cables have slots radiating a portion of the transmitted energy. There are two cables along the railroad—one for inbound and one for outbound. Before the signal goes through the cable and gets weaker, a relay is installed along the railroad every given distance apart to boast up the signal level. The two slot antennas are mounted on the train, one on each side. Therefore, the link budget of the signal is constant.

Because of the fast speed of the train, the received signal suffers the frequency dependence (time-delay spread) and time variance (Doppler spread) of the channel. The Doppler spread directly influences the inter-subcarrier interference of an OFDM system. One approach to reduce Doppler spread and time-delay spread is to use the tracking directional antenna aiming at the base station.[56]

6.7.5   Millimeter Wave Signal

The millimeter wave signal can be only used for indoor communications. In 1973, Lee had installed a 3-mm link between the Empire State and Pan Am Building, using a 15-mW IMPATT diode as a source, two 30-in parabola dish antennas for a link gain of 106 dB, and a Schottky diode for detection. The link was clearly established, but the rainfall attenuation during the heavy rainfall caused the 3-mm wave signal faded.[57] Therefore, we concluded at that time that an outdoor millimeter wave communication

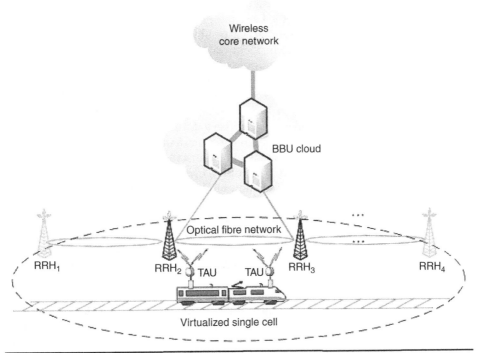

Figure 6.7.4.1   Cloud-based high-speed train communication system architecture.

was not cost effective. An indoor mm-wave communication is applicable in the future. It should be worked under LOS condition or under multi-reflected environment created by the providers. The ray tracing techniques and FDTD method can be used for modeling the millimeter wave signal.

6.8   Summary and Conclusions

This chapter discusses integrating the Lee macro-, micro-, and picocell models into one comprehensive model by combining the different parts of Lee models for different areas. In a macrocell system, each cell is very large. The antenna heights of all base stations are very tall. It is not easy to move any base station around after it is installed. Therefore, the point-to-point prediction model is an important and useful tool for designing macrocell and also microcell systems.

As demands increase for more integrated and more accurate capabilities to provide ubiquitous coverage, throughput and capacity, the Lee models have evolved and developed to meet the capabilities as well as accuracy, speed, required data input, and flexibilities. Furthermore, the measured data have to be collected in order to improve the accuracy of the model. And again, different Lee models will need different input data to improve the accuracy. In the macrocell, we need terrain and clutter data; in the microcell, we need building, terrain, and attribute date; and in the picocell, we need to know wall and building material and the layout of window and room dividers. This chapter discusses design aspects involving relatively new technology, repeaters, special areas, and link budget analysis in designing a system. This chapter also provides a more detailed and insight view of the Lee comprehensive models.

Usually, the ray tracing and FDTD techniques are very accurate for finding radiation patterns from exciting an incident wave on an object of a known shape. However, in a small area with irregular human-made structures, such as buildings, those rays bouncing and reflecting from the buildings based on ray-tracing techniques or FDTD cannot come to give accurate predictions of signal strength at the mobile unit, but rather are statistical in nature. Using both ray-tracing and FDTD techniques is complicated and costly. Therefore, using empirical data to form a statistical prediction model of path loss can allow one to predict the receiving signal in a small area fairly well and is simple. Furthermore, when an area is small, we can install the base stations or access points by using the cut-and-try method. Sometimes, even though we use either the ray-tracing techniques or the statistical prediction model, such as Lee in-building model, to first install the base stations, we still have to use the cut-and-try method to move the base stations around and fine-tune their locations.

Table 6.8.1 comprises the characteristics of the propagation models with their techniques. The Lee model is compared with theoretical and empirical and statistical prediction models on characteristics, data requirements, ease of integration, and speed of calculation with high accuracy.

Form Table 6.8.1, the advantage of using the Lee comprehensive model is that it is a compound of all three model types—theoretical, empirical, and statistical. Also, it integrates three cell size–specific models—macrocell, microcell, and picocell (in-building). The software tool is easy to use in designing a large system with relatively less effort.

In Sec. 6.7, we have explored four different media. Each of them is unique and different from the terrestrial mobile medium. The property of each medium and the

Model Name	Suitable Environment	Complexity	Experimental Data	Details of Environment	Accuracy	Time	Other
Theoretical	Macro, micro, and pico	High	N/A	Low-high (depends on accuracy)	Good	Can be long	Not integrated with limited accuracy
Empirical	Macro, micro, and pico	Medium	Yes	Medium	Good	Short	Not integrated Can fine tune with measurement data
Statistical	Macro, micro, and pico	Low	Yes	Low	Good	Short	Not integrated
Lee(compound of theoretical, statistical, empirical) ...	Macro, micro, and pico	Low	Yes	Low	Good to great	Short	Fully integrated Can fine tune with measurement data

TABLE 6.8.1   A Comparison of Various Models

ways of modeling the medium are different. This section provides new material so that the reader can think of these problems differently.

In conclusion, the Lee comprehensive prediction model is a point-to-point model. It is based on wave propagation theory, statistical theory and empirical data. It is simple to use with a fair degree of accuracy. Since it composes the three Lee models from three different cell sizes, an overall predicted chart in a large system over a large area consisting of all the sizes of cells can be obtained. Of course, every prediction model from other authors has its merit and has been included in this book to give us plenty of leeway for a further improvement in prediction models in the future. Hopefully the propagation prediction models provided in this book can be used in designing the future 5G systems and beyond.

This book has referred to many different papers and books, which are listed in the reference list of each chapter. The authors would like to thank the contributions from these people.

References

1. Fujimoto, K., and James, J. R., eds. *"Mobile Antenna System Handbook."* Artech House, Inc. Norwood, MA 1994, Chap. 2, "Local- Mean Prediction Model – Lee's Model" pp. 28–85.
2. Parsons, J. D. *The Mobile Radio Propagation Channel.* 2nd ed. New York: John Wiley & Sons, 2000.
3. Saunders, S. *Antennas and Propagation for Wireless Communication Systems.* Wiley, & Sons, 2000.
4. Stuber, Gordon L. *Principles of Mobile Communication.* Boston: Kluwer Academic Publishers, 1996.
5. Seybold, John S. *Introduction to RF Propagation.* New york: John Wiley & Sons, 2005.
6. IEEE VTS Committee on Radio Propagation. "Lee's Mobile." *IEEE Transactions on Vehicular Technology*, February 1988, 68–70.
7. Lee, W. C. Y., and Lee, D. J. Y. "Microcell Prediction Enhancement for Terrain." Personal, Indoor and Mobile Radio Communications, PIMRC '96, IEEE International Symposium, vol. 2 (October 1996): 286–90.
8. Lee, D. J. Y., and Lee, W. C. Y. "Propagation Prediction in and through Buildings." *IEEE Transactions on Vehicular Technology*, 49 (2000): 1529–33.
9. Lee, W. C. Y., and Lee, D. J. Y. "Pathloss Prediction from Microcell to Macrocell." Proceedings of IEEE Vehicular Technology Conference, VTC-Spring, Tokyo, May 2000, 1988–92.
10. Lee, W. C. Y., and Lee, D. J. Y. "Microcell Prediction by Street and Terrain Data." Proceedings of IEEE Vehicular Technology Conference, VTC-Spring, Tokyo, May 2000, 2167–71.
11. Lee, W. C. Y., and Lee, D. J. Y. "Microcell Prediction in Dense Urban Area." *IEEE Transactions on Vehicular Technology* 47 (February 1998): 246–53.
12. Lee, W. C. Y., and Lee, D. J. Y. "Fine Tune Lee Model." Proceedings of IEEE Personal Indoor Mobile Radio Conference, PIMRC, London, September 2000, 406–10.
13. Lee, W. C. Y., and Lee, D. J. Y. "Enhanced Lee Model from Rough Terrain Sampling Data Aspect." Proceedings of IEEE Vehicular Technology Conference, VTC-Fall, Ottawa, Sept. 2010.

14. Lee, W. C. Y., and Lee, D. J. Y. "Enhanced Lee In-Building Model." Proceedings of IEEE Vehicular Technology Conference, VTC-Fall, Las Vegas, Sept. 2013.

15. Lee, W. C. Y., and Lee, D. J. Y. "Integrated models and their usage in predicting the signal strength." Proceedings of IEEE Vehicular Technology Conference, VTC-Spring, Seoul, May 2014.

16. Honcharenko, W., Bertoni, H. L., and Dailing, J. "Mechanisms Governing Propagation between Different Floors in Building." *IEEE Transactions on Antennas and Propagation* 41, no. 6 (June 1993): 787–90.

17. Seidel, S. Y., and Rappaport, T. S. "914 MHz Path Loss Prediction Models for Indoor Wireless Communication in Multifloored Buildings." *IEEE Transactions on Antennas and Propagation* 40, no. 2 (February 1992): 207–17.

18. Kouyoumjian, R. G. "The Geometrical Theory of Diffraction and Its Applications." *In Numerical and Asymptotic Techniques in Electromagnetics*, edited by R. Mittra, New York: Springer-Verlag, 1975, 165–215.

19. Molkdar, D. "Review on Radio Propagation into and within Buildings." *IEE Proceedings-H*, February 1991, 61–73.

20. Walfisch, J., and Bertoni, H. L. "A Theoretical Model of UHF Propagation in Urban Environments." *IEEE Transactions on Antennas and Propagation* 36, no. 12 (1988): 1788–96.

21. Nagy, L. "FDTD Field Strength Prediction for Mobile Microcells." ICECom 2005, Dubrovnik (October 2005): 1–3.

22. Lee, W. C. Y., and Lee, D. J. Y. "The Impact of Repeaters on CDMA System Performance." IEEE Vehicular Technology Conference Proceedings vol. 3, VTC 2000-Spring, Tokyo, 2000, 1763–67.

23. Lee, W. C. Y., and Lee, D. J. Y. "The Impact of Front End LNA on Cellular System." IEEE Vehicular Technology Conference, vol. 5, IEEE-VTS, Fall, VTC 2000, 2180–84.

24. Lee, W. C. Y. *Mobile Communications Design Fundamentals.* 2nd ed., Chapter 10, New York: John Wiley & Sons, 1993.

25. Lee, W. C. Y. Six U.S. patents of microcell system listed in Chap. 1, Reference 27.

26. Stuber, Gordon L. *Principles of Mobile Communication.* Boston: Kluwer Academic Publishers, 1996. Ibid. Sec. 9.4.

27. Communications Audit UK Ltd. "CA 2056-21 Masthead Amplifier (LNA)—Overcome Cable Loss." http://info2@commsaudit.com.

28. Lee, W. C. Y. *Mobile Communications Engineering.* 2nd ed., Sec. 16.10, New York: McGraw-Hill, 1998.

29. Lee, W. C. Y. *Wireless and Cellular Telecommunications.* 3rd ed., New York: McGraw-Hill, 2005, Chapter 8 and Chapter 16.

30. Panagopoulos, Athanasios D., Arapoglou, Pantelis-Daniel M., and Cottis, Panayotis G. "Satellite Communications at KU, KA, and V Bands: Propagation Impairments and Mitigation Techniques." The Electronic Magazine of Original Peer-Reviewed Survey Articles. IEEE Communications Surveys and Tutorials, 3rd quarter, 2004, vol. 6, no. 3, 1–14.

31. ITU-R. "Propagation Data and Prediction Methods Required for the Design of Earth-Space Telecommunication Systems." *Propagation in Non-Ionized Media*, Rec, Geneva, 2001, 618–7.

32. Castanet, L., et al. "Comparison of Various Methods for Combining Propagation Effects and Predicting Loss in Low-Availability Systems in the 20-50 GHz Frequency Range." *International Journal of Satellite Communication*, 19, (2001): 317–34.

33. Abdi, A., Lau, C. W., Alouini, M., and Kaveh, M. "A New Simple Model for Land Mobile Satellite Channels: First- and Second-Order Statistics." IEEE Transactions on Wireless Communications 2, no. 3 (2003): 519–28.

34. Mehmood, Asad, and Mohammed, Abbas, "Characterisation and Channel Modelling for Satellite Communication Systems." In *Satellite Communications*, edited by Nazzareno Diodato. Shanghai. InTech, 2010, 133–51.

35. Loo, C. "A Statistical Model for a Land Mobile Satellite Links." *IEEE Transactions on Vehicular Technology* 34, no. 3, (1985): 122–27.

36. Loo, C., and Butterworth, J. S. "Land Mobile Satellite Measurements and Modelling." *IEEE Proceding* 86, no. 7, (1998): 1442–62.

37. Corraza, G. E., and Vatalaro, F. "A Statistical Channel Model for Land Mobile Satellite Channels and, Its Application to Nongeostationary Orbit Systems." *IEEE Transactions on Vehicular Technology* 43, no. 3 (1994): 738–42.

38. Suzuki, H. "A Statistical Model for Urban Radio Propagation." *IEEE Transactions on Communications* 25, no. 7 (1977): 673–80.

39. Xie, Y., and Fang, Y. "A General Statistical Channel Model for Mobile Satellite Systems." *IEEE Transactions on Vehicular Technology* 49, no. 3 (2000): 744–52.

40. Ming, H., Dongya, Y., Yanni, C., Jie, X., Dong, Y., Jie, C., and Anxian, L. "A New Five-State Markov Model for Land Mobile Satellite Channels." International Symposium of Antennas, Propagation and EM Theory, 2008, 1512–15.

41. Karasawa, Y., Kimura, K., and Minamisono, K. "Analysis of Availability Improvement in LMSS by Means of Satellite Diversity Based on Three-State Propagation Channel Model." *IEEE Transations on Vehicular Technology* 46, no. 4 (1997): 1047–56.

42. M. Stojanovic. "Acoustic (Underwater) Communications." In *Encyclopedia of Telecommunications*, edited by John G. Proakis. New York: John Wiley & Sons, 2003.

43. Stojanovic, Milica. "Underwater Wireless Communications: Current Achievements and Research Challenges." Oceans '95. MTS/IEEE. 'Challenges of Our Changing Global Environment' Conference Proceedings, vol. 2 San Diego, CA, October 9–12, 1995, 1189–96.

44. Sozer, E. M., Stojanovic, M., and Proakis, J. G. "Underwater Acoustic Networks." *IEEE Journal of Oceanic Engineering* 25, no. 1 (January 2000): 72–83.

45. Bello, P. A. "Aeronautical Channel Characterization." *IEEE Transactions on Communications* 21, no. 5 (May 1973): 548–63.

46. Dedryvere, Arnaud, Roturier, Benoit, and Chateau, Beatrice. "A General Model for VHF Aeronautical Multipath Propagation Channel." AMCP/WG-D/WP6, Aeronautical Mobile Communications Panel (AMCP), Working Group D, Agenda Item: 2, Honolulu, January 19–28, 1999.

47. Elnoubi, Said M. "A Simplified Stochastic Model for the Aeronautical Mobile Radio channel." IEEE 42nd Vehicular Technology Conference, Denver, CO, May 10–13, 1992, 960–63.

48. Dyer, G., and Gilbert, T. Gaudette. "Channel Sounding Measurements in the VHF A/G Radio Communications Channel." AMCP WG/D8, WP/19, Oberpfafenhoffen, Germany, December 2–11, 1997.

49. Château, B. "VHF Propagation Channel, Modeling for Aeronautical Mobile Communications." PhD Report (in French), Toulouse, October 1998.

50. Hoeher, P., and Haas, E. "Aeronautical Channel Modeling at VHF Band." DLR note, July 22, 1998.

51. Roturier, B., Chateau, B., Souny, B., Combes, P., and Chevalier, H. "Experimental and Theoretical Field Strength Evaluation on VHF Propagation Channel for Aeronautical Mobiles." Aeronautical Mobile Communication Panel WG-D, Madrid, April 7–12, 1997.

52. Miyagaki,Y., Morinaga, N., and Namekawa, T. "Double Symbol Error Rates of M-ary DPSK in a Satellite Aircraft Multipath Channel." *IEEE Transations on Communications* 31, no. 12 (December 1983): 1285–89.

53. Luo, Qinglin, Fang, Wei, Wu, Jinsong, and Chen, Qingchun. "Reliable Broadband Wireless Communication for High Speed Trains Using Baseband Cloud." *EURASIP Journal on Wireless Communications and Networking*, September 2012, 1–12.

54. Zhou, Tao, Tao, Cheng, Liu, Liu, Qiu, Jiahui and Sun, Rongchen. "High-Speed Railway Channel Measurements and Characterizations: A Review." *Journal of Modern Transportation* 20, no. 4 (December 2012): 199–205.

55. Fujimoto, K., and James, J. R., ed. *Mobile Antenna System Handbook.* Ibid. Sec. 5.3.

56. Knorzer, Sandra, Baldauf, A., Fugen, Thomas, and Wiesbeck, Werner. "Channel Medelling for an OFDM Train Communications System Including Different Antenna Types." 64th Vehicular Technology Conference, VTC-2006, September 25–28, 2006, 1–5.

57. Lee, W. C. Y. "Wireless & Cellular Telecommunications" Ibid, pp 750–753.

Index

Note: Page numbers followed by *f* denote figures; page numbers followed by *t* denote tables.